C000151602

# Signal Processing and Networking for Big Data Applications

Help make sense of big data in engineering applications using tools and techniques from signal processing with this unique text. It presents fundamental signal processing theories and software implementations, reviews current research trends and challenges, and describes the techniques used for analysis, design, and optimization. You will learn about key theoretical issues such as data modeling and representation, scalable and low-complexity information processing and optimization, tensor and sublinear algorithms, deep learning and software architecture, and their application to a wide range of engineering scenarios. Applications discussed in detail include wireless networking, smart grid systems, sensor networks, and cloud computing. This is the ideal text for researchers and practicing engineers wanting to solve practical problems involving large amounts of data, and for students looking to grasp the fundamentals of big data analytics.

**Zhu Han** is a Professor in the Department of Electrical and Computer Engineering at the University of Houston and a Fellow of the IEEE. He has co-authored several books, including *Wireless Device-to-Device Communications and Networks* (Cambridge University Press, 2015) and *Game Theory in Wireless and Communication Networks* (Cambridge University Press, 2011).

**Mingyi Hong** is an Assistant Professor and a Black and Veatch Faculty Fellow in the Department of Industrial and Manufacturing Systems Engineering at Iowa State University.

**Dan Wang** is an Associate Professor in the Department of Computing at the Hong Kong Polytechnic University, and a Senior Member of the IEEE.

"A very nice balanced treatment over two large-scale signal processing aspects: mathematical backgrounds vs. big data applications, with a strong flavor of distributed optimization and computation."

**Shuguang Cui** *University of California, Davis*

# Signal Processing and Networking for Big Data Applications

ZHU HAN

University of Houston

MINGYI HONG

Iowa State University

DAN WANG

The Hong Kong Polytechnic University

# CAMBRIDGE
## UNIVERSITY PRESS

University Printing House, Cambridge CB2 8BS, United Kingdom

One Liberty Plaza, 20th Floor, New York, NY 10006, USA

477 Williamstown Road, Port Melbourne, VIC 3207, Australia

4843/24, 2nd Floor, Ansari Road, Daryaganj, Delhi – 110002, India

79 Anson Road, #06–04/06, Singapore 079906

Cambridge University Press is part of the University of Cambridge.

It furthers the University's mission by disseminating knowledge in the pursuit of education, learning, and research at the highest international levels of excellence.

www.cambridge.org
Information on this title: www.cambridge.org/9781107124387
DOI: 10.1017/9781316408032

© Cambridge University Press 2017

This publication is in copyright. Subject to statutory exception and to the provisions of relevant collective licensing agreements, no reproduction of any part may take place without the written permission of Cambridge University Press.

First published 2017

Printed in the United Kingdom by TJ International Ltd. Padstow Cornwall

*A catalogue record for this publication is available from the British Library.*

ISBN 978-1-107-12438-7 Hardback

Cambridge University Press has no responsibility for the persistence or accuracy of URLs for external or third-party Internet websites referred to in this publication and does not guarantee that any content on such websites is, or will remain, accurate or appropriate.

For my family, without whom the book may have been
published earlier, but with much less meaning.

Zhu Han

For my wife, who has been always supportive.

Mingyi Hong

For my wife, who is so patient and understanding.

Dan Wang

# Contents

# Part I

# Overview of Big Data Applications

# 1 Introduction

## 1.1 Background

Today, scientists, engineers, educators, citizens, and decision-makers have unprecedented amounts and types of data available to them. Data come from many disparate sources, including scientific instruments, medical devices, telescopes, microscopes, satellites; digital media including text, video, audio, e-mail, weblogs, twitter feeds, image collections, click streams, and financial transactions; dynamic sensor, social, and other types of networks; scientific simulations, models, and surveys; or computational analysis of observational data. Data can be temporal, spatial, or dynamic; structured or unstructured. Information and knowledge derived from data can differ in representation, complexity, granularity, context, provenance, reliability, trustworthiness, and scope. Data can also differ in the rate at which they are generated and accessed. The phrase "big data" refers to the kinds of data that challenge existing analytical methods due to size, complexity, or rate of availability.

The challenges in managing and analyzing "big data" can require fundamentally new techniques and technologies in order to handle the size, complexity, or rate of availability of these data. At the same time, the advent of big data offers unprecedented opportunities for data-driven discovery and decision-making in virtually every area of human endeavor. A key example of this is the scientific discovery process, which is a cycle involving data analysis, hypothesis generation, the design and execution of new experiments, hypothesis testing, and theory refinement. Realizing the transformative potential of big data requires addressing many challenges in the management of data and knowledge, computational methods for data analysis, and automating many aspects of data-enabled discovery processes. Combinations of computational, mathematical, and statistical techniques, methodologies, and theories are needed to enable these advances.

On March 29, 2012, the White House announced the Big Data Research and Development Initiative to mobilize the research and development toward Big Data analytics for solving many of the nation's most pressing challenges. A great many agencies are involved, spanning from National Science Foundation (NSF) and National Institutes of Health to the Department of Defense and the Department of Energy. Signal processing and systems engineering communities can be important contributors to big data research and development, complementing computer and information science-based efforts in this direction. Big data analytics entail high-dimensional, decentralized, online, and robust statistical signal processing, as well as large, distributed, fault-tolerant, and intelligent systems engineering. There is a need and opportunity for the signals and systems communities to jointly pursue big data research and development.

The aim of this book is to bring together signal processing engineers, computer and information scientists, applied mathematicians and statisticians, as well as systems engineers to carve out the role that analytical and experimental engineering has to play in big data research and development. This book will emphasize signal analytics, networking, computation, optimization, as well as systems engineering aspects of big data.

In this book, we plan to address the challenges from the management of the big data, through the lens of signal processing. It should be noted that the term signal processing here is not limited to the processing of the traditional analog or digital signals, but rather should be understood as a wide range of computational and/or analytical techniques for transformation and interpretation of information. Therefore this book will focus on various theories and techniques that help make sense of the big data, as well as their applications in various engineering domains, such as machine learning, networking, energy systems, and so on. The potential audience of this book includes, for example, graduate or undergraduate engineering students who seek the first introductory course on Big Data Analytics, or researchers in the Signal Processing community who desire to apply their knowledge and expertise for solving big data-related problems, or practitioners who look for tools and theories that can help resolve their practical problems involving large amounts of data.

There are three main objectives of writing this book. The first objective is to provide an introduction to the big data paradigm, from the signal processing perspective. The second objective is to introduce the key techniques to enable signal processing for big data in a comprehensive way. The third objective is to present the state-of-the-art big data applications. This will include classifications of the different schemes and the technical details in each scheme.

## 1.2      Market and Readership

The book will provide the state-of-the-art of research on data analysis for variety of big data applications. It will also include fundamental theories based on which big data signal processing is built. The book will discuss in detail various theoretical issues enabling signal processing for big data, including data modeling and representation; scalable, online, low-complexity, information processing and optimization; deep learning; and software architecture of big data processing. The book will also discuss the application of these theories and techniques to various engineering domain, including wireless networking, smart grid systems, sensor networks, and cloud computing. Other challenges to be discussed in the book include parallel data processing architecture, and distributed/decentralized data processing.

The key features of this book will be:

- a unified view of big data analysis and signal processing
- comprehensive review of the state-of-the-art research and key technologies from big data literature

- coverage of a wide range of techniques for design, analysis, optimization, and applications of bid data
- outlining the key research issues related to signal processing for big data
- outlining major big data applications relevant to the signal processing community
- industrial activities on big data.

To the best of the authors' knowledge, this will be the first book on signal processing on big data. This book intends to provide the background knowledge, discuss the research challenges, review the existing literature, and present the techniques for analysis, design, optimization, and applications under given objectives and constraints for big data applications

The primary audience for the proposed book will consist of:

- researchers and communications engineers interested to study the new paradigm of big data signal processing
- researchers and communications engineers interested in the state-of-the-art research on big data
- engineers in the field of data analysis
- graduate and undergraduate students interested in obtaining comprehensive information on the design, evaluation, and applications of big data and signal processing.

## 1.3 Brief Description of the Chapters

### Part I: Overview of Big Data Applications

This part will present an overview of the book, the angle of signal processing in big data applications, and an introduction on data parallelism and popular software.

1. **Introduction:** Starting with a motivation to big data, the introduction will cover the background, followed by the reasons to write this book. Then the signal processing aspects of big data analysis is illustrated. Finally, the book structure is briefly described.
2. **Data Parallelism: The Supporting Architecture:** To store and process the continuous increasing of the data, past computing framework must undergo a significant change. The key change is data parallelism, leading to a switch from scale up to scale out. More specifically, rather than developing super machines (scale up) to speed up massive data processing, the choice nowadays is to use a massive number of physical machines (scale out) to process data in parallel. Data parallelism greatly differs from computing parallelism. An example of computing parallelism is playing chess, where there is a minimal amount of data, yet the requirement for computation is massive. To facilitate such change, an important data parallelism framework is MapReduce. MapReduce has a map phase and a reduce phase. In the map phase, the data are divided into smaller chunks and sent to different physical machines. In the reduce phase, the outputs from map phase are aggregated for a partial or final

result. In this chapter, we will first present data parallelism, the challenges, and solution approaches in massive parallel data processing. We progressively discuss the enabling technologies, including the file system HDFS, and a series of programming models such as MapReduce, Spark, GraphLab, and Spark Streaming. We then present big data programming models supported by the cloud. Such service is useful for the users who do not have a massive number of physical machines. We present a set of research challenges and advances in this field. Finally, we present how signal processing algorithms are supported by the big data programming models, and give an example implementation of the K-Means algorithm.

## Part II: Methodology and Mathematical Background

This part will present variety of signal processing techniques, which can be applied to the problems of big data analysis.

3. **First-Order Algorithm:** As a fundamental tool for big data analytics, first-order optimization algorithms have recently attracted significant research efforts. This trend is also exemplified in the signal processing community, where researchers are challenged to make sense of a huge amount of data. For example, emerging problems in image processing, social network science, and computational biology can easily exceed millions or even billions of variables. Generally speaking, first-order algorithms are capable of solving problems of huge sizes because: (1) each of their computational steps is simple, closed-form, and easy to perform; and (2) a desired solution can be found using a small number of iterations. Further, first-order algorithms can be implementable in either distributed or parallel fashion so that one can exploit the modern multi-core and cluster computing architectures. In this chapter we are going to focus on a large family of first-order algorithms that are found to be extremely useful for big data optimization. Examples include the popular alternating direction methods of multipliers (ADMM), block coordinate descent methods (BCD), gradient/proximal gradient methods, and their accelerated version. The techniques introduced in this section will be frequently used in the remaining chapters of the book.

4. **Sparse Optimization:** This chapter reviews a collection of sparse optimization models and algorithms. The review focuses on introducing these algorithms along with their motivations and basic properties. A large number of major algorithms for sparse optimization are covered in this chapter, such as classic solvers; shrinkage operation; homotopy algorithms and parametric quadratic programming; continuation, varying stepsizes, and line search; nonconvex approaches for sparse optimization; greedy algorithms; and algorithms for low-rank matrices. Finally, we discuss how to choose a specific algorithm for a certain application.

5. **Sublinear Optimization:** The sublinear algorithm is a random algorithm technique developed in theoretical computer sciences. The initial motivation is that the total amount of data is so massive that they are beyond even linear processing time. The objective of current state-of-the-art sublinear algorithms is to use either $o(n)$ in space,

or o(n) in time, or o(n) in communication to output the results, with quantitative bounds of guarantee. In this chapter, we will first present the foundations and some examples of sublinear algorithms. We then specifically discuss algorithms that are sublinear in time, and sublinear in space and sublinear data streaming algorithms where online processing is required.

6. **Tensor for Big Data:** Tensors are multidimensional generalizations of matrices, which can have non-numeric entries. Extremely large and sparsely coupled tensors arise in numerous important applications that require the analysis of large, diverse, and partially related data. The effective analysis of coupled tensors requires the development of algorithms and associated software that can identify the core relations that exist among the different tensor modes, and scale to extremely large data sets. We will discuss some techniques such as tensor networks, tensor decompositions, big sparse tensors, etc. Then we investigate the current state of art for tensor research specifically targeting big data applications.

7. **Deep Learning and Applications:** Deep neural network (DNN) is a multilayer neural network, whose appeal stems from its strong ability in finding useful patterns from massive amounts of data. Recently it has attracted significant research attention from computer theorists, applied mathematicians, as well as engineers working in various domains. Representative application of DNN includes pattern classification, speech recognition, speech denoising, etc. In this chapter we will first briefly go over the history of neural network, the predecessor of DNN. Then we will describe the architecture of DNN, many of its variants, as well as the state-of-the-art methods in training a large-scale DNN. Finally we will present a few exciting applications of DNN, and discuss the open research directions.

## Part III: Big Data Applications

This part will present various large-scale signal processing applications that involves massive amounts of data.

8. **Compressive Sensing-Based Big Data Analysis:** A new paradigm of signal acquisition and processing, named compressive sensing (CS), has emerged since 2004. The CS theory, which integrates data acquisition, compression, dimensionality reduction, and optimization, has attracted lots of research attention. The CS theory consists of three key components: signal sparsity, incoherent sensing, and signal recovery. It claims that, as long as the signal to be measured is sparse or can become sparse under a certain transform or dictionary, the information in the signal can be sparsely represented in a small number of incoherent measurements, and the signal can be faithfully recovered by tractable computation. CS is a new tool bearing a large number of potential applications in engineering, especially in big data analytics. In this chapter, we will cover several aspects of the recent CS developments. Topics will include a general discussion on its theory as well as several of its applications in data acquisition, communication, imaging, and so on.

9. **Distributed Large-Scale Optimization:** Distributed signal and data processing have long been recognized as an enabling technique for big data analytics. The most obvious, and also the most important reason, is that in the big data era the amount of data to be analyzed will easily exceed the storage and processing capability of a single machine. Therefore pieces of data need to be stored and processed distributively among clusters of machines. The research challenge is how to effectively coordinate the distributed computing nodes so that their individual efforts collectively yield a good interpretation of the data. In this chapter we start by introducing several interesting applications in signal processing, networking, smart grid systems and machine learning. Then we review several state-of-the-art methods for distributed signal processing/computation, including the subgradient based algorithm, the gossip based algorithm, as well as the ADMM-based algorithm. When discussing each family of algorithms, we will focus on both theory and their representative applications.

10. **Optimization of Finite Sums:** In many applications involving massive amounts of data, it is usually preferable to process the data in a sequential manner (as opposed of the batch processing). Several reasons can be attributed to this, all of which have to do with the amounts of the data available. The most obvious reason is that the sheer size of the data prevents any single machine to load the entire data set into its memory. Therefore the data needs to be read sequentially thus processed sequentially. Perhaps an equally important reason is that there are an increasing number of applications in which decisions need to be made at the same pace as the data becoming available. In this chapter, we introduce the state-of-the-art methods in online data processing and optimization. We will start with two basic approaches for such purpose, one based on minimizing the so-called regret function, the other based on the stochastic optimization criteria, where the data is modeled as drawing from certain statistical distribution. After introducing the theory, we will then walk the reader through a few practical applications, including those arriving in machine learning and dynamic network resource management.

11. **Big Data Optimization for Communication Networks:** The insatiable demand for accessing massive amount of data anywhere anytime is the driving force in the current development of next-generation wireless network infrastructure. It is projected that within ten years' time, the wireless cellular network will be able to offer up to 1,000x throughput performance over the current 4G technology. By that time the network should also be able to deliver fiber-like user experience boasting 10 Gb/s individual transmission rate for data intensive cloud-based applications. To move such a huge amount of data from the network to the palms of users hands in real-time, revolutionary infrastructure and highly sophisticated resource management solutions are required. The emerging network architecture with great potential is the so-called cloud-based radio access network, where a large number of base stations (BSs) are deployed for wireless access, while powerful cloud centers are used at the back end to perform centralized network management. Ideally the dense deployment of access nodes should offer significant improvement of spectrum efficiency, the capability of real-time load balancing, and hotspot

coverage. In practice the optimal network provisioning is extremely challenging, and its success depends on the intelligent integration of factors such as backhaul provisioning, physical layer transmit/receive schemes, BS/user cooperation, and so on. In this chapter, we advocate the use of modern first-order, large-scale optimization techniques for massive data delivery over the cloud-based, densely deployed, next-generation wireless networks. We show how to solve many difficult problems in this domain, by using popular optimization algorithms such as the block successive upper-bound minimization (BSUMM) algorithm and the ADMM algorithm.

12. **Big Data Optimization for Smart Grid Systems:** The development of smart grid, impelled by the increasing demand from industrial and residential customers together with the aging power infrastructure, has become an urgent global priority due to its potential economic, environmental, and societal benefits. Smart grid refers to the next-generation electric power system which aims to provide reliable, efficient, secure, and quality energy generation/distribution/consumption using modern information, communications, and electronics technology. In this chapter, the applications of big data processing techniques for smart grid are investigated from several perspectives: how to exploit the inherent structure of the data set, and how to deal with the huge data sets. Specific applications are included such as sparse optimization for false data injection detection, a distributed parallel approach for the security-constrained optimal power flow (SCOPF) problem, smart metering and smart pricing, and a comprehensive literature survey.

13. **Processing Large Data Sets in MapReduce:** Today's lightning-fast data generation from massive sources is calling for efficient big data processing. A state-of-the-art tool is MapReduce. MapReduce provides a general model in processing data in parallel. However, the job completion time can vary greatly, largely depending on how balancing the load is distributed in different machines. MapReduce default configuration emphasizes common cases, and thus is not optimized for various applications. In particular, since MapReduce uses a simple hash function in distributing the data to different machines, if input data have large skew, the performance can be poor. This chapter presents how sublinear algorithms can be used to fast detect data skewness in large data sets, which fits the MapReduce framework. We can thus peek at a small set of data and make a much better decision in distributing loads into machines.

14. **Massive Data Collection using Wireless Sensor Networks:** Wireless sensor networks were initially proposed with a key switch to make the sensors not only a sensing device, but also a computing, storage, and communication device. Enhanced sensors can thus self-organize, self-construct a network, and collectively achieve tasks that were unable to achieve in the past. In this chapter, we will first briefly present wireless sensor networks, its history, and its layering architecture. We will then present various wireless data collection scenarios, because data collection is the primary task for wireless sensor networks. We then discuss the delay issues, storage issues, and communication issues in wireless sensor data collection. Depending on different functional tasks, there needs to be a balance in the factors of delay, storage

and communication. We then present a sublinear approach for wireless sensor data collection in delay-sensitive applications. We then present a sublinear approach for balancing storage and data communication in wireless sensor data collection.

## 1.4    Acknowledgments

Finally, we would like to thank our collaborators for contributions to the full-duplex research: Lanchao Liu, Yun Liao, Tianyu Wang, Kaigui Bian, Radwa Sultan, Karim G. Seddik, Hongyu Cui, Bingli Jiao, Mingxin Zhou, Hongyu Cui, Kun Yang, Boya Di, Kun Yang, Yunxiang Jiang, Francis Lau, He Chen, Xusheng Du, Yanfang Le, Jiangchuan Liu, Yi Yuan, Xiao Ling, Funda Ergun, and Siavash Bayat. We would also like to thank Mr. Xusheng Du for his final editing work.

This work is partially supported by US NSF ECCS-1547201, CCF-1456921, CNS-1443917, ECCS-1405121, NSFC61428101, RGC/GRF PolyU 5264/13E, NSFC 61272464, HK PolyU 4-BCB4, G-YBAG.

# 2 Data Parallelism: The Supporting Architecture

## 2.1 Data Parallelism

Conventionally, the platforms used to store and manage a large data set are relational databases. Examples of leading commercial databases include Oracle, IBM DB2, Microsoft SQL Server, and others. Relational databases are efficient for applications that process a large number of records and transactions that are common in banks, commercial sectors, and so on. The underlying physical supports for advanced relational databases are IBM servers and EMC storage platforms. If there is an increase in data, more powerful machines and larger storage platforms are developed. Such a model to handle increases in data is commonly called a *scale-up* model.

This state of affairs is changing in the face of the recent wave of big data applications. In these applications, the amount of data involved is bigger than ever before, and these applications require more flexible computations. It is common for all data to need processing for the outcome of an application to be obtained. This situation differs significantly from applications that are supported by relational databases.

One notable example is the Google PageRank algorithm. The PageRank algorithm is the foundation of Google's web search. It sorts the ranks of the webpages, so that higher-ranked webpages can be listed during a web search operation. An example of how the PageRank algorithm functions is given below.

**A PageRank Example:** Assume that there are four webpages, $A, B, C, D$. Let $PR(A), PR(B), PR(C), PR(D)$ denote their respective ranks. If $A$ has incoming links from $B, C, D$, and assuming that $D$ has $m$ outgoing links, the rank of $A$ is calculated as $PR(A) = PR(B) + PR(C) + \frac{PR(D)}{m}$. The rank of each individual webpage will be updated periodically. It can be proven in theory that the PageRank algorithm converges.

We can see that to obtain a result from the PageRank algorithm, each piece of data needs to be processed, even though the computation on the data is simple. The data is big and new data (webpages) are generated every day. Applications with similar needs are abundant. For example, companies need to analyze periodically user behaviors and update user profiles to support recommendations.

Google has proposed a new computing architecture called MapReduce. In MapReduce, a job (e.g., a PageRank job) is first divided into tasks. These tasks are assigned to different physical machines, each of which has a chunk of data. The results of these tasks will be aggregated, and a final result will be outputted. The tasks of a MapReduce job run in parallel. For example, the computation of the ranks of webpages $B, C$ will be assigned to one physical machine that stores $B$ and $C$ and the computation of the rank of $D$ will be assigned to another physical machine that stores $D$. These machines can be inexpensive commodity machines. As a result, when the amount of data increases, it is easy to extend the computation by adding more physical machines to handle the additional data. Such a model to handle increases in data is commonly called a *scale-out* model.

The key change in a scale-out model as compared to a scale-up model is data parallelism. More specifically, rather than developing faster machines to process the massive amount of data, the choice of the scale-out model is to use a massive number of inexpensive machines to process data in parallel. It also differs greatly from past high-performance parallel computing architectures. In older parallel computing architecture, data are sent to the computing nodes when necessary. The amount of data is small as compared to current big data applications. Typical applications include playing chess, scientific numerical calculations, and so on. In the MapReduce architecture, tasks are assigned to the physical machines which hold the data.

The MapReduce architecture has rapidly gained popularity since having been proven suitable for use in a great number of applications. In recent years, we have seen a large number of supporting models and techniques developed for the MapReduce architecture, and a large number of research studies conducted on improving the efficiency of the MapReduce architecture under various scenarios.

The MapReduce architecture is not a single technique. It is a set of techniques that supports data parallelism. Figure 2.1 sets out the MapReduce architecture.[1] On the bottom is the supporting file system, the Hadoop Distributed File System (HDFS). On top of the HDFS is a resource management system called Yarn which manages physical resources and requirements from different application jobs. On top of Yarn, there are programming models that can abstract a large set of applications. Typical examples are the MapReduce programming model, which is best suited for independent batch jobs; the Spark programming model, which is best suited for real-time processing jobs; the GraphLab programming model, which is best suited for dependent jobs that can be captured by a graph-based model; the Spark Streaming programming model, which is best suited for streaming applications; and others. There are supporting tools for each of these programming models, such as programming languages, libraries, and so on.

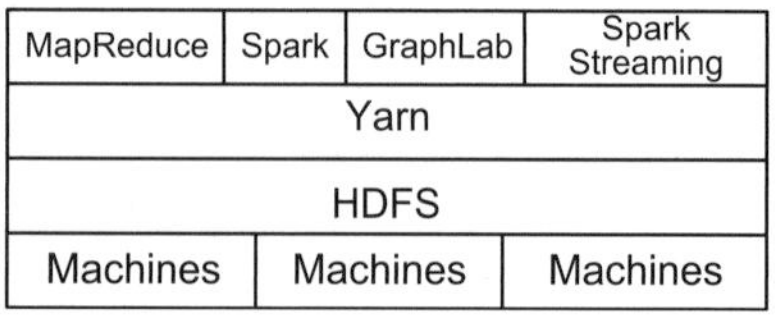

| MapReduce | Spark | GraphLab | Spark Streaming |
|---|---|---|---|
| Yarn | | | |
| HDFS | | | |
| Machines | | Machines | Machines |

**Figure 2.1** The MapReduce architecture.

---

[1] MapReduce is most commonly used to refer to the MapReduce programming model, which is one technique that supports data parallelism. We will discuss the MapReduce programming model later in this chapter. We use the MapReduce architecture to refer to the general data parallel architecture used in the big data applications of today.

In this chapter, we first present the aforementioned set of techniques, which are the core parts in the development of the MapReduce architecture. We present the MapReduce architecture in the cloud environment, which is a suitable solution for grassroots users. We present some commercialized MapReduce-oriented cloud platforms.

We then present recent developments and research studies of the MapReduce architecture. There are two main directions. The first direction is to develop techniques that can achieve faster and more scalable processing of data, and to develop richer tools and supports for the large amount of functionalities needed by various applications. The second direction is to optimize the resource utilization and fairness of the MapReduce architecture. We present resource utilization problems and solution approaches from a system provider's point of view and from a user's point of view. Finally, we discuss problems that are particularly related to iterative operations, which are of interest for signal processing algorithms.

## 2.2    Hadoop Distributed File System (HDFS)

The data storage system of the MapReduce architecture is the Hadoop Distributed File System (HDFS). The HDFS is a distributed file system, yet it differs from previously developed distributed file systems such as AFS [1], NFS [2], and others. In this chapter, we do not specifically present details of the design of the HDFS. A comprehensive study can be found in [3]. We present the assumptions that were made in the design of the HDFS. These assumptions shed light on the choices and trade-offs that were made when designing the HDFS.

In the HDFS, the first assumption is that the system is built on top of a large set (e.g., thousands) of inexpensive commodity machines (e.g., x86). These machines are fault-prone. Therefore, faults are the norm rather than the exception. The files are assumed to be big (e.g., in terms of GB per file), as compared to conventional files. It is assumed that the workloads primarily consist of a large number of streaming reads and a small number of random reads. This assumption essentially means that the files are accessed and processed collectively (to achieve an overall objective), rather than accessed individually. It is also assumed that there is high and sustained bandwidth.

Clearly, the HDFS is designed for big data applications, where large chunks of data are processed to achieve a certain objective in a scale-out model on top of low-end, interconnected physical machines. Nowadays, the HDFS has become the standard file system support for upper-layer systems.

## 2.3    MapReduce and Yarn

### 2.3.1    MapReduce

MapReduce usually refers to the MapReduce programming model. A programming model is how programs are executed by invoking the library calls that are supported

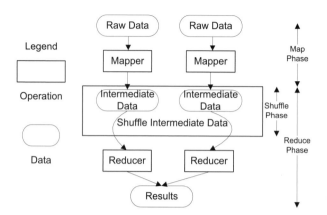

**Figure 2.2** The MapReduce programming model.

by the programming model. The MapReduce programming model has now become the de facto standard model for big data applications. Figure 2.2 illustrates the basic structure of the MapReduce programming model.

A MapReduce job consists of two phases, namely, *map* and *reduce*. Accordingly, there are two program functions: *mapper* and *reducer*. In the map phase, the input data is split into blocks. The mappers then process the data blocks and produce intermediate data. The reduce phase starts from a *shuffle* subphase, during which, run by the reducers, the intermediate data are shuffled and moved to the corresponding nodes for reducing. More specifically, a map function transfers the input raw data into (key, value) pairs, and the reduce function merges all intermediate values associated with the same intermediate key. The reducers then process the shuffled data and generate the final results. For complex problems, multiple rounds of map and reduce are executed.

**A MapReduce Example:** We now use a word count example to illustrate the operations of MapReduce in detail. Assume that we want to develop a WordCount program to count the number of words in a big file. Without loss of generality, assume that we have a file with the following content.

Hello World Hello World

Hello World Hello Hadoop

Hello World Hello Spark

It is easy to see that the output of the WordCount program should be: Hello 6, World 4, Hadoop 1, Spark 1.

Assume that MapReduce performs this file on a per-line basis. As such, the input should be three lines. These three lines are assigned to three machines, A, B, C, and each machine will operate on a line to count the words. After the map phase, the map output will be (key, value) pairs as follows:

Machine A: (Hello, 2), (World, 2)

Machine B: (Hello, 2), (World, 1), (Hadoop, 1)

Machine C: (Hello, 2), (World, 1), (Spark, 1)

The keys will be grouped in the shuffling subphase. The output of the shuffling subphase will be (Hello, 2, 2, 2), (World, 1, 1, 1), (Hadoop, 1), (Spark, 1).

Finally, the reduce phase will be assigned to a machine D, and the intermediate results will be aggregated in the reduce phase based on the keys and the output: (Hello, 6), (World, 3), (Hadoop, 1), (Spark, 1).

The MapReduce programming model is an abstraction on how computation should be done. There are many implementations of the MapReduce programming model. The most widely used implementation is Hadoop, an open source software that is continuously maintained.[2]

## 2.3.2    Yarn

In the Hadoop 1.0 implementation of the MapReduce programming model, a JobTracker module handles the job submitted from the users to the system. As the number of jobs and the amount of resources increase, Yarn[3] is developed from the Hadoop 2.0 implementation, to efficiently manage the jobs and the resources.

The core feature of Yarn is to separate resource management from the application process. Such a separation makes it possible for a Hadoop Yarn cluster to run multiple application jobs or even multiple data parallel programming models, such as MapReduce, Spark, GraphLab, and others, at the same time.

Recall the MapReduce architecture in Figure 2.1. The bottom layer consists of a set of off-the-shelf heterogenous machines, which can easily be scaled out. The HDFS unifies their file systems. Yarn is on top of the HDFS and is responsible for the management of resources such as CPUs, storage, and others. On top of Yarn are applications that run on implementations of data parallel programming models such as MapReduce, Spark, GraphLab, and so on. When running a big data application, the data will first be loaded from a data source (which can be any storage) to the HDFS. This protects the raw data source from being damaged.

We now go on to describe Yarn in more details (see Figure 2.3). Each node (i.e., physical machine) has a *node manager*. There is a *resource manager* running on a

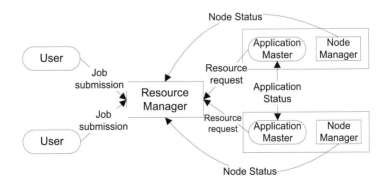

**Figure 2.3**  The Yarn system.

[2]  http://hadoop.apache.org/.
[3]  http://hadoop.apache.org/docs/current/hadoop-yarn/hadoop-yarn-site/YARN.html.

separate master node that monitors the resource status of each node. In the Yarn system, the resource manager and the node managers are not concerned with the specifics of the applications. There are separate *application masters* that handle the specifics of the applications and conduct resource requests to the resource manager.

Many studies have been conducted on the resource management needed to optimize the utilizations of the resources of the entire MapReduce architecture. We will discuss these studies in more detail in Section 2.7.

## 2.4    Spark

The MapReduce programming model is restricted in two ways. First, big data applications have diverse requirements. For example, machine learning algorithms commonly have iterative operations and multi-pass operations. The expressiveness of the MapReduce programming model is limited in supporting these applications. Second, in the MapReduce programming model, the data are stored in hard disks. This design makes the execution time of the MapReduce programming model unacceptable for many applications.

The Spark programming model is designed for near real-time processing, and also with richer functional supports. One key difference between Spark and MapReduce is that in the Spark model, the data are stored in memory, while in the MapReduce model, the data are stored in hard disks. Figures 2.4 and 2.5 illustrate the difference between the Spark model and the MapReduce model in detail. In the Spark model, only the first iteration and the last iteration involve HDFS (disk) reads and writes. This is unavoidable, since we need to load the data from the HDFS to memory. In each of the other iterations, the reads and writes are memory-based. This largely improves the efficiency for real-time applications because the most time-consuming operations for a big data application are the reads and writes of a massive amount of data.

While using memory storage to replace disk storage can substantially speed up operations on data, it also brings in challenges. The key challenge is failure recovery. Recall that the HDFS faces failures as well and the HDFS is designed to be fault-tolerant.

**Figure 2.4**  Operations in the MapReduce programming model.

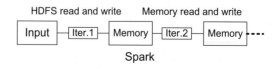

**Figure 2.5**  Operations in the Spark programming model.

The failures of the HDFS and the MapReduce model are on different levels, however. In the HDFS, the failures are on the machine level. The HDFS continuously monitors machine failures and conducts machine-level recovery through logs and redundancy. In MapReduce, the failures are on the job level. More specifically, some tasks of a job (e.g., mappers) can fail. The MapReduce model handles this by simultaneously running multiple mappers of the same task. When a mapper fails, MapReduce will trace another mapper that is conducting the same task on the same data.

In the Spark model, the failures are much less recoverable than the failures in the MapReduce model. This is because disk storage on the HDFS is permanent yet memory storage is transient. The solution approach from the Spark model is a memory abstraction called *Resilient Distributed Dataset* (RDD). An RDD can be regarded as a data structure to store intermediate states. It allows the Spark model to provide fault tolerance through logs that simply record the transformations which can be used to build a data set for recovery instead of using all data. In other words, failure recovery can be started from an intermediate state.

The designs of both the MapReduce and Spark programming model for handling failures illustrate the trade-off made for a typical scale-out data parallelism model. Memory storage and disk storage are assumed to be cheap, since we can scale out by increasing the number of machines to be used.

The design of the RDD also allows the Spark programming model to support a wide range of applications. RDD is a lineage-based data structure. More specifically, it records the operations that build a data set instead of replicating a data set. This design makes *data sharing* efficient across parallel operations since data replicating leads to a very high cost since we run big data applications.

The Spark model is fast and has a rich set of function primitives so that it can support diverse applications. It has much more overhead in memory storage than the MapReduce model because it needs to store a great number of intermediate states, however. MapReduce is thus a choice for batch applications such as back-end log processing, profile updating, and so on.

## 2.5 GraphLab and Spark Streaming

There are other programming models that were designed for specific application types. We discuss GraphLab and Spark Streaming in this section.

### 2.5.1 GraphLab

Big graph is a data structure that is widely used in different applications. Many programming models have been developed for processing big graphs, such as Pregel [4], GraphLab [5, 6], and others.

Both the Pregel programming model and the GraphLab programming model maintain the philosophy of data parallelism. Their abstractions are graph-based. The difference between Pregel and GraphLab is that the Pregel model is a synchronous programming model, while the GraphLab model is an asynchronous programming model. As such, the Pregel model is relatively easy to build in comparison with the GraphLab model, yet it can be less efficient. This is because different parts of the big graph can converge at different speeds as the running time is restricted by the slowest machine.

We shall now discuss the GraphLab programming model in more detail. We again use the PageRank algorithm to illustrate the execution of the GraphLab model. The GraphLab model has three key elements: Data Graph, Update Functions, and Sync Operations. The Data Graph is a graph-based data structure that is used to organize data. For example, each vertex of the Data Graph corresponds to a webpage, each edge of the Data Graph corresponds to a link between two webpages, and the edge data of the Data Graph store the weights of the links. The Update Function is similar to the map function in the MapReduce model. It is a user-defined function that applies to the vertices. For example, a user can define the update of the edge data, that is, the weights of the edges according to the PageRank algorithm as $PR(A) = PR(B) + PR(C) + \frac{PR(D)}{m}$. The Sync Operation is similar to the reduce function in the MapReduce model. For example, a user can implement Sync Operations by defining the global synchronization criteria. The Sync Operations are usually executed in the back end.

### 2.5.2     Spark Streaming

Streaming is another common application. In a streaming application, data are not pre-stored or pre-available. Live data streams arrive continuously and the system may need to output results continuously.

The Spark streaming programming model was developed on top of the Spark programming model. It again applies a data parallel model. A data stream is divided into different intervals according to time. In each time interval, the data in this interval are divided into mini-batches. These batches are constructed into RDDs which are then handled by Spark.

## 2.6     Big Data in the Cloud

At the very beginning, when the MapReduce architecture was proposed, industry leaders such as Google and Facebook ran their big data applications on their dedicated server clusters. Nowadays, a growing number of big data applications are coming not only from the IT industry, but also from civil, environmental, finance, and health industries. For grassroots users or non-computing professionals, the cost of deploying and maintaining large-scale dedicated server clusters can be prohibitive, to say nothing of the technical skills involved. On the other hand, public clouds, which have also emerged in recent years, allow general users to rent computing and storage resources and run their applications in a pay-as-you-go manner. The massive resources available at a public cloud provider's data centers offer ultra-high scalability and yet minimized upfront costs for its users. This new computing paradigm has been tremendously popular, becoming a highly attractive alternative to dedicated server clusters.

A public cloud has very different characteristics from dedicated server clusters. Through machine virtualization, a public cloud effectively hides the many levels of implementation and platform details involved, making shared resources appear exclusive to the end users. In such a virtualized environment, user applications share the underlying hardware by running in virtual machines (VMs). Each VM, during its initial creation (by the users), is provisioned with a certain amount of resources (such as CPU, memory, and I/O). The number and capacity of the VMs can be requested in a pay-as-you-go fashion, or even adjusted in runtime. In other words, the cloud resource allocation is highly elastic, and an application may adjust the resources that it needs for processing its big data at different stages. For example, a greater amount of resources may be required in the mapper phase than in the reduce phase. It is therefore possible to use fewer resources in the reduce phase and more resources in the mapper phase, while the total amount of resources remains the same.

Commercial cloud systems are commonly classified as Infrastructure as a Service (IaaS), Platform as a Service (PaaS), and Software as a Service (SaaS). In IaaS, a cloud service provider provides resources that emulate physical machines. In PaaS, a cloud service provider provides a platform environment to application developers. In SaaS, a cloud service provider provides software and databases.

There are several platforms that support big data applications. The first approach is represented by Google App Engine MapReduce.[4] This is a PaaS approach, where users directly submit their MapReduce jobs to the Google App Engine platform. The advantage of this approach is that the users do not need to worry about the details, yet they also yield control of their applications to the cloud. The second approach is represented by Amazon EMR[5] and Microsoft HDInsight.[6] A user first provisions VMs (Amazon EC2 or Microsoft Azure) from the cloud provider to construct a cluster. The cloud provider then deploys a big data processing system such as Hadoop or Spark on top of the cluster for the user. The user application is then executed. The users are charged according to the infrastructure that is consumed with a certain additional cost for maintaining the system. The advantage of this approach is that it gives the users full control of the system. Users own their cluster (i.e., they do not need to share it with others) and, if needed, it is easy for them to switch their applications (together with the open source Hadoop, Spark, etc.) to another cloud provider. We call this an infrastructure-based PaaS (I-PaaS) approach. The third approach involves the rental of VMs from Amazon EC2, or other sources, and the development of the Hadoop environment entirely by the user. This is a pure IaaS approach. The balance and choice of different approaches depends on the need of the application with the IaaS-oriented approaches enjoying more control but requiring more effort to develop and the PaaS-oriented approaches yielding some control to the cloud providers yet requiring less technical expertise to employ.

To run big data applications in a cloud environment, users need to make decisions about two specific problems. First, given a MapReduce job, it is necessary to find an

---

[4] https://cloud.google.com/appengine/docs/python/dataprocessing/.
[5] https://aws.amazon.com/elasticmapreduce.
[6] http://azure.microsoft.com/enus/services/hdinsight/.

effective way of constructing VMs in the cloud, which balances performance and costs (user bills). We call this the *Resource Provisioning* problem. Second, given the set of VMs that has been provisioned/constructed, it is necessary to find an optimal way to schedule MapReduce jobs/tasks. We call this the *VM-Job Scheduling* problem.

The optimization of resource utilization problems has been heavily studied in the research community in recent years. We discuss them in the next section.

## 2.7        Advances in the Data Parallelism Architecture

There are a lot of development and research studies in the MapReduce Architecture. We can classify them in two broad directions. First, there are studies on how to complete a job or a set of jobs faster. This problem can also be considered as a scalability problem. In other words, the problem is how to run a job with more data (a "bigger" job) fast. Second, there are studies on how to optimize resource utilization of the MapReduce architecture when running big data applications. This direction compensates the first direction. The running time of a job is not the sole emphasis, and there needs trade-offs on a set of criteria. We classify the research directions in Figure 2.6.

### 2.7.1        Fast Job Completion Time

In what follows, we first present the research studies on a data locality problem. Data locality means that a job should process the data in the physical machines that host the data or close to the data as much as possible. This is because for big data applications, massive data movement will lead to high cost. Data locality is a pertinent problem to big data applications. We then present job-task scheduling for fast job completion time, a grand problem that receives tremendous attention. Finally, we discuss the data-computation contention problem. MapReduce jobs are both data intensive and computation intensive. This can mean that executing I/O-intensive operations (e.g., outputting the data to local file systems or to the network) will affect the computation operations. A balance on computation and I/O resources is thus needed.

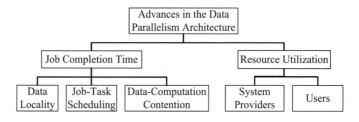

**Figure 2.6**  Optimizing resource utilization in the MapReduce architecture.

## Data Locality

Data parallelism scales out the processing of data to multiple machines and causes data to move from one machine to another. The movement of massive amounts of data, involving slow networking and disk operations, can aggravate resource contention and introduce excessive delays. It is therefore desirable to process data close to local machines.

The design of the MapReduce programming model intrinsically fits data locality. A MapReduce job is divided into tasks and these tasks are dispatched to different machines. It is therefore desirable to distribute the task to the machine with the right data. The Spark programming model is also designed to share data. Instead of replicating data (such replication is expensive in the big data context) Spark records the process that can lead to the transformation of the data; such process is cheap in the big data context.

There are many more specific studies. The data locality of map tasks in scheduling under heavy traffic was investigated in [7]. To achieve a balance between data locality and load balancing while maximizing throughput and minimizing delays, the system was modeled into a queuing architecture. Then, a scheduling algorithm based on the Join the Shortest Queue policy and the Max Weight policy was proposed. The proposed algorithm was proven to be throughput optimal. A stochastic optimization framework was proposed in [8] to optimize the data locality of reduce tasks. Based on the optimal solution under restricted conditions, a receding horizon control policy was proposed. In [9], delayed scheduling was proposed to balance fairness and data locality. The study showed that delaying the launching of jobs/tasks for a short period of time can significantly enhance the throughput.

## Job-Task Scheduling

For a big data job, how to assign its tasks to the appropriate machines (i.e., the Job-Task Scheduling problem) is a grand problem that receives tremendous attention. It is also the core problem to complete a job fast. We first present studies on theoretical advances. These studies analyze the theoretical complexity in various contexts. We then present studies on the data load balancing problem. Intrinsically, balancing the data to be processed on different machines will complete a job faster. Finally, we present the problem on the scalability when the amount of data to be processed increases.

### 1. Theoretical Advances

There are decades of studies on general job scheduling problems on parallel machines [10, 11, 12, 13]. Many approximation algorithms are developed with bounds. The MapReduce architecture has a special structure with map and reduce tasks where mappers proceed reducers. There are studies on general precedence constraints. The best-known result is a four-approximation algorithm for minimizing the total weighted job completion time [14]. However, note that the precedence constraints in these studies are general and cannot be directly applied to the MapReduce architecture.

Recently, many studies have been conducted specifically on job scheduling with the kind of MapReduce sequences depicted in Figure 2.2. In [15], a problem on fast job completion times for MapReduce-like jobs was formulated. The complexity of the problem was shown to be NP-hard. Approximation algorithms were developed based on relaxation to prime-dual linear programming. Online algorithms were further presented. In [16], a problem on precedence constraints between map tasks and reduce tasks in a MapReduce job was investigated. In particular, the shuffle phase was taken into consideration. It was shown that an eight-approximation algorithm can be developed for the problem. A recent episode of this series was [17]. In [15] and [16], tasks and jobs were assumed to be pre-allocated into server machines, and scheduling was conducted on the mapper tasks and reducer tasks of the jobs. In [17], the impact of server assignment was studied. The study showed that scheduling jobs without taking server assignment into consideration may result in less optimal solutions. A problem of the joint scheduling of servers and jobs, called a MapReduce server-job organizer problem was formulated. A three-approximation algorithm and a fast heuristic were developed.

These studies are mostly theoretical. Nevertheless, they provided an in-depth understanding of the origin of the complexity of the problems involved, and insights on how to develop algorithms to overcome such complexities of the problems. Implementation and real-world experiments have validated the effectiveness of the algorithms [17].

## 2. Data Load Balancing

In the MapReduce programming model, mapper tasks are executed before reducer tasks. Since different mappers and reducers are executed on different machines, the completion times of some mapper tasks might be delayed for various reasons. This will have an effect on the execution of reducer tasks. This problem is also called the skewed load problem, or the straggler problem.

Studies first concentrated on the causes of the straggler problem. There have been many studies on the measuring and monitoring of stragglers. An early study that described the straggler problem was [18]. It has been shown that the straggler problem in the MapReduce model is caused by the Zipf distribution of the inputs or intermediate data [19]. TopCluster was developed to monitor data skew [20]. In [21], five types of skews that can arise in MapReduce applications were presented, and five best practices to mitigate skew were proposed.

There are two common approaches to handling the data load balancing problem: the reactive approach and the proactive approach. In a reactive approach, the system monitors the load of different servers and switches the loads from those servers that become congested to those that become idle. One typical example of the reactive approach is SkewTune, developed in [22]. In a proactive approach, models are developed for the performance of the servers or the distribution of data. Load allocation to the servers then follows on these models. One example of the modeling of servers is [23], where MapReduce applications on heterogeneous servers were studied. A Markov chain model was first established to characterize the load of different servers, and the Markov chain was updated from time to time. Placement of the data to servers was based on the Markov chain model. One example of modeling the distribution of data is to learn the distribution of the input data [24]. The argument is that the intrinsic reason for the differences in the server loads is the unbalanced allocation of data. Recall that the MapReduce programming model has a structure of (key, value) pairs. The data are allocated according to their keys, and the default key assignment is through a random hashing. With skewed data distribution, the default hashing function will lead to a skewed key assignment. In [24], a simple online sampling algorithm was developed to learn the data distribution. The algorithm samples a sublinear amount of data, and achieves a $\frac{3}{2}$-competitive ratio with high probability, that is, it is provable that the difference between this online algorithm and the best offline algorithm is at most $\frac{3}{2}$.

There are other studies on fast map execution [25], or on overcoming the barriers of the slow-start-synchronization of reducers [26].

*3. Scalability*

The amount of the data that a big data job needs to process is increasing. In other words, a MapReduce and Spark job is growing bigger. As such, the number of tasks is increasing. In particular, it is advocated that tasks should be small [27]. Otherwise, the job completion time might be heavily affected by some big tasks that are delayed either because of poor scheduling or because of a straggler effect. While the amount of data increases, and the size of each task (i.e., the amount of data each task processes) remains stable or even reduces, the number of tasks becomes even bigger.

The design on the scheduling of tasks needs to be scalable. It is soon realized that a centralized task scheduler itself becomes a bottleneck of the job completion time. Sparrow [28] was proposed. The objective of Sparrow is to better allocate tasks on machines in a distributed way. Sparrow has two core techniques. First, each of the Sparrow distributed schedulers will randomly select two machines to check their loads. The distributed scheduler will assign a task to the low load machine out of these two machines. Second, instead of pushing the task to the selected machine, Sparrow adopts a late binding technique (i.e., the machine will pull tasks from the scheduler). This can avoid estimation error (i.e., straggler may happen on the machine). The techniques are simple, yet the biggest merit of Sparrow is also its simplicity so that it achieves good scalability.

Hawk [29] improved Sparrow as it treats big tasks (which are called elephant tasks in Hawk) and small tasks (which are called mouse tasks in Hawk) differently. Elephant tasks are assigned first and certain resources (machines) are set aside for mouse tasks.

## 2.7.2  Data-Computation Contention

MapReduce jobs are both data intensive and computation intensive. This can mean that executing I/O-intensive operations (e.g., outputting the data to local file systems or to the network) will affect the computation operations. Measurement studies have been conducted [30, 31]. It was reported that when executing a big data job in Amazon EC2, the completion times of those jobs running on two extra-large EC2 instances are constantly more than 10% faster than the completion times of jobs running on 16 small EC2 instances. Note that the total number of cores of the two extra-large EC2 instances and that of the 16 small EC2 instances is the same. When executing non-big data jobs, the job completion times are nearly the same.

Intrinsically, the data-computation contention means that some resources can be idle. As a matter of fact, when I/Os are busy, the CPUs need to wait for the data and cannot perform normal computation operations. When CPUs are busy in computation, the data in I/Os cannot be handled; thus, the I/Os are idle. As a result, the utilization of the resources is not efficient.

Many solutions have been proposed to solve such a problem [31, 32]. It has been observed that the loads of big data applications differ from those of web applications in the sense that they are not entirely dynamic. In addition, an analysis of big data (e.g., the PageRank algorithm) usually runs periodically [33]. In other words, the same job, with the same data size, the same functions to process the data, and so on, will run again and again. A proactive solution is thus considered more suitable than a reactive solution. In [31], network traffic loads of different MapReduce jobs were modeled. Network resource management algorithms were then designed according to these models. In [34], the task completion times and job completion times of different MapReduce jobs were modeled. A simple efficiency turning point (ETP) was introduced to capture the contention between CPUs and I/Os. System dynamic adjustment algorithms were based on ETP. It was shown that such a design made the system simple, distributed, and able to avoid overheads such as synchronization and others.

### 2.7.3    Optimizing Resource Utilization for Big Data Jobs

Along with the establishment of the MapReduce architecture, a large number of research studies have focused on how to best utilize resources to run big data jobs. Resource utilization optimization problems exist in conventional distributed and parallel computing architecture as well. In this section, we study particular problems pertaining to the MapReduce architecture. There can be two ways of looking at the problem of optimizing the utilization of resources for big data applications. First, we look at it from viewpoint of the system providers, where they want to optimize the resource that they utilize in support of their jobs. Second, we look at the problem from the viewpoint of the users. Such a problem commonly exists in the cloud environment, where users want to optimize their job completion times, with the trade-off being the price that they pay to the cloud providers.

#### Resource Utilization Optimization by System Providers

From the viewpoint of the system providers, they have a cluster of machines and they need to support a set of jobs efficiently. When the number of jobs increased and the types of the jobs became diverse (e.g., not only MapReduce jobs, but also Spark jobs, GraphLab jobs, Spark Streaming jobs, and even jobs that do not belong to big data applications) dedicated modules for resource management emerged.

Yarn was the first module to separate resource management and user job requirements. There are many new episodes on middleware after Yarn. Mesos [35] was designed to be a platform for fine-grained resource sharing. Omega [36] built a shared-state approach to provide better resource visibility for very "picky" jobs that are difficult to schedule.

There are studies on sharing clusters by both big data jobs and non-big data jobs, such as online interactive jobs. Production systems include Borg [37] from Google, Apollo [38] from Microsoft, and Bistro [39] from Facebook.

Fairness is one important criterion in optimizing resource utilization. Fair sharing on multiple resources was studied in [40]. Preemption is a commonly used technique to maintain fairness. There are studies on preemption for data-intensive jobs to ensure that no job is unfairly starving [41, 42].

### Resource Utilization Optimization by the Users

From the viewpoint of the users, as discussed, there are two specific problems: the resource provisioning problem and the VM-Job Scheduling problem.

There are studies on running MapReduce applications in the cloud and on the resource provisioning problems [43, 44]. Strategies that optimize the clusters according to different job types and workloads have been presented in [45]. In these studies, different VM types were regarded as containers with given computing capacities.

As we have discussed, in a public cloud, the labeled capacity may not be accurate for data-intensive applications as there can be contention between resources. In [30], the resource provisioning problem in cloud-based big data systems was re-modeled and an interference-aware solution that smartly allocates the MapReduce jobs to different VMs was proposed. The key to the modeling is to capture the interference into a parameter. This parameter is then formulated into the performance of the mapper task and reducer task. Trace-driven evaluations using Facebook data and four types of Amazon EC2 instances (small, media, large, and extra-large) demonstrated that the interference-aware solution can outperform a state-of-the-art cost-aware resource provisioning algorithm.

There are studies on moving big data into the cloud platform. For example, in [46], an online algorithm was developed to efficiently move data from geographically different locations to the cloud.

The VM-Job Scheduling problem is similar to the Job-Task Scheduling problem discussed in the previous section when the VMs and jobs are fixed. Note, however, that the public cloud has capacity far beyond the capacity requirement of one big data application. Together with the elastic nature of the public cloud, this offers opportunities to change the VM capacity during runtime [47] in order to accommodate different data intensities in different stages of the MapReduce programming model. In other words, rather than moving data, it is possible to change the VM capacities at different times, according to the amount of data to be processed in the local VMs. An initial investigation of such an opportunity was reported in [47]. For a certain MapReduce job, and a set of VMs, each of which has a set of CPUs, when a task is planned for execution, the number of CPUs in different VMs is adjusted according to the location of the data to be processed by this task.

Currently, runtime elastic VMs are not available in existing commercial services offered by cloud providers. Yet, in [34] a new I-PaaS platform was proposed. An initial implementation is shown in the construction of runtime elastic VMs by adjusting the number of CPUs in each VM during the runtime of a job. Using a Xen hypervisor, such an adjustment introduced an ignorable delay. With runtime elastic VMs, a 43% earlier finishing time has been observed, while the total number of CPU-hours remains the same. There can be future opportunities for both performance and pricing optimization in such an I-PaaS platform in supporting big data applications in the cloud.

## 2.8    Supports for Signal Processing Algorithms

Signal processing algorithms are iterative in nature and there are various dependencies between data. If a signal processing algorithm can be transformed into a graph-based structure, the Pregel or GraphLab programming model can be applied. There are other programming models proposed for specific signal processing algorithms. For example, MatrixMap [48] was developed for Matrix operations, where Matrix operations are the foundation for signal processing algorithms. Nevertheless, MatrixMap is research exploratory; thus, while it optimizes performance in Matrix operations, it lacks supporting tools, such as programming language, libraries, version updates, and so on. There is a trade-off between making a programming model easy to use and optimized for an application, and making the model general and expressive enough that it is worthwhile for industry to develop of an entire set of supports.

Currently, the most supported toolkit for signal processing algorithms is Machine Learning Library (MLlib) [49]. MLlib is based on the Spark programming model, where data sharing is more efficient as compared to the MapReduce programming model. More importantly, it currently provides a supporting library of a large number of algorithms [50], such as basic statistics, classification and regression, collaborative filtering, clustering, dimensionality reduction, feature extraction and transformation, and frequent pattern mining.

**An Example on the K-Means algorithm:** We now give an example of the K-Means algorithm implemented by MLlib. K-means is a clustering algorithm. Given a set of points in a space, and $K$, the K-means algorithm iteratively clusters the points into $K$ clusters in which each point belongs to the cluster with the nearest mean. From another point of view, the space will be partitioned into Voronoi cells (clusters). The detailed algorithm is in Algorithm 1.

---

**Algorithm 1** K-Means
1: Randomly select $K$ out of $N$ points. Set these points as cluster centers.
2: For each of the rest points $i$, compute its distance to the cluster centers. Clustering point $i$ to the cluster center with the smallest distance.
3: Recompute cluster centers.
4: Repeat steps 2 and 3 until the distance between the new clustering center and the previous clustering center is smaller than a predefined threshold.

---

The K-Means algorithm is supported by MLlib. We show how it can be used in the following Java program:

```java
1  import java.util.List;
2  import java.util.regex.Pattern;
3  import org.apache.spark.SparkConf;
4  import org.apache.spark.api.java.JavaRDD;
5  import org.apache.spark.api.java.JavaSparkContext;
6  import org.apache.spark.api.java.function.Function;
7  import org.apache.spark.mllib.clustering.KMeans;
8  import org.apache.spark.mllib.clustering.KMeansModel;
9  import org.apache.spark.mllib.linalg.Vector;
10 import org.apache.spark.mllib.linalg.Vectors;
11
12 public final class JavaKMeans {
13
14   public static void main(String[] args) {
15     if (args.length < 3) {
16       System.err.println(
17         "Usage: JavaKMeans <input_file> <k> <max_iterations> [<runs>]");
18       System.exit(1);
19 }
20
21 // Parses the arguments
22     String inputFile = args[0];
23     int k = Integer.parseInt(args[1]);
24     int iterations = Integer.parseInt(args[2]);
25     int runs = 1;
26     if (args.length >= 4) {
27       runs = Integer.parseInt(args[3]);
28     }
29     SparkConf sparkConf = new SparkConf().setAppName("JavaKMeans");
30     JavaSparkContext sc = new JavaSparkContext(sparkConf);
31
32     // Loads data
33     JavaRDD<String> lines = sc.textFile(inputFile);
34     JavaRDD<Vector> points = lines.map(new ParsePoint());
35
36     // Trains the k-means model
37     final KMeansModel model = KMeans.train(points.rdd(), k, iterations, runs,
           KMeans.K_MEANS_PARALLEL());
38
39     //Prints each point belongs to which cluster
40     List<String> mapresult = points.map(new Function<Vector, String>() {
41                 @Override
42                 public String call(Vector v) throws Exception {
43                         // TODO Auto-generated method stub
44                         return v.toString()
45                                 + " belong to cluster :" +
                                    model.predict(v);
46                 }
47         }).collect();
48     for(String item:mapresult)
49     {
50         System.out.println(item);
51     }
52
53     //Prints the cluster centers
54     System.out.println("Cluster centers:");
55     for (Vector center : model.clusterCenters()) {
56       System.out.println(" " + center);
57     }
58
59     //Prints the cost
60     double cost = model.computeCost(points.rdd());
61     System.out.println("Within Set Sum of Squared Errors = " + cost);
62
63     sc.stop();
64   }
```

```
65
66    //Parses the point
67    private static class ParsePoint implements Function<String, Vector> {
68            private static final Pattern SPACE = Pattern.compile(" ");
69
70        public Vector call(String line) {
71            String[] tok = SPACE.split(line);
72            double[] point = new double[tok.length];
73            for (int i = 0; i < tok.length; ++i) {
74                point[i] = Double.parseDouble(tok[i]);
75            }
76            return Vectors.dense(point);
77        }
78    }
79 }
```

Since the K-Means is supported by MLlib, we can directly call the KMeans package, see line 37.

We look into the details of how MLlib supports the K-Means algorithm in a data parallelism style. The implementation of the K-Means algorithm in MLlib is shown as follows, which is written in the Scala programming language. Line 5 shows the data partitioning by a map function and line 29 shows data aggregation by a reduce function:

```
1  private def runAlgorithm(data: RDD[VectorWithNorm]): KMeansModel = {
2  ......
3
4  // Find the sum and count of points mapping to each center
5          val totalContribs = data.mapPartitions { points =>
6          val thisActiveCenters = bcActiveCenters.value
7          val runs = thisActiveCenters.length
8          val k = thisActiveCenters(0).length
9          val dims = thisActiveCenters(0)(0).vector.size
10
11          val sums = Array.fill(runs, k)(Vectors.zeros(dims))
12          val counts = Array.fill(runs, k)(0L)
13
14          points.foreach { point =>
15            (0 until runs).foreach { i =>
16              val (bestCenter, cost) =
17                  KMeans.findClosest(thisActiveCenters(i), point)
18              costAccums(i) += cost
19              val sum = sums(i)(bestCenter)
20              axpy(1.0, point.vector, sum)
21              counts(i)(bestCenter) += 1
22            }
23          }
24
25          val contribs = for (i <- 0 until runs; j <- 0 until k) yield {
26            ((i, j), (sums(i)(j), counts(i)(j)))
27          }
28          contribs.iterator
29        }.reduceByKey(mergeContribs).collectAsMap()
30
31  ......
32
33 }
```

## 2.9     Summary

In this chapter, we presented the history of development of the MapReduce architecture. Intrinsically, MapReduce is a data parallelism architecture that deviates from relational database architecture to organize massive data. The MapReduce architecture is suitable for big data applications where the data is big and the output results need to be obtained by the processing of all or a significant part of the data set.

We presented the MapReduce architecture in detail. At the very bottom is a new file organization system, the HDFS. On top of the HDFS, Yarn was developed to separate the processes of resource management and application requirements. On top of Yarn are the programming models, the most fundamental of which is the MapReduce programming model. The MapReduce model consists of mapper tasks and reducer tasks. The data are divided into chunks and processed by the tasks in parallel. We then presented the Spark programming model. It has much better performance since it is memory-based. Its data-sharing design makes it easy to get extended into various application-specific models, and many extensions based on the Spark programming model are available. We then presented other programming models such as GraphLab and Spark Streaming, each of which is an abstraction that puts emphasis on a set of big data applications. GraphLab is designed for graph-based iterative applications, and Spark Streaming is designed for real-time live streaming applications.

We then presented big data applications in the cloud. We have seen that more and more big applications are from grassroots users or non-IT professionals, who cannot afford large-scale dedicated server clusters. The cloud environment thus becomes a viable, if not the only, solution.

There are a lot of advances in data parallelism architecture. We classified the advances in two directions: fast job completion time and resource utilization optimization. For job completion time, we discussed three problems: (1) the data locality problem, where data should be processed in or close to the local machine, and moving data from one machine to another should be avoided as much as possible. This reflects the essential idea in the design of processing big data jobs (i.e., moving data is considered high cost); (2) the Job-Task Scheduling problem, where tasks are assigned to the appropriate machines for processing; and (3) the data-computation contention problem. As for resource utilization optimization, we discussed this from the viewpoint of system providers and from the viewpoint of users.

Finally, we presented the supports for signal processing algorithms. Signal processing algorithms are iterative in nature and there are various dependencies between data. There are many projects focusing on various data-dependent structures, such as Graph processing and Matrix processing. Currently, the most supported toolkit for signal processing algorithms is MLlib [49]. We present a concrete example of how a typical signal processing algorithm, the K-means, can be implemented by MLlib.

# Part II

## Methodology and Mathematical Background

# 3 First-Order Methods

In this chapter, we introduce a few popular first-order algorithms for solving modern big data problems. These algorithms are particularly suitable for big data optimization mainly because of the following three features:

1. *Decomposability*: The algorithms are able to decompose a huge problem, having millions or billions of variables and/or data points, to small-sized problems. Such decomposition can be done either across different variables, or across different data points. Each type of decomposition gives rise to small-dimensional subproblems that have simple (often closed-form) solutions.
2. *Cheap Update*: Most of the algorithms only use first-order gradient information to perform the computation (hence the name of this family of the algorithms), which is far cheaper to obtain compared with the second-order information such as Hessian matrices.
3. *Parallel Implementability*: Most of the algorithms are amendable for parallel implementation on modern multi-core clusters, making them scale well with the increased problem size.

This chapter is organized as follows. To begin with, we first introduce some fundamental notions of continuous optimization, which serve as the basis for the subsequent discussion. We then present a variety of first-order methods including the classical (projected) gradient descent (GD) methods, the coordinate descent methods and the augmented Lagrangian-based methods. For each family of methods, we will introduce its most basic version, analyze its theoretical performance, and discuss its advanced variants. Note that the main purpose here is not to survey all first-order methods (which would be impossible), but to introduce to the reader some necessary tools and some basic algorithms that will be repeatedly used in later chapters.

## 3.1 A Brief Introduction to Optimization

### 3.1.1 Definition and Notations

Let us consider the following optimization problem:

$$
\begin{aligned}
\min_{\mathbf{x}} \quad & f_0(\mathbf{x}) \\
\text{s.t.} \quad & f_i(\mathbf{x}) \le 0, \ i = 1, \ldots, m, \\
& h_j(\mathbf{x}) = 0, \ j = 1, \ldots, p.
\end{aligned}
\tag{3.1}
$$

Here $\mathbf{x} = (x_1, \ldots, x_n)^T \in \mathbb{R}^n$ are the optimization variables; the continuous function $f_0 : \mathbb{R}^n \rightarrow \mathbb{R}$ is the objective function; the continuous functions $f_i : \mathbb{R}^n \rightarrow \mathbb{R}$, and $h_j : \mathbb{R}^n \rightarrow \mathbb{R}$ are the functions defining the inequality and the equality constraints, respectively. The *feasible set* of problem (3.1) is given by

$$\mathcal{X} := \left\{ x \mid f_i(\mathbf{x}) \leq 0, \ i = 1, \ldots, m, \quad h_j(\mathbf{x}) = 0, \ j = 1, \ldots, p \right\}.$$

Using this notation, problem (3.1) can be equivalently written as

$$\min_{\mathbf{x}} \quad f_0(\mathbf{x}), \quad \text{s.t. } \mathbf{x} \in \mathcal{X}. \tag{3.2}$$

A feasible solution $\mathbf{x}^* \in \mathcal{X}$ that achieves the minimum objective value is called a *globally optimal solution*. That is

$$f_0(\mathbf{x}^*) \leq f_0(\mathbf{x}), \quad \forall x \in \mathcal{X}.$$

If a solution $\hat{\mathbf{x}} \in \mathcal{X}$ has the minimal objective value among a neighborhood around $\hat{\mathbf{x}}$, that is,

$$f_0(\hat{\mathbf{x}}) \leq f_0(\mathbf{x}), \quad \forall \mathbf{x} \in \{\|\mathbf{x} - \hat{\mathbf{x}}\| \leq \delta\} \cap \mathcal{X}, \quad \text{for some } \delta > 0,$$

then $\hat{\mathbf{x}}$ is called a *locally optimal solution*.

A set $\mathcal{X}$ is called a *convex set* if the line segment between any two points in $\mathcal{X}$ lies within $\mathcal{X}$ (i.e., for any $\mathbf{x}_1, \mathbf{x}_2 \in \mathcal{X}$ and $\alpha \in [0, 1]$) we must have $\alpha\mathbf{x}_1 + (1 - \alpha)\mathbf{x}_2 \in \mathcal{X}$. A function $f : \mathbb{R}^n \rightarrow \mathbb{R}$ is called a *convex function* if dom $f$ is a convex set, and for any $\mathbf{x}_1, \mathbf{x}_2 \in \text{dom} f$ and for all $\alpha \in [0, 1]$, we have

$$f(\alpha\mathbf{x}_1 + (1 - \alpha)\mathbf{x}_2) \leq \alpha f(\mathbf{x}_1) + (1 - \alpha)f(\mathbf{x}_2).$$

Note that a function that does not satisfy the above definition is often referred to as a *nonconvex function*. Also note that a convex function $f(\mathbf{x})$ is not necessarily continuous, but it must be continuous over any *open set* $(\mathbf{a}_1, \mathbf{a}_2) \in \text{dom} f$.

A function is called a *strongly convex function* with modulus $\sigma > 0$ if dom $f$ is a convex set, and for any $\mathbf{x}_1, \mathbf{x}_2 \in \text{dom} f$ and for all $\alpha \in [0, 1]$, we have

$$f(\alpha\mathbf{x}_1 + (1 - \alpha)\mathbf{x}_2) + \frac{\sigma\alpha(1 - \alpha)}{2}\|\mathbf{x}_1 - \mathbf{x}_2\|^2 \leq \alpha f(\mathbf{x}_1) + (1 - \alpha)f(\mathbf{x}_2).$$

The optimization problem (3.1) is called a *convex problem* if $f_0$ is continuous, and that $f_0, \ldots, f_m$ are convex and $h_1, \ldots, h_p$ are affine functions (3.1) (or equivalently the feasible set $\mathcal{X}$ is convex). For convex problems, all locally optimal solutions are globally optimal. It is possible to have multiple global optimal solutions, but the convex combination of any two of them must also be a globally optimal solution. That is, the optimal solution set, denoted as $\mathcal{X}^*$, must be a convex set. If $f_0$ is further assumed to be strongly convex, then there is at most a single globally optimal solution.

Next we discuss explicit characterization of optimality conditions for problem (3.1). We will start with the simplest *unconstrained problem* where $f_i \equiv 0$, $i = 1, \cdots, m$, and $h_j \equiv 0$, $j = 1, \cdots p$, which is then followed by the discussion of problems with constrains. The optimality condition developed here will be useful in developing various algorithms in the subsequent sections. Before we delve into the detailed discussion, we mention that every continuous function $f$ attains its infimum over a compact feasible set $\mathcal{X}$. This result, known as the Weierstrass theorem, ensures the existence of the globally optimal solution.

## 3.1.2 Unconstrained Optimization

Let us first assume that $f_0$ is a twice continuously differentiable function. Then the following condition is necessary for a feasible $\hat{\mathbf{x}} \in \mathcal{X}$ to be a locally optimal solution:

$$\nabla f_0(\hat{\mathbf{x}}) = 0, \quad \nabla^2 f_0(\hat{\mathbf{x}}) \succeq 0. \tag{3.3}$$

On the other hand, the sufficient condition for local optimality, expressed below, is slightly more restrictive:

$$\nabla f_0(\hat{\mathbf{x}}) = 0, \quad \nabla^2 f_0(\hat{\mathbf{x}}) \succ 0. \tag{3.4}$$

The proof of the sufficiency is a simple application of the Mean Value Theorem. Interestingly, for a class of *quadratic* optimization problems where $f_0(\mathbf{x}) = \mathbf{x}^T \mathbf{A} \mathbf{x}$ with $\mathbf{A} \in \mathbb{R}^{n \times n}$ being a symmetric matrix, the necessary condition is also sufficient. Further, each locally optimal solution is a globally optimal solution; see [51, Section 1.1.1] for detailed argument. If one further assumes that problem (3.1) is a convex problem, then $\mathbf{x}^*$ is globally optimal if and only if $\nabla f_0(\mathbf{x}^*) = 0$.

Next let us consider problems with $f_0$ being convex but not necessarily differentiable. In this case the optimality condition is characterized using the so-called *subdifferential*, a set denoted as $\partial f(\mathbf{x})$:

$$\partial f(\mathbf{x}) = \{\mathbf{y} : f(\mathbf{x}') \geq f(\mathbf{x}) + \langle \mathbf{y}, \mathbf{x}' - \mathbf{x} \rangle, \ \forall \, \mathbf{x}' \in \mathrm{dom}\, f\}. \tag{3.5}$$

If $f$ is differentiable at $\mathbf{x}$, then $\partial f(\mathbf{x}) = \{\nabla f(\mathbf{x})\}$. Any $\mathbf{y} \in \partial f(\mathbf{x})$ is called a *subgradient*. See Figure 3.1 for an illustration. Utilizing the definition of subdifferential, we can define the optimal solution for minimizing a convex but non-differentiable function $f_0$ as any $\mathbf{x}^* \in \mathcal{X}$ that satisfies

$$\mathbf{0} \in \partial f_0(\mathbf{x}^*).$$

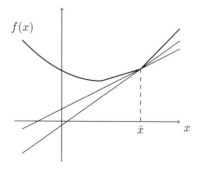

**Figure 3.1** Two subgradients of a nonsmooth function at $\bar{x}$.

## 3.1.3    Constrained Optimization

When constraints are present, the solutions that satisfy the necessary or sufficient conditions given in the previous subsection may no longer be feasible. To deal with this type of problem, a different set of optimality conditions are needed. Assume that the feasible set $\mathcal{X}$ in problem (3.2) is convex, and $f_0$ is differentiable. Then if $\hat{\mathbf{x}}$ is a local minimum, the following condition must be satisfied (first-order necessary condition):

$$\langle \nabla f_0(\hat{\mathbf{x}}), \mathbf{x} - \hat{\mathbf{x}} \rangle \geq 0, \ \forall \, \mathbf{x} \in \mathcal{X}. \tag{3.6}$$

Further, if $f_0$ is convex, then this condition is sufficient and necessary for $\hat{\mathbf{x}}$ to be a globally optimal solution. We note that condition (3.6) is equivalent to the following fixed-point equation:

$$\hat{\mathbf{x}} = \text{proj}_{\mathcal{X}} \left[ \hat{\mathbf{x}} - \alpha \nabla f(\hat{\mathbf{x}}) \right], \tag{3.7}$$

where $\alpha > 0$ is any positive constant; the *projection operator* $\text{proj}_X : \mathbb{R}^n \to \mathbb{R}^n$ is defined below:

$$\text{proj}_{\mathcal{X}}[\mathbf{z}] := \arg \min_{\mathbf{x}} \|\mathbf{z} - \mathbf{x}\|^2, \quad \text{s.t. } \mathbf{x} \in \mathcal{X}. \tag{3.8}$$

The equivalence of (3.7) and (3.6) can be verified by simply checking the optimality condition of the convex problem (3.8). Later we will see that condition (3.7) will be useful in designing first-order algorithms.

Similarly, if $f_0$ is nonsmooth and is given by

$$f_0(\mathbf{x}) := \ell(\mathbf{x}) + h(\mathbf{x})$$

with $\ell(\mathbf{x})$ being continuously differentiable and $h(\mathbf{x})$ being convex but possibly nonsmooth, then the condition (3.6) can be extended to

$$\langle \nabla \ell(\hat{\mathbf{x}}) + \xi, \mathbf{x} - \hat{\mathbf{x}} \rangle \geq 0, \quad \text{for some } \xi \in \partial h(\hat{\mathbf{x}}), \ \forall \, \mathbf{x} \in \mathcal{X}. \tag{3.9}$$

This condition is equivalent to

$$\hat{\mathbf{x}} = \text{prox}_h^\alpha \left[ \hat{\mathbf{x}} - \alpha \nabla \ell(\hat{\mathbf{x}}) \right], \tag{3.10}$$

where $\alpha > 0$ is any positive constant; the *proximity operator* $\text{prox}_h^\alpha : \mathbb{R}^n \to \mathbb{R}^n$ is defined below:

$$\text{prox}_h^\alpha[z] := \arg \min \ h(x) + \frac{\alpha}{2} \|z - x\|^2, \quad \text{s.t. } x \in \mathcal{X}. \tag{3.11}$$

An alternative way to characterize the (local) optimality of (3.1) is to use the so-called *Lagrangian function* $L : \mathbb{R}^n \times \mathbb{R}^m \times \mathbb{R}^p \to \mathbb{R}$, defined below:

$$L(\mathbf{x}, \lambda, \nu) := f_0(\mathbf{x}) + \sum_{i=1}^m \lambda_i f_i(\mathbf{x}) + \sum_{j=1}^p \nu_j h_j(\mathbf{x}), \tag{3.12}$$

where $\lambda := \{\lambda_i\}_{i=1}^m$, $\nu := \{\nu_j\}_{j=1}^p$; $\lambda_i \geq 0$ and $\nu_j$ are the Lagrange multipliers associated with the $i$th inequality constraint $f_i(\mathbf{x}) \leq 0$ and the $j$th equality constraint $h_j(\mathbf{x}) = 0$, respectively. The variables $\mathbf{x}$ and $(\lambda, \nu)$ in (3.12) are also known as the *primal variable* and the *dual variables*, respectively.

The *Lagrange dual function* $g(\lambda, \nu) : \mathbb{R}^m \times \mathbb{R}^p \to \mathbb{R}$ is defined as

$$g(\lambda, \nu) = \inf_{\mathbf{x}} L(\mathbf{x}, \lambda, \nu). \qquad (3.13)$$

The *dual problem* is given by

$$\max_{\lambda, \nu} \ g(\lambda, \nu) \quad \text{s.t. } \lambda_i \geq 0, \ i = 1, \ldots, m. \qquad (3.14)$$

It is easy to verify that the function $g(\lambda, \nu)$ is concave w.r.t. $\lambda$ and $\nu$, as it is the infimum of a collection of linear functions (cf. (3.13)). Therefore the dual problem is always a convex problem (maximizing a concave objective is equivalent to minimizing a convex one).

Regardless of the convexity of problem (3.1), the following *weak duality* always holds:

$$g(\lambda, \nu) \leq f_0(\mathbf{x}), \ \forall \, \mathbf{x} \in \mathcal{X}, \quad \lambda \geq \mathbf{0}.$$

If for all $\mathbf{x} \in \mathcal{X}$ and all $\lambda \geq \mathbf{0}$, the equality is always achieved, and then we say the *strong duality* holds. For convex problems, such strong duality holds, for example, if the problem is a linear program, or certain constraint qualifications such as the Slater's condition holds true, which requires the existence of $\mathbf{x} \in \mathcal{X}$ satisfying $f_i(\mathbf{x}) < 0, \ i = 1, \ldots, m$.

Utilizing the preceding definitions, one can derive an alternative first-order necessary optimality condition using both the primal and dual variables, and such condition is known as the Karush–Kuhn–Tucker (KKT) condition. Let $\mathbf{x}^*$ be any locally optimal solution of problems (3.1). Assume that $f_i$'s $h_j$'s are continuously differentiable at $\mathbf{x}^*$, and assume that certain constraint qualification is satisfied. Then there must exist a solution $(\lambda^*, \nu^*)$ feasible to (3.14) such that the following system of equations are satisfied:

$$\nabla f_0(\mathbf{x}^*) + \sum_{i=1}^{m} \lambda_i^* \nabla f_i(\mathbf{x}^*) + \sum_{j=1}^{p} \nu_j^* \nabla h_j(\mathbf{x}^*) = 0, \qquad (3.15)$$

$$f_i(\mathbf{x}^*) \leq 0, \ i = 1, \ldots, m, \qquad (3.16)$$

$$h_j(\mathbf{x}^*) = 0, \ j = 1, \ldots, p, \qquad (3.17)$$

$$\lambda_i^* \geq 0, \ i = 1, \ldots, m, \qquad (3.18)$$

$$\lambda_i^* f_i(\mathbf{x}^*) = 0, \ i = 1, \ldots, m. \qquad (3.19)$$

The last condition is called the *complementarity condition*, which says that if $f_i(\mathbf{x}^*)$ is not active (i.e., $f_i(\mathbf{x}^*) < 0$), then the corresponding $\lambda_i^*$ must be zero. It is important to note that when problem (3.1) is convex, $f_i$'s are continuous differentiable, and $h_j$'s are all affine, then the KKT condition is also sufficient for global optimality.

## 3.2 Gradient Descent Method

First let us consider problem (3.2) with a differentiable objective function and with $\mathcal{X} = \mathbb{R}^n$ (i.e., the constrained problem). Suppose that for any given point $\mathbf{x} \in \mathcal{X}$, only

**Table 3.1** Common stepsize selection rules for GD

- *Constant Rule*: Let $\alpha^t = \alpha$ for all $t$
- *Limited Minimization Rule*: Pick $\alpha^t$ according to

$$\alpha^t = \arg \min_{\alpha \in [0,\, 1]} f_0(\mathbf{x}^t + \alpha \mathbf{d}^t)$$

- *Diminishing Rule*: Pick $\alpha^t$ satisfying

$$\sum_{t=1}^{\infty} \alpha^t = \infty, \quad \alpha^t \to 0. \qquad (3.22)$$

- *Amijo Rule*: Let $\sigma \in (0, \frac{1}{2})$. Fix a constant $s$ and $\beta \in (0, 1)$. Keep shrinking $\alpha$ by $s$, $s\beta$, $s\beta^2$ until the following is satisfied

$$f_0(\mathbf{x}^t + \alpha \mathbf{d}^r) - f_0(\mathbf{x}^t) \le \sigma \alpha \langle \nabla f_0(\mathbf{x}^t), \mathbf{d}^t \rangle.$$

the gradient information $\nabla f_0(\mathbf{x})$ is available, and we are interested in finding a *stationary solution* $\hat{\mathbf{x}} \in \mathcal{X}$ that satisfies the first-order necessary optimality condition $\nabla f_0(\hat{\mathbf{x}}) = 0$.

The well-known GD method generates a sequence $\{\mathbf{x}^t\}$ by the following iteration:

$$\mathbf{x}^{t+1} = \mathbf{x}^t + \alpha^t \mathbf{d}^t \qquad (3.20)$$

where $\alpha^t > 0$ is some properly selected stepsize; $\mathbf{d}^t$ is some direction. To ensure the convergence of the GD algorithm, one needs to be careful about picking the direction as well as the stepsize. A good choice of the direction $\mathbf{d}^t$ should satisfy the following condition:

$$\langle \mathbf{d}^t, \nabla f_0(\mathbf{x}^t) \rangle < 0, \quad \forall\, t \text{ such that } \mathbf{x}^t \text{ is not stationary.} \qquad (3.21)$$

If we pick $\mathbf{d}^t = -\nabla f_0(\mathbf{x}^t)$, then the algorithm is known as the steepest descent method. In Table 3.1, a few commonly used stepsize selection rules are listed.

To characterize the convergence behavior of GD methods with different direction and/or stepsize rule, we need the following Lipschitzian assumption on the gradient of the objective:

$$\|\nabla f_0(\mathbf{x}) - \nabla f_0(\mathbf{z})\| \le L \|\mathbf{x} - \mathbf{z}\|, \quad \forall\, \mathbf{x}, \mathbf{z} \in \mathcal{X}. \qquad (3.23)$$

Here the constant $L > 0$ is essentially the upper bound on the size of $\nabla f_0(\mathbf{x})$ over $\mathcal{X}$. Further assume that (3.21) is true, then every limit point generated by the GD method is a stationary solution if one of the following conditions holds [51, Proposition 1.2.1–1.2.4]:

1. There exits a scalar $\epsilon > 0$ such that for all $t$

$$0 < \epsilon < \alpha^t \le -\frac{(2 - \epsilon)\langle \nabla f_0(\mathbf{x}^t), \mathbf{d}^t \rangle}{L \|\mathbf{d}^t\|^2}.$$

2. The limited minimization rule is used.

3. The diminishing rule (3.22) is used, and there exist positive scalars $c_1$ and $c_2$ such that for all $t$ the following are true:

$$c_1\|\nabla f_0(\mathbf{x}^t)\| \leq -\langle\nabla f_0(\mathbf{x}^t), \mathbf{d}^t\rangle, \quad \|\mathbf{d}^t\|^2 \leq c_2\|\nabla f_0(\mathbf{x}^t)\|^2$$

4. The Armijo rule is used.

When $\mathcal{X}$ is not the whole space $\mathbb{R}^n$, the GD method no longer works because it cannot ensure the feasibility of the iterates. Below we introduce the *gradient projection* (GP) method, which slightly generalizes the GD method. Recall that when there are constraints, the optimality condition is given by (3.7). The GP method, described below, generates a sequence that attempts to satisfy such optimality condition:

$$\mathbf{x}^{t+1} = \mathbf{x}^t + \alpha^t \underbrace{(\bar{\mathbf{x}}^{t+1} - \mathbf{x}^t)}_{:=\mathbf{d}^t} \tag{3.24}$$

$$\bar{\mathbf{x}}^{t+1} = \text{proj}_X\left[\mathbf{x}^t - s^t \nabla f_0(\mathbf{x}^t)\right] \tag{3.25}$$

where $\alpha^t \in (0, 1]$ and $s^t > 0$ are stepsizes. The GP method is closely related to the GD method: when viewing $\bar{\mathbf{x}}^{t+1} - \mathbf{x}^t := \mathbf{d}^t$ as a direction, the iterate (3.24) is exactly the GD iteration (3.20), just that the particular combination of direction and stepsize ensures the feasibility (as will be shown shortly). Alternatively, if we pick $\alpha^t = 1$, then $\mathbf{x}^{t+1}$ is generated according to (3.25), which is just the GD iteration with an additional projection step.

To analyze the GP method, first it is easy to see that $\mathbf{x}^{t+1}$ is always feasible, as it is a convex combination of two feasible points $\mathbf{x}^t$ and $\bar{\mathbf{x}}^{t+1}$. Second, observe that the direction $\mathbf{d}^t = \bar{\mathbf{x}}^{t+1} - \mathbf{x}^t$ satisfies the condition (3.21). This is because by the optimality condition of (3.25), we have

$$\langle\mathbf{x}^t - s^t\nabla f_0(\mathbf{x}^t) - \bar{\mathbf{x}}^{t+1}, \mathbf{x} - \bar{\mathbf{x}}^{t+1}\rangle \leq 0, \forall x \in \mathcal{X}. \tag{3.26}$$

Let $\mathbf{x} = \mathbf{x}^t$, one obtains

$$\langle\nabla f_0(\mathbf{x}^t), \bar{\mathbf{x}}^{t+1} - \mathbf{x}^t\rangle \leq -\frac{1}{s^t}\|\mathbf{x}^t - \bar{\mathbf{x}}^{t+1}\|^2. \tag{3.27}$$

The GP method converges to the stationary solution (i.e., those satisfying (3.7)) under the following stepsize selection rules [51, Section 2.3.2]:

1. $s^t = s$, and $\alpha^t$ is chosen either by the Armijo rule or the limited minimization rule.
2. $\alpha^t = 1$, and $s^t$ is chosen either by the constant rule, the diminishing rule or by the Armijo rule. If the constant rule is chosen, then one should pick $s^t \in (0, \frac{2}{L})$.

Now that we have briefly reviewed the convergence of both GD and GP, a closely related question is: Can we say more about the behavior of these methods? After all, achieving asymptotic convergence is the most basic requirement for any sensible algorithm. Below we briefly discuss the *convergence rate* for these methods. For simplicity of presentation, we will focus on the GD method, but the reader should note that the GP method would have similar behavior.

The rate of convergence analysis addresses the following question: How many iterations are sufficient to achieve a good enough solution? To make the statement precise, let us define an $\epsilon$-optimal solution as a feasible $\mathbf{x}^\epsilon \in \mathcal{X}$ that satisfies

$$f_0(\mathbf{x}^\epsilon) - f_0(\mathbf{x}^*) \le \epsilon, \quad \text{where } x^* \in \mathcal{X}^*. \tag{3.28}$$

Define the constant $R$ as the distance between the initial solution $\mathbf{x}^0$ to the optimal solution set, that is,

$$R := \min_{\mathbf{x}^* \in \mathcal{X}^*} \|\mathbf{x}^0 - \mathbf{x}^*\|.$$

For general convex problem in the form of (3.2), we can show that the GD with constant stepsize $\alpha^t = \frac{1}{L}$ achieves a *sublinear* convergence rate [52, Corollary 2.1.2]

$$\Delta^t := f_0(\mathbf{x}^t) - f_0(\mathbf{x}^*) \le \frac{2LR^2}{t+4}. \tag{3.29}$$

This means that to achieve an $\epsilon$-optimal solution requires about $\frac{2LR^2}{\epsilon}$ number of iterations. If we further assume that $f_0$ is strongly convex with modulus $\sigma$, then the GD with constant stepsize achieves a *linear* convergence

$$\Delta^t := f_0(\mathbf{x}^t) - f_0(\mathbf{x}^*) \le \frac{LR^2}{2} \left( \frac{\kappa - 1}{\kappa + 1} \right)^{2t} \tag{3.30}$$

where $\kappa$ is the condition number defined as $\kappa := L/\sigma$. Clearly in this case to achieve an $\epsilon$-optimal solution requires about $\kappa \log\left(\frac{c}{\epsilon}\right)$ number of iterations, where $c > 0$ is some constant.

We also mention here that GD is capable of achieving linear convergence *without* assuming that the objective is strongly convex; see, for example [53, 54]. More precisely, if the objective can be written as the following *composite form*:

$$f_0(x) := g_0(Ax),$$

where $A \in \mathbb{R}^{\ell \times n}$ is not necessarily a full rank matrix, and $g_0(\cdot)$ is a strongly convex function. Clearly $f_0(x)$ is not necessarily a convex function, but somewhat surprisingly, by utilizing certain error-bound condition the linear convergence of GD can be shown; see [53].

The GD-based algorithms play an important role in modern big data optimization. These types of method only require first-order information about the function, and they enjoy nice convergence properties. Further, it is closely related to many algorithms that we are going to introduce later, such as coordinated descent methods and prox-linear methods.

A fundamental question at this point is, is it possible to improve the convergence rate of the GD, but still only use the gradient information? The answer is affirmative, and it is indeed possible to improve the rate of GD (for general convex problem) from $\mathcal{O}(1/\epsilon)$ to $\mathcal{O}(1/\sqrt{\epsilon})$. Below we introduce a simple version of the Nesterov's method for solving problem (3.1) without constraints.

**Table 3.2** The Accelerated Gradient Method

- *Initialization*: Let $\mathbf{x}^0 = \mathbf{x}^{-1}$
- *At iteration t, do:*

$$\mathbf{y}^t = \mathbf{x}^{t-1} + \frac{t-2}{t+1}(\mathbf{x}^{t-1} - \mathbf{x}^{t-2})$$

$$\mathbf{x}^t = \mathbf{y}^{t-1} - \alpha \nabla f_0(\mathbf{y}^t).$$

It can be shown that if $\alpha \leq \frac{1}{L}$, then the algorithm converges by the following rate:

$$\Delta^t := f_0(\mathbf{x}^t) - f_0(\mathbf{x}^*) \leq \frac{2R^2}{\alpha(t+1)^2}. \tag{3.31}$$

Also such rate is optimal, in the sense that there exists a problem such that if only the gradient information is used, then the best rate that can be achieved by a certain family of algorithms that only use gradient information is $\mathcal{O}(1/\sqrt{\epsilon})$; for the precise statement and the construction of the example, please see the textbook by Nesterov [52].

## 3.3 Block Coordinate Descent Method

In big data optimization problems, the entire dimension of the optimization variable can be huge. The block coordinate descent (BCD) method is very popular for solving this type of problem, as it represents an effective divide and conquer strategy in which each time only a small subset of variables are optimized, while fixing the rest of the variables.

More specifically, consider the following problem where the decision variable $\mathbf{x}$ is naturally decomposed into $K$ non-overlapping blocks $\mathbf{x}_k \in \mathbb{R}^{n_k}$, $k = 1, \cdots, K$:

$$\begin{aligned} \min_{\mathbf{x}} \quad & f_0(\mathbf{x}) \equiv f_0(\mathbf{x}_1, \mathbf{x}_2, \ldots, \mathbf{x}_K), \\ \text{s.t.} \quad & \mathbf{x}_k \in \mathcal{X}_k, \ \forall \, k. \end{aligned} \tag{3.32}$$

In each step the BCD updates only one block of variable while fixing the rest of the blocks. Here the blocks can be updated either by performing an exact minimization, that is,

$$\mathbf{x}_k^{t+1} \in \arg \min_{\mathbf{x}_k \in \mathcal{X}_k} f_0(\mathbf{x}_k, \mathbf{x}_{-k}^t) \tag{3.33}$$

or by performing a GP step

$$\mathbf{x}_k^{t+1} = \text{proj}_{\mathcal{X}_k} \left[ \mathbf{x}_k^t - s^t \nabla f_0(\mathbf{x}^t) \right]. \tag{3.34}$$

In the above two equations, we have defined

$$\mathbf{x}_{-k}^t := [\mathbf{x}_1^t, \cdots, \mathbf{x}_{k-1}^t, \mathbf{x}_{k+1}^t, \cdots, \mathbf{x}_K^t]^T, \ \forall \, k.$$

For simplicity of presentation, in the rest of this section we will mainly consider the exact minimization version of the BCD (3.33). But most of the conclusions hold for the gradient-based BCD as well.

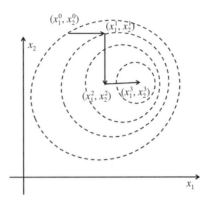

**Figure 3.2** Illustration of BCD method with $K = 2$.

There are various different rules for block selection (see Table 3.3 for a summary). In most schemes listed in this table, a single block is updated at each iteration, therefore the objective function is guaranteed to be at least monotonically non-increasing. The only exception is the Jacobian scheme, where the entire set of variables are updated. A naive implementation of such block update rule just updates $\mathbf{x}^{t+1}$ by $\mathbf{x}^{t+1} = \hat{\mathbf{x}}^{t+1}$, but such scheme does not guarantee convergence, as the blocks are not optimized consistently. However it can be shown that $\hat{\mathbf{x}}^{t+1} - \mathbf{x}^t$ indeed is a descent direction for $f_0(\mathbf{x}^t)$ (i.e., $\langle \hat{\mathbf{x}}^{t+1} - \mathbf{x}^t, \nabla f_0(\mathbf{x}^t) \rangle \le 0$), therefore a common strategy is to use a sequence of stepsizes and perform the following iteration [56]:

$$\mathbf{x}^{t+1} = \mathbf{x}^t + \alpha^t(\hat{\mathbf{x}}^{t+1} - \mathbf{x}^t), \tag{3.36}$$

where the stepsizes satisfy

$$\lim_t \alpha^t = 0, \quad \lim_t \sum_t \alpha^t = \infty. \tag{3.37}$$

Although the stepsizes need to be tuned carefully, it has been shown that this type of Jacobian scheme is suitable for parallel and distributed implementation, therefore it has become quite popular for solving big data problems; see [57, 58, 59, 60] for detailed discussions and applications. We also note that for the MBI and G-So rule given in Table 3.3, $K$ computational steps need to be performed before *a single* coordinate is finally selected and updated. This can cause underutilization of computational resources, unless $K$ computational steps can be carried out in parallel.

The main strength of the BCD method is its simplicity: updating the block variables one at a time results in simple subproblems with significantly reduced dimensions; see (3.33) and (3.34). This is precisely the reason that for a long time BCD has been popular for solving a wide variety of convex and nonconvex problems. The BCD methods can be traced back to algorithms developed in [61] for solving systems of equations, and to [62, 63, 64, 65] which analyze smooth and convex optimization problems. When the objective of the problem is nonconvex, Powell has provided a series of examples showing that BCD method can get stuck at non-interesting points [66]. If the objective function is not differentiable, BCD can be again trapped at non-stationary points; see, for example, [64] and [67]. Therefore the key theoretical question is whether the BCD even converges, and when it does, what is the convergence rate. Next we analyze various theoretical issues of the BCD methods. We will focus our attention on problems with continuously differentiable $f_0$. We will defer the discussion of nonsmooth problems to Chapter 4.

First of all, it is known that the BCD converges when the objective is convex and the feasible sets $\mathcal{X}_k$'s are all convex sets (regardless of the uniqueness of the subproblem (3.33)). However, when the feasible set becomes *coupled*, the BCD does not work, regardless of the convexity of the feasible set or the objective function. For example if applying the BCD to the following problem with $\mathbf{x}^0 = (-1, 1)$:

$$\begin{aligned} \min \quad & 2x_1^2 + 3x_2^2 \\ \text{s.t.} \quad & x_1 + x_2 = 0 \end{aligned} \tag{3.38}$$

then the algorithm will be stuck at the non-interesting point $(-1, 1)$, while the optimal solution is obviously $\mathbf{x}^* = (0, 0)$. Interestingly, a few recent results demonstrate that if one can update *two* out of $n$ variables at each iteration, then it is possible to deal with linear constraints of the form [68]

$$\sum_{k=1}^{K} a_k x_k = a_0.$$

Other methods that deal with linearly or nonlinearly coupling constraints will be discussed in later sections.

Second, when the objective $f_0$ is not convex, but the feasible sets $\mathcal{X}_k$'s are all convex sets, the convergence of BCD becomes delicate. By [51, Proposition 2.7.1], we have that if the following sets of conditions are satisfied, then every limit point generated by the cyclic version of the BCD is a stationary solution of problem (3.32):

1. $f_0$ continuously differentiable, and $\mathcal{X}_k$ are closed and convex sets.
2. Each subproblem (3.33) has a unique solution.
3. The function $f(\mathbf{x}_k, \mathbf{x}_{-k})$, when viewed as a function of $\mathbf{x}_k$, is monotonically non-increasing in the interval from $\mathbf{x}_k$ to its minimizer $\mathbf{x}_k^*$.

It is important to note that the uniqueness of each subproblem is essential for the convergence of BCD, without which the algorithm can diverge, see the counterexample provided in [66]. However, when assuming that each subproblem is also convex, the above conditions can be significantly weakened. For example, Tseng has shown in [69, Theorem 4.1] that the following condition is sufficient to guarantee the convergence of the cyclic version of the BCD (to the set of stationary solutions):

1. $f_0$ is continuously differentiable, and $\mathcal{X}_k$'s are closed and convex sets.
2. The set $\{\mathbf{x} \mid f(\mathbf{x}) \leq f(\mathbf{x}^0)\}$ is compact.
3. $f_0(\mathbf{x}_1, \cdots, \mathbf{x}_K)$ has at most one minimum in $\mathbf{x}_k$ for $k = 2, \cdots, K - 1$.

Similarly conditions are also needed for convergence of the E-C rule and the G-So rule in Table 3.3 as well. Somewhat surprisingly, for the MBI update rule and randomized update rule given in Table 3.3, the uniqueness condition is no longer required. See reference [70] for results related to the MBI rule and reference [59] for results related to the randomized block update rule. In particular, the following condition guarantees the convergence of randomized BCD to the set of stationary solutions, with probability one:

1. $f_0$ is continuously differentiable, and $\mathcal{X}_k$'s are closed and convex sets.
2. The set $\{\mathbf{x} \mid f(\mathbf{x}) \leq f(\mathbf{x}^0)\}$ is compact.

Similar conditions guarantee that the MBI converges to the set of stationary solutions, see [70, Theorem 3.1].

Let us then proceed to analyze the rate of convergence of BCD-related problems. Analyzing the rate of convergence of BCD-type method is a relatively new research direction which has been under extensive study in the last couple of years. Similarly as in the gradient-based method, here the difference of the objective value and the optimal solution, given by $\Delta^t := f_0(\mathbf{x}^t) - f_0(x^*)$, is usually used to measure the optimality gap. It is known that when the objective $f_0$ is strongly convex with respect to all the block variables $\mathbf{x}_k$, $k = 1, \cdots, K$, then BCD converges linearly [53]. Luo and Tseng in [53] have shown that such strong convexity condition can be relaxed, that is, when the following conditions are satisfied, the BCD converges linearly without requiring that the entire problem to be strongly convex:

1. The objective function is a composite function, that is,

$$f_0 = g(Ax) = g \left( \sum_{k=1}^{K} A_k x_k \right)$$

where $g(\cdot)$ is a strongly convex function, and $A$ is a matrix that is not necessarily full rank.
2. Each feasible set $\mathcal{X}_k$ is a ployhedral set.
3. Each matrix $A_k$, which consists of a few columns of $A$, has full column rank.

It is possible to generalize this linear convergence result to certain nonconvex quadratic optimization problems as well (i.e., to problems with nonconvex quadratic objective and polyhedral feasible sets); see [71] for a review of these results.

When the problem is a general convex problem, analyzing the convergence of BCD with *deterministic rules* (i.e., all the step selection rules in Table 3.3 except the randomized rule) is challenging. In fact, Nesterov has stated in one of his recent paper that "The simplest variant of the coordinate descent method is based on a cyclic coordinate search. However, for this strategy it is difficult to prove convergence, and almost impossible to estimate the rate of convergence" [72]. Nevertheless, a few encouraging progresses have been made toward this direction. Beck et al. in [73] have shown that if $f_0$ is smooth and each block is updated using the GP method (3.34), then the cyclic version of the BCD converges in a sublinear rate $\mathcal{O}(1/t)$. Recently, the authors in [74] extended this result to problems with nonsmooth objectives as well as many other coordinate selection rules, such as the G-So rule, the E-C rule, and the MBI rule. Further, they have shown that the BCD with the exact minimization step (3.33) converges in a sublinear rate $\mathcal{O}(1/t)$, *regardless* of the uniqueness of each subproblem. More precisely, the BCD with exact minimization step (3.33) converges sublinearly under the following conditions:

1. The set $\{\mathbf{x} \mid f(\mathbf{x}) \leq f(\mathbf{x}^0)\}$ is compact.
2. The objective can be written as

$$\min \quad f_0(\mathbf{x}) := \ell(\mathbf{x}_1, \cdots, \mathbf{x}_K) + \sum_{k=1}^{K} h_k(\mathbf{x}_k) \tag{3.39}$$

where $\ell$ is a smooth function with Lipschitz continuous gradient; $h_k$ is a convex but not necessarily smooth function.
3. Either the cyclic, the MBI, or the E-C rule is used.

The analysis used there consists of three simple main steps, outlined below:

(S1) *Sufficient descent*: Show that the successive difference of the objective $f_0(\mathbf{x})$ between two iterations is large enough.
(S2) *Estimating cost-to-go*: Show that the difference between $f_0(\mathbf{x}^t)$ and the optimal objective is close enough.
(S3) *Obtain the bounds*: Match these two bounds and obtain the desired result.

These three steps can also be generalized to show the sublinear convergence rate for the Jacobian block selection scheme; see [60].

When the blocks are selected *randomly*, the convergence rate of BCD has been widely studied. It is shown in [72, 75, 76] that for either smooth problems or the nonsmooth problems of the form (3.39), the BCD with GP step (3.34) converges sublinearly in expectation (or with certain high probability). More precisely, we have the following:

$$\mathbb{E}[f_0(\mathbf{x}^{t+1})] - f_0(\mathbf{x}^*) \leq \frac{C}{t+1}$$

where $C$ is some constant that can be explicitly computed using the problem data; the expectation is taken over the randomization of the coordinates. Also there are related high probability convergence results associated with the randomized BCD. We note here that this type of result is different from the previous deterministic analysis. The

---

**Algorithm 2** The BSUM Algorithm [67]

---

Find a feasible solution $\mathbf{x}^0 \in \mathcal{X}$ and set $t = 0$
**For** $r = 1, \cdots$
Pick index set $\mathcal{I}^t$
For all $i \in \mathcal{I}^t$, compute

$$\mathbf{x}_k^t \in \arg \min_{\mathbf{x}_k \in \mathcal{X}_k} u_k(\mathbf{x}_k, \mathbf{x}^{t-1}) \tag{3.40}$$

Set $\mathbf{x}_j^t = \mathbf{x}_j^{t-1}, \quad \forall j \notin \mathcal{I}^t$
$t = t + 1$,
**until** some convergence criterion is met.

---

sublinear convergence analysis for randomized rule represents the rate for the "average" case, meaning if one runs the algorithm for a large number of times with the same data set and with randomized coordinate selection rule, then the average objective value will be close to the global optimal. This, however, does not guarantee that such rate will be achieved for any given realization of the algorithm. In contrast, the rate derived by the deterministic block selection rule represents a "worst case" rate – every realization of the algorithm is guaranteed to achieve that given rate.

## 3.4        Block Successive Upper-Bound Minimization Method

In this section, we describe a framework that significantly generalizes the applicability of the BCD-type methods discussed in the previous section. The generalization comes from the key observation that, in many applications, it is beneficial to solve the per-block subproblem (3.32) *inexactly*.

The resulting framework, termed the Block Successive Upper-bound Minimization (BSUM) method [67], is described in the following table.

In BSUM, the rules that determine the index set $\mathcal{I}^t$ is the same as those given in Table 3.3, so the only difference with BCD is in (3.40) where the objective to be minimized is a function $u_k(\mathbf{x}_k, \mathbf{x}^{t-1}) : \mathbb{R}^{n_k} \to \mathbb{R}$, rather than $f_0(\mathbf{x}_k, \mathbf{x}^t_{-k})$. Below we will again assume that the original objective $f_0$ is smooth, and describe the conditions in which $u_k$ needs to be satisfied. The more general case for nonsmooth problems will be discussed in later chapters.

We have the following assumption on $u_k$:

$$u_k(\mathbf{x}_k, \mathbf{x}) = f_0(\mathbf{x}), \quad \forall \, x \in \mathcal{X}, \, \forall \, k \tag{3.41a}$$

$$u_k(\mathbf{x}_k, \mathbf{z}) \geq f_0(\mathbf{x}_k, \mathbf{z}_{-k}), \quad \forall \, \mathbf{x}_k \in \mathcal{X}_k, \, \forall \, \mathbf{z} \in \mathcal{X}, \forall \, k \tag{3.41b}$$

$$\nabla u_k(\mathbf{x}_k, \mathbf{y})\Big|_{\mathbf{x}_k = \mathbf{y}_k} = \nabla_k f_0(\mathbf{y}), \quad \forall \, \mathbf{x}_k, \mathbf{y}_k \in \mathcal{X}_k, \forall \, k, \, \forall \, \mathbf{y} \in \mathcal{X} \tag{3.41c}$$

$$u_k(\mathbf{x}_k, \mathbf{z}) \text{ is continuous in } (\mathbf{x}_k, \mathbf{z}) \quad \forall \, k. \tag{3.41d}$$

where $\nabla_k f_0(\mathbf{y})$ means the partial derivative w.r.t. the $k$th block. Clearly $u_k(\cdot)$ is an locally tight upper bound of the original function $f_0(\cdot)$, hence the name of the framework. Commonly used upper bounds include the quadratic upper bound, the proximal upper bound, and so on; see Table 3.4.

BSUM converges under similar condition as the BCD-type methods we have seen in the previous section. More specifically, if either one of the following two conditions are satisfied, then every limit point of the BSUM is a stationary solution [67, Theorem 2]:

1. For each $k$, $u_k(\mathbf{x}_k, \mathbf{y})$ is quasi-convex in $\mathbf{x}_k$, and each subproblem (3.40) has a unique solution for any point $\mathbf{x}^{t-1} \in \mathcal{X}$.
2. The level set $\mathcal{X}^0 = \{\mathbf{x} \mid f_0(\mathbf{x}) \leq f_0(\mathbf{x}^0)\}$ is compact, and the subproblem (3.40) has a unique solution for any point $\mathbf{x}^{t-1} \in \mathcal{X}$, for at least $K - 1$ blocks.

Further, one can also use the three-step approach outlined in the previous section to analyze the convergence rate of the BSUM; see [74]. Specifically, if we further assume that each upper-bound function has Lipschitz continuous gradient, and the entire function is convex, then the BSUM algorithm with various block selection rules converges in a sublinear manner.

To see the applicability of BSUM, the first observation is that the block coordinate GP iteration described in (3.34) is precisely the BSUM method with the quadratic upper bound (cf. Table 3.4). Other popular methods that are special cases of BSUM includes the Expectation-Maximization (EM) algorithm [77] in statistics, the Convex-Concave Procedure (CCCP) in machine learning [78], the multiplicative update for solving non-negative matrix factorization (NMF) problem [79], the weighted minimum mean square error (WMMSE) in wireless communication [80], just to name a few. Below we demonstrate why EM method is a special case of BSUM. For more detailed discussion and other applications, we refer the reader to [67, 55].

Let $\mathbf{w}$ be the observation from which we would like to estimate $\mathbf{x}$. The maximum likelihood (ML) estimate of $\mathbf{x}$ is given by

$$\hat{\mathbf{x}}_{\text{ML}} = \arg\max_{\mathbf{x}}\ \ln p(\mathbf{w}|\mathbf{x}) = \min_{\mathbf{x}}\ -\ln p(\mathbf{w}|\mathbf{x}). \qquad (3.42)$$

Let the random vector $\mathbf{z}$ be some hidden/unobserved variable. The EM generates a sequence of estimates $\{\mathbf{x}^r\}$ by

1. E-Step: Calculate $f_0(\mathbf{x}, \mathbf{x}^t) := \mathbb{E}_{\mathbf{z}|\mathbf{w},\mathbf{x}^t}\{\ln p(\mathbf{w}, \mathbf{z}|\mathbf{x})\}$.
2. M-Step: $\mathbf{x}^{t+1} = \arg\min_{\mathbf{x}} -f_0(\mathbf{x}, \mathbf{x}^t)$.

Below we show that the EM iteration in fact can be interpreted as successively minimizing certain upper-bound function, hence a special case of BSUM. The objective of (3.42) can be bounded by

$$-\ln p(\mathbf{w}|\mathbf{x}) = -\ln\ \mathbb{E}_{\mathbf{z}|\mathbf{x}}\ p(\mathbf{w}|\mathbf{z}, \mathbf{x})$$

$$= -\ln\ \mathbb{E}_{\mathbf{z}|\mathbf{x}}\left[\frac{p(\mathbf{z}|\mathbf{w}, \mathbf{x}^t)p(\mathbf{w}|\mathbf{z}, \mathbf{x})}{p(\mathbf{z}|\mathbf{w}, \mathbf{x}^t)}\right]$$

$$= -\ln\ \mathbb{E}_{\mathbf{z}|\mathbf{w},\mathbf{x}^t}\left[\frac{p(\mathbf{z}|\mathbf{x})p(\mathbf{w}|\mathbf{z}, \mathbf{x})}{p(\mathbf{z}|\mathbf{w}, \mathbf{x}^t)}\right]$$

$$\leq -\mathbb{E}_{\mathbf{z}|\mathbf{w},\mathbf{x}^t}\ln\left[\frac{p(\mathbf{z}|\mathbf{x})p(\mathbf{w}|\mathbf{z}, \mathbf{x})}{p(\mathbf{z}|\mathbf{w}, \mathbf{x}^t)}\right]$$

$$= -\mathbb{E}_{\mathbf{z}|\mathbf{w},\mathbf{x}^t}\ln p(\mathbf{w}, \mathbf{z}|\mathbf{x}) + \mathbb{E}_{\mathbf{z}|\mathbf{w},\mathbf{x}^t}\ln p(\mathbf{z}|\mathbf{w}, \mathbf{x}^t)$$

$$:= u(\mathbf{x}, \mathbf{x}^t),$$

Clearly the M-step can be written as

$$\mathbf{x}^{t+1} = \arg\min_{\mathbf{x}} u(\mathbf{x}, \mathbf{x}^t).$$

Furthermore, it is not hard to see that the conditions (3.41) are all satisfied, therefore we conclude that the EM is in fact the BSUM method with a single block of variable (i.e., $K = 1$). More applications of the BSUM algorithm will be provided in Chapter 4 when we discuss sparse optimization algorithms.

## 3.5        The Augmented Lagrangian Method

As we have mentioned in the previous section, generally speaking neither BCD nor BSUM is able to deal with problems that have coupling constraints. The augmented Lagrangian method (also known as the method of multipliers) is one of the most well-known algorithms that address such issue.

Consider the following equality constrained problem:

$$\begin{aligned} \min_{\mathbf{x}} \quad & f_0(\mathbf{x}) \\ \text{s.t.} \quad & h_j(\mathbf{x}) = 0, \ j = 1, \ldots, p. \end{aligned} \tag{3.43}$$

Note that inequality constraints of the form $f_i(\mathbf{x}) \leq 0$ can be easily converted to an equality constraint by adding a new variable $z_i$ and replace the inequality constraint by

$$f_i(\mathbf{x}) + z_i^2 = 0.$$

Construct the *augmented Lagrangian function* $L_\rho : \mathbb{R}^{n \times p} \to \mathbb{R}$, given by

$$L_\rho(\mathbf{x}, \mathbf{y}) := f_0(\mathbf{x}) + \sum_{j=1}^{p} y_j h_j(\mathbf{x}) + \rho \sum_{j=1}^{p} \|h_j(\mathbf{x})\|^2. \tag{3.44}$$

The main idea here is to find a correct multiplier $\mathbf{y} := \{y_j\}_{j=1}^{p}$ such that minimizing the augmented Lagrangian yields the optimal solution of (3.43).

The augmented Lagrangian iteration [81, 82] for problem (3.43) is

$$\mathbf{x}^{t+1} = \arg\min L_\rho(\mathbf{x}, \mathbf{y}^t), \tag{3.45a}$$

$$y_j^{t+1} = y_j^t + \rho h_j(\mathbf{x}^{t+1}), \ j = 1, \cdots, p. \tag{3.45b}$$

The convergence analysis of the augmented Lagrangian method is somewhat complicated for general nonconvex problems; for example, it may require that $\rho$ be a very large number, or may need to impose the assumption that the multipliers are bounded. We refer the reader to [56, Chapter 4] for detailed discussion.

Below we consider the simpler but nevertheless very useful case where $f_0(\mathbf{x})$ is convex, and the equality constraints collectively is written as the affine constraint $\mathbf{Ax} - \mathbf{b} = \mathbf{0}$, with $\mathbf{A} \in \mathbb{R}^{m \times n}$ and $\mathbf{b} \in \mathbb{R}^m$. In this case, the KKT condition of (3.43) is given by

$$\text{(primal feasibility)} \quad \mathbf{0} = \mathbf{Ax}^* - \mathbf{b}, \tag{3.46a}$$

$$\text{(dual feasibility)} \quad \mathbf{0} = \nabla f_0(\mathbf{x}^*) - \mathbf{A}^T \mathbf{y}^*. \tag{3.46b}$$

At every iteration, $\mathbf{x}^{t+1}$ and $\mathbf{y}^{t+1}$ maintain condition (3.46b), because from the optimality condition of (3.45a) we have

$$\mathbf{0} = \nabla f_0(\mathbf{x}^{t+1}) - \mathbf{A}^T(\mathbf{y}^t + \rho(\mathbf{b} - \mathbf{Ax}^{t+1})) = \nabla f_0(\mathbf{x}^{t+1}) - \mathbf{A}^T \mathbf{y}^{t+1}. \tag{3.47}$$

Therefore if the correct multipliers are identified in the limit, then the primal feasibility (3.46a) will also be satisfied. Please see textbook [83] for detailed discussion.

## 3.6    Alternating Direction Method of Multipliers

Now we know that the augmented Lagrangian method can deal with equality constraints for general nonconvex problems and linear constraints for convex problems. The question is, does it also apply to problems with block structures, for example those described in (3.32)? In this section we introduce the alternating direction method of multipliers (ADMM), which can be viewed as a hybrid of the augmented Lagrangian and the BCD method. Once again, like the BCD and BSUM, utilizing the block structure here is beneficial because doing so oftentimes leads to simple subproblems with closed-form solutions.

### 3.6.1    The Algorithm

Let us consider the following problem, where both $f_0$ and $g_0$ are convex functions, and $\mathcal{X}$ and $\mathcal{Z}$ are closed and convex sets, and $\mathbf{A} \in \mathbb{R}^{m \times n_x}$, $\mathbf{B} \in \mathbb{R}^{m \times n_z}$, and $\mathbf{b} \in \mathbb{R}^m$:

$$\min_{\mathbf{x}} \quad f_0(\mathbf{x}) + g_0(\mathbf{z}) \tag{3.48}$$

$$\text{s.t.} \quad \mathbf{A}\mathbf{x} + \mathbf{B}\mathbf{z} = \mathbf{b}$$

$$\mathbf{x} \in \mathcal{X}, \ \mathbf{z} \in \mathcal{Z}. \tag{3.49}$$

Clearly there are two blocks of variables, and they are separable in the objective but coupled by a linear constraint. As in the previous section, we can write the augmented Lagrangian for (3.48) as

$$L_\rho(\mathbf{x}, \mathbf{z}; \mathbf{y}) = f_0(\mathbf{x}) + g_0(\mathbf{z}) - \mathbf{y}^T(\mathbf{A}\mathbf{x} + \mathbf{B}\mathbf{z} - \mathbf{b}) + \frac{\rho}{2}\|\mathbf{A}\mathbf{x} + \mathbf{B}\mathbf{z} - \mathbf{b}\|_2^2.$$

The ADMM algorithm can be viewed as an inexact version of the augmented Lagrangian update (3.45), where the primal problem is not required to be solved exactly. Formally the algorithm is described as follows:

$$\mathbf{x}^{t+1} = \arg\min_{\mathbf{x}} L_\rho(\mathbf{x}, \mathbf{z}^t; \mathbf{y}^t), \tag{3.50a}$$

$$\mathbf{z}^{t+1} = \arg\min_{\mathbf{y}} L_\rho(\mathbf{x}^{t+1}, \mathbf{z}; \mathbf{y}^t), \tag{3.50b}$$

$$\mathbf{y}^{t+1} = \mathbf{y}^t - \gamma\rho(\mathbf{A}\mathbf{x}^{t+1} + \mathbf{B}\mathbf{z}^{t+1} - \mathbf{b}), \tag{3.50c}$$

where $\gamma > 0$ is some constant. Each iteration of the algorithm includes two steps of primal computation and a single step of dual update.

The basic ADMM iteration can be extended in several ways. For instance, if one of the subproblems has no closed-form solution, an inexact solution is often sufficient to keep the algorithm going. Below is one popular way to generate such inexact solution for $\mathbf{x}$ [84]:

$$\mathbf{x}^{t+1} = \arg\min_{\mathbf{x}} f_0(\mathbf{x}) + \rho\langle \mathbf{A}^T(\mathbf{A}\mathbf{x}^t + \mathbf{B}\mathbf{z}^t - \mathbf{b} - \mathbf{y}^t/\rho), \mathbf{x} - \mathbf{x}^t\rangle$$

$$+ \frac{1}{2P}\|\mathbf{x} - \mathbf{x}^t\|_2^2, \tag{3.51}$$

where $P > 0$ is some appropriately chosen constant. Clearly, when $P$ is small enough, the objective function is simply a quadratic upper bound (cf. Table 3.4) for the augmented Lagrangian. However, it is important to note that generally speaking the original function $f_0$ is not allowed to be approximated, unless in certain special cases; see [85]. Also one can perform the over-relaxation by replacing $\mathbf{A}\mathbf{x}^{t+1}$ in (3.50b) and (3.50c) by the following quantity:

$$\alpha^t \mathbf{A}\mathbf{x}^{t+1} - (1 - \alpha^t)(\mathbf{B}\mathbf{z}^t - \mathbf{b}) \tag{3.52}$$

where $\alpha^t \in (0, 2)$ is some constant; for more discussions see [86, Chapter 3] and the references therein.

## 3.6.2    Convergence Analysis

Generally speaking, the ADMM iteration (3.50) converges under quite mild conditions. Below we provide a sufficient condition for the convergence of the algorithm (for more refined conditions, the reader is referred to [87]):

1. Problem (3.48) is feasible.
2. Both $f_0$ and $g_0$ are proper convex and lower semi-continuous functions.
3. $\mathbf{A}$ and $\mathbf{B}$ both have full column rank.
4. The constant $\gamma \in (0, (\sqrt{5} + 1)/2)$.

Then the sequence $(\mathbf{x}^t, \mathbf{z}^t, \mathbf{y}^t)$ is bounded and every limit point of $(\mathbf{x}^t, \mathbf{z}^t)$ is an optimal solution for problem (3.50).

Further, it is also possible to establish the rate of convergence for ADMM. It is worth noting that differently from the BCD and gradient-based method in which the optimality gap can be simply measured using the objective value, for ADMM *both* the objective value and the constraint violation need to be taken into consideration. Using the variational inequality characterization, it is possible to show that an optimal primal-dual solution $(\mathbf{x}^*, \mathbf{z}^*, \mathbf{y}^*)$ must satisfy [88]

$$f_0(\mathbf{x}) + g_0(\mathbf{z}) - \big(f_0(\mathbf{x}^*) + g_0(\mathbf{z}^*)\big) + \langle \mathbf{w} - \mathbf{w}^*, F(\mathbf{w})\rangle \geq 0, \quad \forall \mathbf{w} \in \Omega \tag{3.53}$$

where we have defined

$$\mathbf{w} = \begin{bmatrix} \mathbf{x} \\ \mathbf{z} \\ \mathbf{y} \end{bmatrix}, \quad F(\mathbf{w}) = \begin{bmatrix} -\mathbf{A}^T\mathbf{y} \\ -\mathbf{B}^T\mathbf{y} \\ \mathbf{A}\mathbf{x} + \mathbf{B}\mathbf{z} - \mathbf{b} \end{bmatrix}, \quad \Omega = \mathcal{X} \times \mathcal{Z} \times \mathbb{R}^m. \tag{3.54}$$

This implies that the $\epsilon$-optimal solution can be defined as a solution $(\hat{\mathbf{x}}, \hat{\mathbf{z}}, \hat{\mathbf{y}})$ that satisfies the following condition:

$$f_0(\mathbf{x}) + g_0(\mathbf{z}) - \big(f_0(\hat{\mathbf{x}}) + g_0(\hat{\mathbf{z}})\big) + \langle \mathbf{w} - \hat{\mathbf{w}}, F(\mathbf{w})\rangle \geq -\epsilon, \quad \forall \mathbf{w} \in \Omega. \tag{3.55}$$

Let us define the average of the iterates at a given iteration $T$ as

$$\tilde{\mathbf{w}}^T = \frac{1}{T}\sum_{t=1}^{T} \mathbf{w}^t. \tag{3.56}$$

Then it is shown in [88, Theorem 4.1] that the ADMM achieves a sublinear convergence, in the following sense:

$$f_0(\mathbf{x}) + g_0(\mathbf{z}) - \left(f_0(\tilde{\mathbf{x}}^T) + g_0(\tilde{\mathbf{z}}^T)\right) + \langle \mathbf{w} - \tilde{\mathbf{w}}^T, F(\mathbf{w}) \rangle \geq -\frac{C}{T}, \quad \forall \mathbf{w} \in \Omega$$

where $C > 0$ is some constant. Further, it is also possible to show that the iterates converge sublinearly in the non-ergodic sense (i.e., without taking the average in (3.56)) [89]. More specifically, one can show that the ADMM iteration generates a primal-dual sequence that satisfies

$$\|\mathbf{w}^t - \mathbf{w}^{t+1}\|_{\mathbf{H}}^2 \leq \frac{1}{t+1}\|\mathbf{w}^0 - \mathbf{w}^*\|_{\mathbf{H}}^2, \ \forall \ \mathbf{w}^* \in \Omega^* \tag{3.57}$$

where $\mathbf{H}$ is some positive semidefinite matrix; $\Omega^*$ is the optimal primal-dual solution set. It is possible to show that the quantity $\|\mathbf{w}^t - \mathbf{w}^{t+1}\|_{\mathbf{H}}^2$ can be used to measure the optimality, therefore implying that the ADMM converges sublinearly.

We also mention here that under some more stringent conditions, the ADMM iteration converges linearly. Once again one needs to first properly define the measure of optimality, and then show that under certain conditions such optimality measure shrinks by a constant factor after each iteration. The work [90] shows that for a Jacobi version of the ADMM applied to smooth functions with Lipschitz continuous gradients, the objective value descends at the rate of $O(1/t)$ and that of an accelerated version descends at $O(1/t^2)$. Then, [90] establishes the same rates on a Gauss–Seidel version and requires only one of the two objective functions to be smooth with Lipschitz continuous gradient. Lately, [89] shows that $\|\mathbf{w}^t - \mathbf{w}^{t+1}\|$ converges at $O(1/t)$ assuming at least one of the subproblems is exactly solved. Reference [91] proves that the dual objective value of a modification to the ADMM descends at $O(1/t^2)$ under the assumption that the objective functions are strongly convex, one of them is quadratic, and both subproblems are solved exactly. Reference [92] shows the linear rate of convergence under a variety of scenarios in which at least one of the two objective functions is strongly convex and has Lipschitz continuous gradient. On the other hand, sublinear-rate results [90, 89] do not require any strongly convex functions. In addition, the ADM applied to linear programming is known to converge at a globally linear rate [93].

For example, reference [94] shows that under a few conditions such as (see [94, Table 1.1] for more detailed conditions)

1. at least one of $f_0$ and $g_0$ has Lipschitz continuous gradient,
2. at least one of $f_0$ and $g_0$ is strongly convex,

the ADMM (and certain inexact version of ADMM) generates a sequence $\mathbf{w}^t$ that satisfies

$$\|\mathbf{w}^t - \mathbf{w}^*\|_{\mathbf{G}}^2 \leq \frac{1}{\delta+1}\|\mathbf{w}^{t+1} - \mathbf{w}^*\|_{\mathbf{G}}^2 \tag{3.58}$$

for some $\delta > 0$, and for some $\mathbf{w}^* \in \Omega^*$, and some $\mathbf{G} \succeq 0$, implying that the quantity $\|\mathbf{w}^t - \mathbf{w}^*\|_{\mathbf{G}}^2$ converges linearly.

In [95], the authors show that the ADMM iteration (3.50) is capable of linear convergence without either $f_0$ or $g_0$ to be strongly convex. More precisely, assume that the following conditions hold:

1. $f_0$ and $g_0$ can be expressed as

$$f_0(\mathbf{x}) = \ell_0(\mathbf{Cx}) + h_0(\mathbf{x}), \quad g_0(\mathbf{x}) = \ell_1(\mathbf{Dz}) + h_1(\mathbf{z}),$$

   where $\ell_0$, $\ell_1$ and $h_0$, $h_1$ are all convex and continuous relative to their domains, and $\mathbf{C}, \mathbf{D}$ are some given matrices not necessarily full column rank.
2. Each $\ell_k(\cdot)$ is strictly convex and continuously differentiable with a uniform Lipschitz continuous gradient.
3. Each $h_k$ satisfies either one of the following conditions:
   (a) The epigraph of $h_k(\cdot)$ is a polyhedral set.
   (b) $h_k(\cdot) = \lambda_k \|\cdot\|_1$.
   (c) Each $h_k(\cdot)$ is the sum of the functions described in the previous two items.
4. Both $\mathbf{B}$ and $\mathbf{A}$ have full column rank.
5. The feasible sets $\mathcal{X}$ and $\mathcal{Z}$ are compact polyhedral sets.

Next we need to define the optimality gap. Let us use $d(\mathbf{y}^t)$ to denote the optimal objective of the following problem:

$$\min_{\mathbf{x},\mathbf{z}} L_\rho(\mathbf{x}, \mathbf{z}; \mathbf{y}^t).$$

Let $d^*$ denote the optimal objective for the original problem (3.50), and define the *dual optimality gap* as

$$\Delta_d^t = d^* - d(\mathbf{y}^t) \geq 0.$$

The last inequality is due to weak duality. Define the *primal optimality gap* as

$$\Delta_p^t = L_\rho(\mathbf{x}^{t+1}, \mathbf{z}^{t+1}; \mathbf{y}^t) - d(\mathbf{y}^t) \geq 0.$$

Then [95] shows that if $\gamma$ in (3.50c) is chosen small enough, then the *combined* primal and dual optimality gaps shrinks geometrically, that is,

$$\Delta_p^{t+1} + \Delta_d^{t+1} \leq \frac{1}{1+\hat{\delta}}\left(\Delta_p^t + \Delta_d^t\right) \tag{3.59}$$

for some $\hat{\delta} > 0$. Interestingly, such analysis also works for the inexact version of the ADMM, in which the entire augmented Lagrangian can be approximated. This is quite different than the approximation mentioned in (3.51), where the objective function $f_0$ has to remain the same.

Through the previous discussion, we would like to make the point that the analysis of the rate of ADMM is quite different than that of the BCD/GD/PG/BSUM methods. For different problems or when showing different results, one usually needs to pick a special optimality gap to work with, and there is no general agreement on which optimality gap should be used. For more details about the rate analysis of ADMM, we refer the reader to [94, 95, 96, 88].

## 3.6.3    Solving Multiblock Problems and Beyond

A natural question at this point is that whether the ADMM method can be generalized to problems with multiple-block variables, problems with coupled objectives, or problems with nonconvex objectives. Unfortunately, extending ADMM to either one of these scenarios is proven to be very difficult. In this section, we briefly discuss the extension of the ADMM to these more general scenarios.

First let us consider the following multiple-block problem:

$$
\begin{aligned}
&\min_{\mathbf{x}} && \sum_{k=1}^{K} f_k(\mathbf{x}_k) && (3.60)\\
&\text{s.t.} && \sum_{k=1}^{K} A_k \mathbf{x}_k = \mathbf{b} \\
& && \mathbf{x}_k \in \mathcal{X}_k, \ k = 1, \cdots, K.
\end{aligned}
$$

One can directly extend the ADMM iteration as the following

$$
\mathbf{x}_k^{t+1} = \arg\min_{\mathbf{x}_k} L_\rho(\mathbf{x}_1^{t+1}, \cdots, \mathbf{x}_{k-1}^{t+1}, \mathbf{x}_k, \mathbf{x}_{k+1}^t, \mathbf{x}_K^t; \mathbf{y}^t), \tag{3.61a}
$$

$$
\mathbf{y}^{t+1} = \mathbf{y}^t - \gamma\rho\left(\sum_{k=1}^{K} A_k \mathbf{x}_k^{t+1} - \mathbf{b}\right). \tag{3.61b}
$$

One would expect that such extension should at least converge. Unfortunately, this is by no means the case. Recently, it is shown in [97] the $K$-block ADMM iteration (3.61) can indeed diverge. A counterexample is developed there, where the ADMM is used to solve the following linear systems of equations (which has a unique solution $x_1 = x_2 = x_3 = 0$):

$$
E_1 x_1 + E_2 x_2 + E_3 x_3 = 0, \tag{3.62}
$$

$$
\text{with} \quad [E_1 \ E_2 \ E_3] = \begin{bmatrix} 1 & 1 & 1 \\ 1 & 1 & 2 \\ 1 & 2 & 2 \end{bmatrix}. \tag{3.63}
$$

It is shown in [97] that regardless of the starting point, the 3-block ADMM algorithm always diverges. Motivated by this somewhat surprising result, there have been numerous efforts that attempt to develop a modified version of the ADMM iteration so that it can work for $K$-block problems. We refer the reader to the following references for more detailed discussion: [98, 95, 99, 100, 7]. Interestingly, in [101], the authors have shown that to guarantee the convergence for the counterexample (3.62), one only needs to use an iteration-dependent, *diminishing* stepsize $\gamma^t$. More specifically, they show that for the counterexample (in fact for a much wider problem class), if $\gamma^t$ satisfies

$$
\lim_{t\to\infty} \gamma^t = 0, \quad \lim_{T\to\infty} \sum_{t=1}^{T} \gamma^t = \infty, \tag{3.64}
$$

then the ADMM iteration, explicitly shown below, converges to the global optimal solution for problem (3.62):

$$y^{l+1} = y^l + \gamma^l \rho \left( E_1 x_1^l + E_2 x_2^l + E_3 x_3^l \right)$$
$$x_1^{l+1} = (E_1^T E_1)^{-1} \left( -E_1^T E_2 x_2^l - E_1^T E_3 x_3^l - E_1^T y^{l+1} / \rho \right)$$
$$x_2^{l+1} = (E_2^T E_2)^{-1} \left( -E_2^T E_1 x_1^{l+1} - E_2^T E_3 x_3^l - E_2^T y^{l+1} / \rho \right)$$
$$x_3^{l+1} = (E_3^T E_3)^{-1} \left( -E_3^T E_1 x_1^{l+1} - E_3^T E_2 x_2^{l+1} - E_3^T y^{l+1} / \rho \right).$$

There is very little work on ADMM with convex but coupled objective function. Reference [101] considers the following multiblock, coupled objective problem:

$$
\begin{array}{ll}
\min\limits_{\mathbf{x}} & f_0(\mathbf{x}_1, \cdots, \mathbf{x}_K) \qquad\qquad (3.65)\\
\text{s.t.} & \sum\limits_{k=1}^{K} \mathbf{A}_k \mathbf{x}_k = \mathbf{b}\\
& \mathbf{x}_k \in \mathcal{X}_k, \ k = 1, \cdots, K.
\end{array}
$$

The authors show that under the following conditions (which is similar to those in [95] for the linear convergence case):

1. $f_0$ can be expressed as

$$f_0(\mathbf{x}) = \ell(\mathbf{C}_1 \mathbf{x}_1, \cdots, \mathbf{C}_K \mathbf{x}_K) + \sum_{k=1}^{K} h_k(\mathbf{x}_k)$$

where $\ell$, $h$ are both all convex and continuous relative to their domains, and $\mathbf{C}_k$'s are some given matrices not necessarily full column rank.
2. The function $\ell(\cdot)$ is strictly convex and continuously differentiable with a uniform Lipschitz continuous gradient.
3. Each $h_k$ satisfies either one of the following conditions:
   (a) The epigraph of $h_k(\cdot)$ is a polyhedral set.
   (b) $h_k(\cdot) = \lambda_k \| \cdot \|_1$.
   (c) Each $h_k(\cdot)$ is the sum of the functions described in the previous two items.
4. Each $\mathbf{A}_k$ has full column rank.
5. The feasible sets $\mathcal{X}_k$ are compact polyhedral sets.

Then ADMM with $\gamma^l$ being the diminishing stepsize (3.64) converges to the set of global optimal primal-dual solutions. Some other recent results on using ADMM for solving problems with coupled objective can be found in [102].

Finally, we remark that there is no general theoretical results on using ADMM for solving nonconvex problems, although they have become quite popular recently; see [103, 104, 105, 106, 107, 108, 109] for applications in phase retrieval, distributed matrix factorization, distributed clustering, sparse zero variance discriminant analysis, and so on. Below we present one special nonconvex problem in which

ADMM is capable of computing stationary solutions [110]. Consider the following problem:

$$
\min \quad f_0(\mathbf{x}) := \sum_{k=1}^{K} g_k(\mathbf{x}) + h(\mathbf{x}) \tag{3.66}
$$

$$
\text{s.t.} \quad \mathbf{x} \in \mathcal{X} \tag{3.67}
$$

where $g_k$'s are a set of smooth, possibly nonconvex functions, while $h$ is a convex nonsmooth regularization term. Suppose there are $K$ agents in the system, and each of them is associated with a single function $f_k$. Then we can restate the problem by introducing $K$ auxiliary variables $\{\mathbf{x}_k\}_{k=1}^{K}$:

$$
\min \quad f_0(\mathbf{x}_k) := \sum_{k=1}^{K} g_k(\mathbf{x}_k) + h(\mathbf{x}) \tag{3.68}
$$

$$
\text{s.t.} \quad \mathbf{x}_k = \mathbf{x}, \ \forall\, k, \quad \mathbf{x} \in \mathcal{X} \tag{3.69}
$$

This problem is closely related to the convex global consensus problem [86, Section 7], but with the key difference that $g_k$'s can be nonconvex. The augmented Lagrangian is given by

$$
L_\rho(\{\mathbf{x}_k\}, \mathbf{x}; \mathbf{y}) = \sum_{k=1}^{K} g_k(\mathbf{x}_k) + h(\mathbf{x}) + \sum_{k=1}^{K} \langle \mathbf{y}_k, \mathbf{x}_k - \mathbf{x} \rangle + \sum_{k=1}^{K} \frac{\rho_k}{2} \|\mathbf{x}_k - \mathbf{x}\|^2. \tag{3.70}
$$

Note that here a set of penalty coefficient $\{\rho_k\}$ are used. Consider the following ADMM iteration:

$$
\mathbf{x}^{t+1} = \underset{\mathbf{x} \in \mathcal{X}}{\operatorname{argmin}}\, L_\rho(\{\mathbf{x}_k^t\}, \mathbf{x}; \mathbf{y}^t) \tag{3.71}
$$

Each node $k$ computes $\mathbf{x}_k$ by solving:

$$
\mathbf{x}_k^{t+1} = \underset{\mathbf{x}_k}{\arg\min}\, g_k(\mathbf{x}_k) + \langle \mathbf{y}_k^t, \mathbf{x}_k - \mathbf{x}^{t+1} \rangle + \frac{\rho_k}{2} \|\mathbf{x}_k - \mathbf{x}^{t+1}\|^2. \tag{3.72}
$$

Each node $k$ updates the dual variable:

$$
\mathbf{y}_k^{t+1} = \mathbf{y}_k^t + \rho_k \left( \mathbf{x}_k^{t+1} - \mathbf{x}^{t+1} \right). \tag{3.73}
$$

The following conditions guarantee that the ADMM iteration, when applied to solve problem (3.68), will converge to a stationary solution [110]:

1. Each $g_k$ can be nonconvex, and has Lipschitz continuous gradient with constant $L_k$. Moreover, $h$ is convex (possible nonsmooth); $\mathcal{X}$ is a closed convex set.
2. For all $k$, the stepsize $\rho_k$ is chosen large enough such that:
   (a) For all $k$, the $\mathbf{x}_k$ subproblem (3.72) is strongly convex with modulus $\beta_k(\rho_k)$.
   (b) For all $k$, $\rho_k \beta_k(\rho_k) > 2L_k^2$ and $\rho_k \geq L_k$.
3. $f_0(\mathbf{x})$ is bounded from below over $\mathcal{X}$.

## 3.7     Conclusion

In this section, we have discussed a few basic first-order methods for solving relatively large-scale problems. We have first introduced some basic notions of continuous optimization, and then presented a variety of first-order methods including the classical (projected) GD method, the coordinate descent method, the block successive upper-bound minimization method and the augmented Lagrangian-based methods. Some basic properties of these methods are summarized in the following table:

**Table 3.3** Commonly Used Coordinate Selection Rules [55]

At each iteration $t + 1$, define a set of new variables $\{\hat{\mathbf{x}}_k^{t+1}\}_{k=1}^K$ by:

$$\hat{\mathbf{x}}_k^{t+1} \in \min_{\mathbf{x}_k \in \mathcal{X}_k} f_0\left(\mathbf{x}_k, \mathbf{x}_{-k}^t\right), \quad k = 1, \cdots, K. \qquad (3.35)$$

Let $k(t + 1)$ denote the index picked at iteration $t + 1$.
The following are common block selection rules.

- *Cyclic rule*: The coordinates are chosen cyclically (in a Gauss–Seidel manner), that is,

$$k(t + 1) = \{(t \bmod K) + 1\}.$$

- *Essentially cyclic (E-C) rule*: Each block is updated at least once in $T$ iterations, that is,

$$\bigcup_{i=1}^T k(t + 1) = \{1, \cdots, K\}, \ \forall \, t.$$

- *Jacobian rule*: All blocks are selected at the same time.
- *Gauss–Southwell (G-So) rule*: At each iteration $t$, for some $q \in (0, \, 1]$

$$k(t + 1) \in \left\{ k \ \middle| \ \|\hat{\mathbf{x}}_k^{t+1} - \mathbf{x}_k^t\| \geq q \max_j \|\hat{\mathbf{x}}_j^{t+1} - \mathbf{x}_j^t\| \right\}.$$

- *Maximum block improvement (MBI) rule*: At each iteration $r$, we have

$$k(t) \in \arg\max_k f_0(\hat{\mathbf{x}}_k^{t+1}, \mathbf{x}_{-k}^t).$$

- *Randomized rule*: Define $p_{\min} \in (0, 1)$ to be a constant. At each iteration $r$, use a probability vector $(p_1^{t+1}, \ldots, p_K^{t+1})$ in which $\sum_{k=1}^K p_k^{t+1} = 1$ and $p_k^t > p_{\min}, \forall k$; then we have

$$k(t + 1) \sim \mathrm{multi}(p_1^{t+1}, \cdots, p_K^{t+1})$$

where "multi" here represents a multinomial distribution.

**Table 3.4** Commonly Used Upper Bounds [55]

- *Proximal Upper Bound*: Given a constant $\gamma > 0$, one can construct a bound by adding a quadratic penalization (i.e., the proximal term)

$$u_k(\mathbf{x}_k, \mathbf{z}) := f_0(\mathbf{x}_k, \mathbf{z}_{-k}) + \frac{\gamma}{2}\|\mathbf{x}_k - \mathbf{z}_k\|^2.$$

- *Quadratic Upper Bound*: Suppose $f_0(\mathbf{x}) = f_0(\mathbf{x}_1, \cdots, \mathbf{x}_K)$, where $f_0$ is smooth with $\mathbf{H}_k$ as the Hessian matrix for the $k$th block. Then one can construct the following bound:

$$u_k(\mathbf{x}_k, \mathbf{z}) := f_0(\mathbf{z}) + \langle \nabla_k f_0(\mathbf{z}), \mathbf{x}_k - \mathbf{z}_k \rangle + \frac{1}{2}(\mathbf{x}_k - \mathbf{z}_k)^T \Phi_k (\mathbf{x}_k - \mathbf{z}_k)$$

  where $\Phi_k \succeq \mathbf{H}_k$ is a positive semidefinite matrix.

- *Linear Upper Bound*: Suppose $f_0$ is differentiable and concave, then one can construct

$$u_k(\mathbf{x}_k, \mathbf{z}) := f_0(\mathbf{z}) + \langle \nabla_k f_0(\mathbf{z}), \mathbf{x}_k - \mathbf{z}_k \rangle.$$

- *Jensen's Upper Bound*: Suppose $f_0(\mathbf{x}) := f_0(\mathbf{a}_1^T \mathbf{x}_1, \cdots, \mathbf{a}_K^T \mathbf{x}_K)$ where $\mathbf{a}_k \in \mathbb{R}^{n_k}$ is a coefficient vector, and $f_0$ is convex with respect to each $\mathbf{a}_k^T \mathbf{x}_k$. Let $\mathbf{w}_k \in \mathbb{R}_+^{n_k}$ denote a weight vector with $\|\mathbf{w}_k\|_1 = 1$. Then one can use Jensen's inequality and construct

$$u_k(\mathbf{x}_k, \mathbf{z}) := \sum_{j=1}^{n_k} \mathbf{w}_k(j) f_0 \left( \frac{\mathbf{x}_k(j)}{\mathbf{w}_k(j)} (\mathbf{x}_k(j) - \mathbf{z}_k(j)) + \mathbf{a}_k^T \mathbf{z}_k, \mathbf{z}_{-k} \right)$$

  where $\mathbf{w}_k(j)$ represents the $j$th element in vector $\mathbf{w}_k$.

**Table 3.5** Comparison of Algorithms.

| Algorithm | Problem Types | Properties |
|---|---|---|
| GD | Nonconvex, nonsmooth | Converges to KKT; stepsize rule (Table 3.1) |
| BCD | Nonconvex, nonsmooth, multiblock | Converges to KKT under suitable condition; Common coordinate rules (Table 3.3) |
| BSUM | Nonconvex, nonsmooth, multiblock | Convergence to KKT; Common upper-bound selection (Table 3.4) |
| AL | Nonconvex, general coupling constraints | Convergence to KKT with proper $\rho$ |
| ADMM | Convex, linear coupling constraints | Convergence to optimal primal-dual solution Commonly 2-block variables, separable objective |

# 4  Sparse Optimization

This chapter reviews a collection of sparse optimization models and algorithms. The review focuses on introducing these algorithms along with their motivations and basic properties. The mathematical tools from the previous section will be heavily replied upon.

This chapter is organized as follows. Section 4.1 reviews a list of (convex) sparse optimization models, which deal with different types of signals and different kinds of noise; they can also include additional features as objectives and constraints that arise in practical problems. Section 4.2 demonstrates that the convex sparse optimization problems can be transformed in equivalent cone programs and solved by off-the-shelf algorithms; yet it argues that these algorithms are usually inefficient for large-scale problems. Sections 4.3–4.12 cover a large variety of algorithms for sparse optimization. These algorithms have different strengths and fit different applications. The efficiency of these algorithms, for the most part, comes from the use of the often closed-form shrinkage-like operations, which is introduced in Section 4.4. Then, Section 4.5 presents a prox-linear framework and gives several algorithms under this framework that are based on gradient descent and take advantages of shrinkage-like operations. These algorithms can often be linked with the BSUM framework discussed in Chapter 3. Duality is a very powerful tool in modern convex optimization; sparse optimization is not an exception. Section 4.6 derives a few dual models and algorithms for sparse optimization and discusses their properties. One class of dual algorithms (i.e., the ADMM algorithm introduced in Chapter 3) is very efficient and extremely versatile, applicable to nearly all convex sparse optimization problems. The framework and applications of these algorithms are given in Section 4.8. Unlike other algorithms, the homotopy algorithms in Section 4.9 produces not just one solution but a path of solutions for the LASSO model (which is given in (4.2b) below); not only are they efficient at producing multiple solutions corresponding to different parameter values, they are especially fast if the solution is extremely sparse. All the above algorithms can be (often significantly) accelerated by appropriately setting their stepsize parameters that determine the amount of progress at each iteration. Such techniques are reviewed in Section 4.10. Unlike all previous algorithms, greedy algorithms do not necessarily correspond to an optimization model; instead of systematically searching for solutions that minimize objective functions, they build sparse solutions by constructing their supports step by step in a greedy manner. They have very good performance on fast-decaying signals. A few greedy algorithms are discussed in Section 4.11. Most of the algorithms and techniques in Sections 4.4–4.11 can be extended from sparse vector recovery to low-rank matrix recovery. However, for the latter problem, there is also a class of algorithms based on decomposing a low-rank matrix into the product of a skinny matrix and a fat one; algorithms in this class require relatively less memory and run very efficiently; they are briefly discussed in Section 4.12. Finally, Section 4.13 summarizes these algorithms and compares their advantages and suitable applications.

## 4.1 Sparse Optimization Models

In the next few sections, we provide an overview for a set of widely used algorithms for solving problems in the form of

$$\min_{\mathbf{x}}\{r(\mathbf{x}) : \mathbf{A}\mathbf{x} = \mathbf{b}\}, \tag{4.1a}$$

$$\min_{\mathbf{x}} r(\mathbf{x}) + \mu h(\mathbf{A}\mathbf{x} - \mathbf{b}), \tag{4.1b}$$

$$\min_{\mathbf{x}}\{r(\mathbf{x}) : \bar{h}(\mathbf{A}\mathbf{x} - \mathbf{b}) \le \sigma\}, \tag{4.1c}$$

where $r(\mathbf{x})$, $h(\mathbf{x})$, and $\bar{h}(\mathbf{x})$ are convex functions; $\sigma$ and $\mu$ are some positive scalars.

Function $r(\mathbf{x})$ is a regularizer and is typically non-differentiable in sparse optimization problems; functions $h(\mathbf{x})$ and $\bar{h}(\mathbf{x})$ are data fidelity measures, which are often differentiable though there are exceptions such as $\ell_1$ and $\ell_\infty$ fidelity measures (or loss functions), which are non-differentiable.

When introducing algorithms, our *first* focus is given to the $\ell_1$ problems

$$\min_{\mathbf{x}}\{\|\mathbf{x}\|_1 : \mathbf{A}\mathbf{x} = \mathbf{b}\}, \tag{4.2a}$$

$$\min_{\mathbf{x}} \|\mathbf{x}\|_1 + \frac{\mu}{2}\|\mathbf{A}\mathbf{x} - \mathbf{b}\|_2^2, \tag{4.2b}$$

$$\min_{\mathbf{x}}\{\|\mathbf{x}\|_1 : \|\mathbf{A}\mathbf{x} - \mathbf{b}\|_2 \le \sigma\}. \tag{4.2c}$$

We also consider the variants in which the $\ell_1$ norm of $\mathbf{x}$ is replaced by regularization functions corresponding to transform sparsity, joint sparsity, low rankness, as well as those involving more than one regularization terms. These problems are convex and nonsmooth. For example, as variants of problem (4.2a), most algorithms in this chapter can be extended to solving

$$\min_{\mathbf{s}}\{\|\mathbf{s}\|_1 : \mathbf{A}\Psi\mathbf{s} = \mathbf{b}\}, \tag{4.3a}$$

$$\min_{\mathbf{x}}\{\|\Upsilon\mathbf{x}\|_1 : \mathbf{A}\mathbf{x} = \mathbf{b}\}, \tag{4.3b}$$

$$\min_{\mathbf{u}}\{\mathrm{TV}(\mathbf{u}) : \mathcal{A}(\mathbf{u}) = \mathbf{b}\}, \tag{4.3c}$$

$$\min_{\mathbf{X}}\{\|\mathbf{X}\|_{2,1} : \mathbf{A}\mathbf{X} = \mathbf{b}\}, \tag{4.3d}$$

$$\min_{\mathbf{X}}\{\|\mathbf{X}\|_* : \mathcal{A}(\mathbf{X}) = \mathbf{b}\}, \tag{4.3e}$$

where $\mathcal{A}(\cdot)$ is a given linear operator, $\Upsilon$ is a linear transform, TV(**u**) is a certain discretization of total variation $\int |\nabla \mathbf{u}(x)| dx$ (also, see [111] for a rigorous definition in the space of generalized functions of bounded variation), the $\ell_{2,1}$-norm is defined as

$$\|\mathbf{X}\|_{2,1} := \sum_{i=1}^{m} \left\| [x_{i1} \ x_{i,2} \cdots x_{in}] \right\|_2 , \tag{4.4}$$

and the nuclear norm $\|\mathbf{X}\|_*$ equals the sum of singular values of $\mathbf{X}$. Models (4.3a)–(4.3e) are widely used to obtain, respectively, a vector that is sparse under $\Psi$, a vector that become sparse by transform $\Upsilon$, a piece-wise constant signal, a matrix consisted of joint-sparse vectors, and a low-rank matrix.

Partition the set $\{1, \ldots, n\}$ into $S$ subsets by the following $\{1, \ldots, n\} = \mathcal{G}_1 \cup \mathcal{G}_2 \cup \cdots \cup \mathcal{G}_S$ where the $\mathcal{G}_i$'s are certain sets of coordinates and $\mathcal{G}_i \cap \mathcal{G}_j = \emptyset$, $\forall i \neq j$. Define the (weighted) group $\ell_{2,1}$-norm as

$$\|\mathbf{x}\|_{\mathcal{G},2,1} = \sum_{s=1}^{S} w_s \|\mathbf{x}_{\mathcal{G}_s}\|_2,$$

where $w_s \geq 0$, $s = 1, \ldots, S$, are a given set of weights and $\mathbf{x}_{\mathcal{G}_s}$ is a subvector of $\mathbf{x}$ corresponding to the coordinates in $\mathcal{G}_s$. The model

$$\min_{\mathbf{x}} \{ \|\mathbf{x}\|_{\mathcal{G},2,1} : \mathbf{A}\mathbf{x} = \mathbf{b} \} \tag{4.5}$$

tends to give solutions with all but a few nonzero subvectors $\mathbf{x}_{\mathcal{G}_s}$.

When the equality constraints of the problems in (4.3) and (4.5) do not hold for the target solutions (e.g., when observation error exists), one can apply relaxation and penalty to the constraints like those in (4.2b) and (4.2c).

Depending on the type of noise, one may replace the $\ell_2$-norm in (4.2b) and (4.2c) by other distance measures or loss functions such as the Kullback–Leibler divergence [112, 113] of two probability distributions $p$ and $q$ and the logistic loss function [114].

Practical problems sometimes have additional constraints such as the non-negativity constraints $\mathbf{x} \geq 0$ or bound constraints $\mathbf{l} \leq \mathbf{x} \leq \mathbf{u}$.

The $\ell_{2,1}$ norm in (4.4) can be generalized to the $\ell_{p,q}$-norm

$$\|\mathbf{X}\|_{p,q} = \left( \sum_i (\sum_j |x_{i,j}|^p)^{q/p} \right)^{1/q} ,$$

which is convex if $1 \leq p, q \leq \infty$. For these problems and extensions, we study their algorithms extending from those for $\ell_1$ minimization.

Furthermore, it is not rare to see problems with *complex-valued* variables and data. We discuss how to handle them in Section 4.7.3.

This chapter does not cover the model

$$\min_{\mathbf{x}} \{ \|\mathbf{A}\mathbf{x} - \mathbf{b}\|_2 : \|\mathbf{x}\|_1 \leq k \},$$

as well as its variants. Solving such models may require projection to a polyhedral set such as the $\ell_1$-ball $\{ \mathbf{x} \in \mathbb{R}^n : \|\mathbf{x}\|_1 \leq 1 \}$, which is more expensive than solving subproblems of minimizing $\ell_1$-norm. However, there are efficient projection methods [115, 116, 117], as well as algorithms [118, 119] for $\ell_1$-constrainted sparse optimization problems.

## 4.2          Classic Solvers

Most nonsmooth convex problems covered in this chapter can be transformed to optimization problems of standard forms, for which off-the-shelf algorithms and codes are available. In particular, using the identity

$$\|\mathbf{x}\|_1 = \min_{\mathbf{x}_1,\mathbf{x}_2}\{\mathbf{1}^T(\mathbf{x}_1 + \mathbf{x}_2) : \mathbf{x}_1 - \mathbf{x}_2 = \mathbf{x}, \mathbf{x}_1 \geq \mathbf{0}, \mathbf{x}_2 \geq \mathbf{0}\}, \tag{4.6}$$

where the minimizers $\mathbf{x}_1$ and $\mathbf{x}_2$ represent the positive and negative parts of $\mathbf{x}$, respectively, and $\mathbf{1} = [1, 1, \ldots, 1]^T$, one can transform problem (4.2a) to

$$\min_{\mathbf{x}_1,\mathbf{x}_2}\{\mathbf{1}^T(\mathbf{x}_1 + \mathbf{x}_2) : \mathbf{A}(\mathbf{x}_1 - \mathbf{x}_2) = \mathbf{b}, \mathbf{x}_1 \geq \mathbf{0}, \mathbf{x}_2 \geq \mathbf{0}\}. \tag{4.7}$$

From a solution $(\mathbf{x}_1^*, \mathbf{x}_2^*)$ to (4.7), one obtains a solution $\mathbf{x}^* = \mathbf{x}_1^* - \mathbf{x}_2^*$ to (4.2a).

Problem (4.7) is a standard-form linear program and can be solved by the simplex method, active-set method, as well as interior-point method, which have reliable academic and/or commercial solver packages.

Based on identity (4.6), problem (4.2b) can be transformed to a linearly constrained quadratic program (QP) and problem (4.2c) to a QP with quadratic and linear constraints, both of which can be further transformed to second-order cone programs (SOCPs) and solved by an interior-point method.

Likewise, we can model various nuclear norm-based, low-rank matrix recovery models as semidefinite programs (SDP) based on the identity [120]

$$\|\mathbf{X}\|_* = \min_{\mathbf{Y},\mathbf{Z}}\left\{\frac{1}{2}(\mathrm{tr}(\mathbf{Y}) + \mathrm{tr}(\mathbf{Z})) : \begin{bmatrix} \mathbf{Y} & \mathbf{X} \\ \mathbf{X}^T & \mathbf{Z} \end{bmatrix} \succeq \mathbf{0}\right\} \tag{4.8}$$

where $\mathrm{tr}(\mathbf{M})$ is the trace of $\mathbf{M}$, and $\mathbf{M} \succeq \mathbf{0}$ means that $\mathbf{M}$ is symmetric and positive semidefinite. One can show (4.8) by considering the singular value decomposition $\mathbf{X} = \mathbf{U}_{m\times m}\Sigma_{m\times n}\mathbf{V}_{n\times n}^T$ and noticing that due to the constraint in (4.8),

$$0 \leq \frac{1}{2}\mathrm{tr}\left(\begin{bmatrix} \mathbf{U}^T & -\mathbf{V}^T \end{bmatrix}\begin{bmatrix} \mathbf{Y} & \mathbf{X} \\ \mathbf{X}^T & \mathbf{Z} \end{bmatrix}\begin{bmatrix} \mathbf{U} \\ -\mathbf{V} \end{bmatrix}\right)$$

$$= \frac{1}{2}\left(\mathrm{tr}(\mathbf{U}^T\mathbf{Y}\mathbf{U}) + \mathrm{tr}(\mathbf{V}^T\mathbf{Z}\mathbf{V})\right) - \mathrm{tr}(\Sigma)$$

$$= \frac{1}{2}(\mathrm{tr}(\mathbf{Y}) + \mathrm{tr}(\mathbf{Z})) - \|\mathbf{X}\|_*,$$

which equals 0 if we let $\mathbf{Y} = \mathbf{U}\Sigma_{m\times m}\mathbf{U}^T$ and $\mathbf{Z} = \mathbf{V}\Sigma_{n\times n}\mathbf{V}^T$.

While linear programming, SOCP, and SDP are well known and have very reliable solvers, many sparse optimization problems – especially those involving images and high-dimensional data or variables – are often too large in size for their solvers. For example, to recover an image of 10 million pixels, there will be at least 10 million variables along with, say, 1 million constraints in the problem, and even worse, in sparse optimization problems the constraints tend to have *dense* coefficient matrices. Such data scale is far beyond the limit of the most standard solvers of linear programming, SOCP, and SDP. As a result, these solvers either run very slowly or report out of memory.

In the simplex and active-set methods, a matrix containing $\mathbf{A}$ must be inverted or factorized to compute the next point. This becomes very difficult when $\mathbf{A}$ is large and dense – which is often the case in sparse optimization applications. Each interior-point iteration approximately solves a Newton system and thus also factorizes a matrix involving $\mathbf{A}$.

Even when the factorization of a large and dense matrix can be done, it is still very disadvantageous to omit the facts that the solution will be sparse and that $\mathbf{A}$ may have structures permitting fast multiplications of $\mathbf{A}$ and $\mathbf{A}^T$ to vectors.

Nevertheless, simplex and interior-point methods are *reliable* and can return *highly accurate* solutions. They are good benchmark solvers for small-sized sparse optimization problems, and they can be integrated with fast but less-accurate solvers and in charge of returning accurate solutions to smaller subproblems.

For convex and nonsmooth optimization problems, it is also natural to consider subgradient methods, namely, applying subgradient iterations to the unconstrained model (4.1b) and projected subgradient iterations to the constrained models (4.1a) and (4.1c). These methods are typically simple to implement, and the subgradient formulas for many nonsmooth functions in this chapter are easy to derive. On the down side, the convergence guarantee and practical convergence speed of (projected) subgradient are somewhat weak. The convergence typically requires careful choice of stepsizes, for example, those that diminish in the limit but diverge in sum (see [51]).

Unlike the above off-the-shelf methods, the algorithms we overview in this chapter do not necessarily invert or factorize large matrices or rely on the subgradients of non-smooth functions. Many of them are based on multiplying $\mathbf{A}$ and $\mathbf{A}^T$ to vectors, and exactly solving simpler subproblems involving the nonsmooth functions. For problems with sparse solutions, such algorithms are more CPU and memory efficient than the off-the-shelf solutions above. Certain algorithms can even return highly accurate solutions.

## 4.3    The BSUM Method for Sparse Optimization

The BSUM algorithm introduced in the previous chapter (see Section 3.4) is a generalization of the classical BCD method. In this section we show its applications in sparse optimization.

Consider the following $\ell_{2,1}$ penalized problem

$$\min \sum_{s=1}^{S} \|\mathbf{x}_s\|_2 + f(\mathbf{x}), \tag{4.9}$$

where $f$ is a differentiable function. This problem is a slight generalization of (4.2b), where in the latter problem each scalar $x_i$ is a block variable.

The recent work [121] describes a very efficient method for these problems, which is called the block coordinate gradient descent (BCGD) method. For each block $s$, define

$$\mathbf{d}(\bar{\mathbf{x}}; s) = \arg\min_{\mathbf{d}} \|\bar{\mathbf{x}}_s + \mathbf{d}\|_2 + \nabla_{\mathbf{x}_s} f(\bar{\mathbf{x}})^T \mathbf{d} + \frac{1}{\delta} \|\mathbf{d}\|_2^2, \tag{4.10}$$

which can be viewed as the prox-linear step of block $s$. Given a current point $\mathbf{x}^k$, the standard prox-linear step is

$$\mathbf{x}^{k+1} = \arg\min_{\mathbf{x}} \sum_s \left( \|\mathbf{x}_s\|_2 + \nabla_{\mathbf{x}_s} f(\mathbf{x}^k)^T (\mathbf{x}_s - \mathbf{x}_s^k) + \frac{1}{\delta} \|\mathbf{x}_s - \mathbf{x}_s^k\|_2^2 \right). \tag{4.11}$$

If we focus just on block $s$, it is easy to see $\mathbf{x}_s^{k+1} - \mathbf{x}_s^k = \mathbf{d}(\mathbf{x}^k, s)$.

The algorithm BCGD is based on iterating the following steps

1. choose a block $s$
2. compute $\mathbf{d}(\mathbf{x}^k; s)$
3. set a stepsize $\alpha_k > 0$
4. set $\mathbf{x}_s^{k+1} \leftarrow \mathbf{x}_s^k + \alpha_k \mathbf{d}(\mathbf{x}^k; s)$ and $\mathbf{x}_t^{k+1} \leftarrow \mathbf{x}_t^k$ for $t \neq s$.

Clearly, when $\alpha_k = 0$, each block variable $\mathbf{x}_s$ is obtained by optimizing the following problem:

$$\mathbf{x}_s^{k+1} = \arg\min_{\mathbf{x}_s} \left( \|\mathbf{x}_s\|_2 + \nabla_{\mathbf{x}_s} f(\mathbf{x}^k)^T (\mathbf{x}_s - \mathbf{x}_s^k) + \frac{1}{\delta} \|\mathbf{x}_s - \mathbf{x}_s^k\|_2^2 \right)$$

$$:= u_s(\mathbf{x}_s; \mathbf{x}^k).$$

For small enough $\delta$, $u_s(\mathbf{x}_s; \mathbf{x}^k)$ can be shown as an upper bound for $f(\mathbf{x}_s, \mathbf{x}_{-s})$ (i.e., satisfying conditions (3.41)), hence a BSUM algorithm.

Among the three block selection rules introduced in [121], the Gauss–Seidel rule, the Generalized Gauss–Seidel rule and the Gauss–Southwell rule, the Gauss–Southwell rule has been shown to be very effective for sparse optimization. In such block selection rules, a block index $s$ is picked if it maximizes certain merit value, which is set in one of the following two ways. The first way is referred to as the *Gauss–Southwell-r rule*, in which the merit is given as $m_s := \|\mathbf{d}(\mathbf{x}^k; s)\|_\infty$. That is, the block that potentially can travel the longest distance (measured by the $\ell_\infty$ norm) will be picked for optimization. The second way is referred to as the *Gauss–Southwell-q rule*, in which the merit function is set as

$$m_s := \|\mathbf{x}_s^k\|_2 - (\|\mathbf{x}_s^k + \mathbf{d}\|_2 + \nabla_{\mathbf{x}_s} f(\mathbf{x}^k)^T \mathbf{d} + \frac{1}{\delta} \|\mathbf{d}\|_2^2)$$

$$= f(\mathbf{x}^k) + \|\mathbf{x}_s^k\|_2 - (f(\mathbf{x}^k) + \|\mathbf{x}_s^k + \mathbf{d}\|_2 + \nabla_{\mathbf{x}_s} f(\mathbf{x}^k)^T \mathbf{d} + \frac{1}{\delta} \|\mathbf{d}\|_2^2)$$

$$= f(\mathbf{x}^k) + \|\mathbf{x}_s^k\|_2 - (u_s(\mathbf{d}(\mathbf{x}^k, s) + \mathbf{x}_s^k, \mathbf{x}^k)).$$

Clearly, this rule will pick the block that achieves the largest gap between its previous objective value $f(\mathbf{x}^k) + \|\mathbf{x}_s^k\|_2$ and the minimum value of its upper-bound function.

Roughly speaking, the relatively superior performance of the Gauss–Southwell rule on sparse optimization over the other two rules is due to the built-in greediness –

it selects the block that has the most "potential" to update (according to two different criteria), which tends to be the true nonzero blocks in the solution. Take $f(\mathbf{x}) = \frac{\mu}{2}\|\mathbf{Ax} - \mathbf{b}\|_2^2$ as an example. Let $\mathcal{S}^o$ denote the set of nonzero blocks in $\mathbf{x}^o$, the original group-sparse vector. Suppose $\mathbf{b} = \mathbf{Ax}^o + \mathbf{w}$, where $\mathbf{w}$ is the noise, and $\{s : \mathbf{x}_s^k \neq \mathbf{0}\} \subset \mathcal{S}$, namely, the current active blocks are among the nonzero blocks and thus $\mathbf{Ax}^k = \mathbf{A}_{\mathcal{S}^o}\mathbf{x}_{\mathcal{S}^o}^k$. Then, $\nabla_{\mathbf{x}_s}f(\mathbf{x}^k) = \mu\mathbf{A}_s^T(\mathbf{Ax}^k - \mathbf{b}) = \mu\mathbf{A}_s^T(\mathbf{A}_{\mathcal{S}^o}\mathbf{x}_{\mathcal{S}^o}^k - \mathbf{A}_{\mathcal{S}^o}\mathbf{x}_{\mathcal{S}^o}^o - \mathbf{w}) = \mu\mathbf{A}_s^T\mathbf{A}_{\mathcal{S}^o}(\mathbf{x}_{\mathcal{S}^o}^k - \mathbf{x}_{\mathcal{S}^o}^o) - \mu\mathbf{A}_s^T\mathbf{w}$. When the matrices satisfying certain restricted isometry property (RIP), it tends to have orthogonal columns when restricted to sparse vectors. If $s \in \mathcal{S}^o$, then $\mathbf{A}_s^T\mathbf{A}_{\mathcal{S}^o}$ tends to have large entries; otherwise, $\mathbf{A}_s^T\mathbf{A}_{\mathcal{S}^o}$ tends to have smaller entries. As a result, $\nabla_{\mathbf{x}_s}f(\mathbf{x}^k)$ tends to be larger for $s \in \mathcal{S}^o$, and the same blocks tend to give rise to larger $m_s$. In short, the stronger correlation between the residual $\mathbf{Ax}^k - \mathbf{b}$ and the submatrices $\mathbf{A}_s$ over $s \in \mathcal{S}^o$ is taken advantage of by greedy selections of blocks, and this leads to having a good chance of picking blocks $s \in \mathcal{S}^o$ for update. Hence, blocks in $(\mathcal{S}^o)^c$ are seldom chosen for update; under some conditions, one can even show that they are never chosen. Of course, we have assumed $f(\mathbf{x}) = \frac{\mu}{2}\|\mathbf{Ax} - \mathbf{b}\|_2^2$ and $\mathbf{A}$ to have good RIP, which is not the most general case. However, in many sparse optimization problems, greed is good to varying extents. Some interesting algorithms and analysis can be found in [122, 123].

## 4.4    Shrinkage Operation

Shrinkage operation and its extensions are often used in sparse optimization algorithms. While $\ell_1$-norm is non-differentiable, it has the excellent property of being component-wise separable (in the sense $\|\mathbf{x}\|_1 = \sum_{i=1}^n |x_i|$). It is very easy to minimize $\|\mathbf{x}\|_1$ together with other component-wise separable functions on $\mathbf{x}$ since it reduces to a sequence of independent subproblems. For example, the problem

$$\min_{\mathbf{x}} \|\mathbf{x}\|_1 + \frac{1}{2\tau}\|\mathbf{x} - \mathbf{z}\|_2^2, \tag{4.12}$$

where $\tau > 0$, is equivalent to solving $\min_{x_i} |x_i| + \frac{1}{2\tau}|x_i - z_i|^2$ over each $i$. Applying an analysis over the three cases of solution $(x_{\mathrm{opt}})_i$ – positive, zero, and negative – one can obtain the closed-form solution

$$(x_{\mathrm{opt}})_i = \begin{cases} z_i - \tau, & z_i > \tau \\ 0, & -\tau \leq z_i \leq \tau, \\ z_i + \tau, & z_i < -\tau. \end{cases} \tag{4.13}$$

The solution is illustrated in Figure 4.1. Visually, the problem takes the input vector $\mathbf{z}$ and shrinks it to zero component-wise. Therefore, $(x_{\mathrm{opt}})_i$ is called the *soft-thresholding* or *shrinkage* of $z_i$, written as $(x_{\mathrm{opt}})_i = \mathrm{shrink}(z_i, \tau)$. The solution of (4.12) is given by

$$\mathbf{x}_{\mathrm{opt}} = \mathrm{shrink}(\mathbf{z}, \tau). \tag{4.14}$$

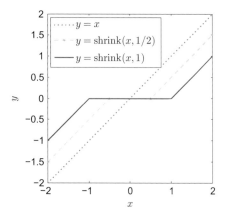

**Figure 4.1** Illustration of shrink$(x, \tau)$.

If we slightly abuse the notation and treat $0/0 = 0$, we get shrink$(z_i, \tau) = \max\{|z_i| - \tau, 0\} \cdot (z_i/|z_i|)$ and thus

$$\text{shrink}(\mathbf{z}, \tau) = \max\{|\mathbf{z}| - \tau, \mathbf{0}\} \cdot \text{sign}(\mathbf{z}), \quad \text{component-wise,} \tag{4.15}$$

where

$$\text{sign}(x) = \begin{cases} 1, & x > 0, \\ 0, & x = 0, \\ -1, & x < 0. \end{cases}$$

Operation (4.15) has been applied to image denoising [124] (applied to the wavelet coefficients of images) and have been found in earlier literature. It is now underpinning a large number of sparse optimization algorithms and solvers.

If problem (4.12) is subject to constraints $\mathbf{x} \geq \mathbf{0}$, the solution becomes

$$\mathbf{x}_{\text{opt}} = \max\{\mathbf{z} - \tau, \mathbf{0}\}, \quad \text{component-wise.} \tag{4.16}$$

Unlike general nonsmooth optimization problems, sparse optimization problems generally have rather simple nonsmooth terms $r(\mathbf{x})$. One kind of simplicity is that $r(\mathbf{x})$ is easy to minimize together with simple functions such as $\|\mathbf{x} - \mathbf{z}\|_2^2$. As a side note, the general problem of minimizing $r(\mathbf{x}) + \frac{1}{2}\|\mathbf{x} - \mathbf{z}\|_2^2$ is called the Moreau–Yosida regularization of $r$, which has a rich literature.

### 4.4.1 Generalizations of Shrinkage

When the second term in (4.12) is *convex* and *component-wise separable*, the solution of (4.12) is often easy to obtain. We next study a few useful generalizations to (4.12), which are not component-wise separable but still easy to handle. They are used in algorithms for obtaining, for example, transform, group, or jointly sparse vectors, as well as low-rank matrices.

If $\ell_1$ is replaced by $\ell_2$ in (4.12), we obtain the problem

$$\min_{\mathbf{x}} \|\mathbf{x}\|_2 + \frac{1}{2\tau}\|\mathbf{x} - \mathbf{z}\|_2^2, \tag{4.17}$$

whose solution is given by

$$\max\{\|\mathbf{x}\|_2 - \tau, 0\} \cdot (\mathbf{x}/\|\mathbf{x}\|_2). \tag{4.18}$$

Unlike component-wise shrinkage (4.15), (4.18) shrinks the entire vector toward the origin along the radial line. Problem (4.18) may appear as subproblems in joint/group-sparse recovery algorithms.

When $\mathbf{x}$ and $\mathbf{z}$ are *complex-valued*, (4.15) and (4.18) remain correct under the definition $\|\mathbf{y}\|_2 = \sum_i |\mathrm{r}(y_i)^2 + \mathrm{i}(y_i)^2|^{1/2}$, where $\mathrm{r}(\cdot)$ and $\mathrm{i}(\cdot)$ are the real and imaginary parts of the input, respectively.

Let $\mathbf{Z} = \mathbf{U}\Sigma\mathbf{V}^T$ be the singular value decomposition of matrix $\mathbf{Z}$ and $\hat{\Sigma}$ be the diagonal matrix with diagonal entries

$$\mathrm{diag}(\hat{\Sigma}) = \mathrm{shrink}(\mathrm{diag}(\Sigma), \tau), \tag{4.19}$$

then we have from [125]:

$$\mathbf{U}\hat{\Sigma}\mathbf{V}^T = \arg\min_{\mathbf{X}} \|\mathbf{X}\|_* + \frac{1}{2\tau}\|\mathbf{X} - \mathbf{Z}\|_F^2, \tag{4.20}$$

where we recall that $\|\cdot\|_*$ and $\|\cdot\|_F^2$ equal the sum of the singular values and squared singular values of the input matrix, respectively. This result follows essentially from the unitary invariant properties: $\|\mathbf{X}\|_* \equiv \|\mathbf{AXB}\|_*$ and $\|\mathbf{X}\|_F \equiv \|\mathbf{AXB}\|_F$ for any matrix $\mathbf{X}$ and unitary matrices $\mathbf{A}$ and $\mathbf{B}$. In general, matrix problems with only unitary-invariant functions (e.g., $\|\cdot\|_*$, $\|\cdot\|_F$, spectral norm, trace, all Schatten (semi)-norms) and constraints (e.g., positive or negative (semi)-definiteness) typically reduce to vector problems regarding singular values. Computationally, singular value decomposition is expensive for large matrices and, if it is needed every iteration, it can become the performance bottleneck. Therefore, algorithms that iteratively call (4.20) often use its approximate; see [126, 127], for example.

If we generalize (4.12) to

$$\min_{\mathbf{x}} \|\Psi\mathbf{x}\|_1 + \frac{1}{2\tau}\|\mathbf{x} - \mathbf{z}\|_2^2, \tag{4.21}$$

where $\Psi$ is an orthogonal linear transform[1] then by introducing variable $\mathbf{y} = \Psi\mathbf{x}$, we have $\mathbf{x} = \Psi^T\mathbf{y}$ and $\|\mathbf{x} - \mathbf{z}\|_2 = \|\Psi(\mathbf{x} - \mathbf{z})\|_2 = \|\mathbf{y} - \Psi\mathbf{z}\|_2$ and obtain the new problem

$$\min_{\mathbf{y}} \|\mathbf{y}\|_1 + \frac{1}{2\tau}\|\mathbf{y} - \Psi\mathbf{z}\|_2^2. \tag{4.22}$$

Given the solution $\mathbf{y}_{\mathrm{opt}}$ of (4.22), applying $\mathbf{x}_{\mathrm{opt}} = \Psi^T\mathbf{y}_{\mathrm{opt}}$, we recover the solution of (4.21):

$$\mathbf{x}_{\mathrm{opt}} = \Psi^T \mathrm{shrink}(\Psi\mathbf{z}, \tau). \tag{4.23}$$

---

[1] $\Psi$ can be represented by a matrix with orthonormal columns.

Another generalization involves total variation, which is often used in image-processing tasks [128] and models for piece-wise constant signals:

$$\min_{\mathbf{u}} TV(\mathbf{u}) + \frac{1}{2\tau}\|\mathbf{u} - \mathbf{z}\|_2^2. \tag{4.24}$$

There is no known closed-form solution for (4.24) except for $\mathbf{z}$ containing simple geometries such as disks. However, problem (4.24) has very efficient solvers that take advantage of the "local" structure of total variation such as graph-cut [129, 130], primal-dual algorithm [131], as well as [132, 133, 134]. The alternating direction method in Section 4.8 below is also efficient, and on total variation problems, it can be applied to models that are more complicated than (4.24); see [135, 136].

A challenge arises when the regularization function consists of more than one term, for example, $r(\mathbf{u}) = \|\Psi\mathbf{u}\|_1 + TV(\mathbf{u})$. Another challenge is $r(\mathbf{u}) = \|\Psi\mathbf{u}\|_1$, where $\Psi$ is *not* orthogonal. Even if $\Psi$ is a tight frame, which satisfies $\Psi^T\Psi = \mathbf{I}$ and enjoys Parseval's identity, the same trick of changing variable no longer guarantees the correctness of (4.23). In these cases, the Moreau–Yosida regularization of $r$ is not trivial. Typically solutions are variable splitting and decomposition; see Section 4.8 for an alternative method based on variable splitting. Also see [137, 138] for novel ways of dealing multiple regularizers.

## 4.5 Prox-Linear Algorithms

The problems in Section 4.4 with simple solutions appear as subproblems in the so-called prox-linear method. To introduce this method, we consider a general form. Problems (4.1b) and (4.2b) can be written in the general form

$$\min_{\mathbf{x}} r(\mathbf{x}) + f(\mathbf{x}), \tag{4.25}$$

where $r$ is the regularization function and $f$ is the data fidelity function. The prox-linear algorithm has a long history in the literature; see [139, 140, 141] for example. It is based on *linearizing* the second term $f(\mathbf{x})$ at each $\mathbf{x}^k$ and adding a *proximal term*, giving rise to the iteration

$$\mathbf{x}^{k+1} = \arg\min_{\mathbf{x}} r(\mathbf{x}) + f(\mathbf{x}^k) + \langle \nabla f(\mathbf{x}^k), \mathbf{x} - \mathbf{x}^k \rangle + \frac{1}{2\delta_k}\|\mathbf{x} - \mathbf{x}^k\|_2^2. \tag{4.26}$$

The last term keeps $\mathbf{x}^{k+1}$ close to $\mathbf{x}^k$, and the parameter $\delta_k$ determines the stepsize. If $r$ is $\ell_1$-norm and the last term of (4.26) was not there, the solution would be either 0 or unbounded. Note that the second term is independent of the decision variable $\mathbf{x}$, and the last two terms can be combined to a complete square plus a constant as follows

$$\langle \nabla f(\mathbf{x}), \mathbf{x} - \mathbf{x}^k \rangle + \frac{1}{2\delta_k}\|\mathbf{x} - \mathbf{x}^k\|_2^2 = \frac{1}{2\delta_k}\left\|\mathbf{x} - \left(\mathbf{x}^k - \delta_k\nabla f(\mathbf{x}^k)\right)\right\|_2^2 + c,$$

where $c = -\delta_k\|\nabla f(\mathbf{x}^k)\|_2^2$ is independent of the decision variable $\mathbf{x}$. Therefore, iteration (4.26) is equivalent to

$$\mathbf{x}^{k+1} = \arg\min_{\mathbf{x}} \; r(\mathbf{x}) + \frac{1}{2\delta_k} \left\| \mathbf{x} - \left( \mathbf{x}^k - \delta_k \nabla f(\mathbf{x}^k) \right) \right\|_2^2. \tag{4.27}$$

It is now clear that the motivation of linearizing just function $f$ is that iteration (4.27) is easy to compute for the aforementioned examples of regularization function $r$.

It is worth mentioning that when $\frac{1}{\delta_k}$ is chosen greater than the Lipschitz constant of $\nabla f(\mathbf{x})$, then problem (4.26) can be viewed as a single-block BSUM algorithm, with the upper-bound function given by

$$u(\mathbf{x}, \mathbf{x}^k) := r(\mathbf{x}) + f(\mathbf{x}^k) + \langle \nabla f(\mathbf{x}^k), \mathbf{x} - \mathbf{x}^k \rangle + \frac{1}{2\delta_k} \|\mathbf{x} - \mathbf{x}^k\|_2^2.$$

## 4.5.1   Forward-Backward Operator Splitting

Besides the above prox-linear interpretation, there are other alternative approaches to obtain (4.26) such as forward-backward operator splitting and optimization transfers. We briefly review forward-backward operator splitting since it is general and very elegant. Various operator splitting schemes have been developed since the 1950s. Well-known ones are included in [142, 143, 144, 145, 146]; a good reference for forward-backward splitting and its applications in signal processing is [147]. It is based on manipulating subdifferential operators.

The first-order optimality condition of unconstrained, convex, nonsmooth minimization is given based on the subdifferential of the objective function.

LEMMA 1 ([148]).  *Point $\mathbf{x}_{\text{opt}}$ is the minimizer of a convex function $f$ if and only if*

$$\mathbf{0} \in \partial f(\mathbf{x}_{\text{opt}}), \tag{4.28}$$

*where we have ignored the technical trivial to ensure the existence of $\partial f$.*

The proof is trivial since $f(\mathbf{x}) \geq f(\mathbf{x}_{\text{opt}}) = f(\mathbf{x}_{\text{opt}}) + \mathbf{0}^T(\mathbf{x} - \mathbf{x}_{\text{opt}})$ if and only if (4.28) holds.

For (4.25), introduce operators $T_1$ and $T_2$:

$$T_1(\mathbf{x}) = \partial r(\mathbf{x}) \quad \text{and} \quad T_2(\mathbf{x}) = \partial f(\mathbf{x}).$$

The subdifferential obeys the rule

$$f(\mathbf{x}) = \alpha_1 f_1(\mathbf{x}) + \alpha_2 f_2(\mathbf{x}) \implies \partial f(\mathbf{x}) = \alpha_1 \partial f_1(\mathbf{x}) + \alpha_2 \partial f_2(\mathbf{x})$$

as long as $\alpha_1, \alpha_2 \geq 0$. Therefore, the optimality condition (4.28) for model (4.25) is $\mathbf{0} \in \partial(r + f)(\mathbf{x}_{\text{opt}}) = T_1(\mathbf{x}_{\text{opt}}) + T_2(\mathbf{x}_{\text{opt}})$, which also holds after multiplying a factor $\tau = \delta_k^{-1} > 0$. Hence, we have

$$\mathbf{x}_{\text{opt}} \text{ solves } (4.25) \Leftrightarrow \mathbf{0} \in \tau T_1(\mathbf{x}_{\text{opt}}) + \tau T_2(\mathbf{x}_{\text{opt}})$$

$$\Leftrightarrow \mathbf{0} \in (\mathbf{x}_{\text{opt}} + \tau T_1(\mathbf{x}_{\text{opt}})) - (\mathbf{x}_{\text{opt}} - \tau T_2(\mathbf{x}_{\text{opt}}))$$

$$\Leftrightarrow (I - \tau T_2)(\mathbf{x}_{\text{opt}}) \in (I + \tau T_1)(\mathbf{x}_{\text{opt}})$$

$$\overset{(*)}{\Leftrightarrow} \mathbf{x}_{\text{opt}} = (I + \tau T_1)^{-1}(I - \tau T_2)(\mathbf{x}_{\text{opt}}), \tag{4.29}$$

where for $(\ast)$ to hold, we have assumed that $\mathbf{z} \in (I + \tau T_1)(\mathbf{x})$ determines a unique $\mathbf{x}$ for a given $\mathbf{z}$. Following from the definition $T_1 = \partial r$ and Lemma 1, the solution $\mathbf{x}$ to $\mathbf{z} \in (I + \tau T_1)(\mathbf{x})$ is the minimizer of $\tau r(\mathbf{x}) + \frac{1}{2}\|\mathbf{x} - \mathbf{z}\|_2^2$. In other words,

$$(I + \tau T_1)^{-1}(\mathbf{z}) = \arg\min_{\mathbf{x}} \tau r(\mathbf{x}) + \frac{1}{2}\|\mathbf{x} - \mathbf{z}\|_2^2. \tag{4.30}$$

If we let $\mathbf{z} = (I - \tau T_2)(\mathbf{x}_{\text{opt}})$, we obtain (4.29); on the other hand, letting $\tau = \frac{1}{2\delta_k}$ and $\mathbf{z} = (I - \tau T_2)(\mathbf{x}^k) = \mathbf{x}^k - \frac{1}{2\delta_k}\nabla f(\mathbf{x}^k)$, we obtain

$$(I + \tau T_1)^{-1}(\mathbf{z}) = \arg\min_{\mathbf{x}} \tau r(\mathbf{x}) + \frac{1}{2}\|\mathbf{x} - (\mathbf{x}^k - \frac{1}{2\delta_k}\nabla f(\mathbf{x}^k))\|_2^2, \tag{4.31}$$

which only differs from the right-hand side of (4.26) by a quantity independent of $\mathbf{x}$. Hence, (4.26) and (4.31) yield the same solution, and thus (4.26) is precisely a fixed point iteration of (4.29):

$$\mathbf{x}^{k+1} = (I + \tau T_1)^{-1}(I - \tau T_2)(\mathbf{x}^k). \tag{4.32}$$

The operation $(I - \tau T_2)$ is called the forward step, or the gradient descent step. The operation $(I + \tau T_1)^{-1}$ is called the backward step, or the minimization step. Such splitting makes (4.29) and (4.32) easy to carry out for various interesting examples of $r$ and $f$.

## 4.5.2  Examples

Since the forward step is straightforward for any differentiable function $f$, we focus on different choices of the function $r$.

---

**Example 1** [Basis pursuit denoising].  Let $r(\mathbf{x}) = \|\mathbf{x}\|_1$ and $f(\mathbf{x})$ be a differentiable function (e.g., $\frac{\mu}{2}\|\mathbf{Ax} - \mathbf{y}\|_2$ corresponding to (4.2b)). The backward step (4.30) is given by (4.14). Hence, (4.32) becomes

$$\mathbf{x}^{k+1} = \text{shrink}(\mathbf{x}^k - \delta_k \nabla f(\mathbf{x}^k), \delta_k). \tag{4.33}$$

The main computation at each iteration $k$ is $\nabla f(\mathbf{x}^k)$. For problem (4.2b), it becomes $\mathbf{A}^T \mathbf{A}\mathbf{x}^k$.

If we generalize to $r(\mathbf{x}) = \|\Psi\mathbf{x}\|_1$ for an orthogonal linear transform, then following (4.23), (4.32) is given by

$$\mathbf{x}^{k+1} = \Psi^T \text{shrink}(\Psi(\mathbf{x}^k - \delta_k \nabla f(\mathbf{x}^k)), \delta_k). \tag{4.34}$$

**Example 2** [Total variation image reconstruction].  Let $\mathbf{x}$ denote an image and $\text{TV}(\mathbf{x})$ denotes its total variation. The backward step (4.30) solves the so-called Rudin–Osher–Fatemi model [128]

$$\min_{\mathbf{x}} \tau \text{TV}(\mathbf{x}) + \frac{1}{2}\|\mathbf{x} - \mathbf{z}\|_2^2. \tag{4.35}$$

Since no closed-form solution is known, the iteration (4.32) is carried out in two steps:

$$\text{forward step:} \quad \mathbf{z}^k = \mathbf{x}^k - \delta_k \nabla f(\mathbf{x}^k),$$

$$\text{backward step:} \quad \mathbf{x}^{k+1} = \arg\min_{\mathbf{x}} \text{TV}(\mathbf{x}) + \frac{1}{2\delta_k} \|\mathbf{x} - \mathbf{z}\|_2^2.$$

**Example 3** [Joint-sparse recovery for inexact measurements]. Suppose we seek a solution $\mathbf{X} \in \mathbb{R}^{N \times L}$, with rows $\mathbf{x}_1, \ldots, \mathbf{x}_n$, that has all but a few rows containing nonzero entries. One approach is to set $r(\mathbf{X}) = \sum_{i=1}^{N} \|\mathbf{x}_i\|_p$ for some $p \geq 1$. The backward step (4.30) is separable across rows of $\mathbf{X}$ and, if $p > 1$, it is not further separable. The iteration (4.32) is carried out in two steps:

$$\text{forward step:} \quad \mathbf{Z}^k = \mathbf{X}^k - \delta_k \nabla f(\mathbf{X}^k),$$

$$\text{backward step:} \quad \mathbf{x}_i^{k+1} = \arg\min_{\mathbf{x}} \|\mathbf{x}\|_p + \frac{1}{2\delta_k} \|\mathbf{x} - \mathbf{z}_i^k\|_2^2, \ i = 1, \ldots, N,$$

where $\mathbf{z}_i^k$ denotes row $i$ of $\mathbf{Z}^k$. For $p = 2$, the backward step is solved by (4.18).

**Example 4** [Low-rank matrix recovery]. In this case, we let $r(\mathbf{X}) = \|\mathbf{X}\|_*$ and consider the model

$$\min_{\mathbf{X}} \ r(\mathbf{X}) + f(\mathbf{X}), \tag{4.36}$$

where $f(\mathbf{X})$ is a convex function of $\mathbf{X}$. Problem (4.36) is a generalization of the unconstrained low-rank matrix recovery model

$$\min \|\mathbf{X}\|_* + \frac{\mu}{2} \|\mathcal{A}(\mathbf{X}) - \mathbf{b}\|_2^2, \tag{4.37}$$

where $\mathcal{A}$ is a linear operator and $\mu > 0$ is a penalty parameter. The iteration (4.32) gives

$$\mathbf{X}^{k+1} = \text{shrink}_*(\mathbf{X}^k - \delta_k \nabla f(\mathbf{X}^k), \delta_k), \tag{4.38}$$

where $\text{shrink}_*$ is given by (4.19) and (4.20). Algorithm FPCA [126] is based on (4.38).

**Example 5** [Composite TV+$\ell_1$ regularization]. In this example, the model is

$$\min_{\mathbf{x}} r(\mathbf{x}) + \frac{\mu}{2} \|\mathbf{A}\mathbf{x} - \mathbf{b}\|_2^2,$$

where

$$r(\mathbf{x}) = \text{TV}(\mathbf{x}) + \alpha \|\Psi \mathbf{x}\|_1.$$

The regularizer arises in image processing [149] to take advantages of both total variation and a sparsifying transform $\Psi$ and also in signal processing and statistics [150] with $\Psi = \mathbf{I}$ to obtain a sparse solution with clustered nonzero entries. Sparsity and clustering are results of minimizing $\|\mathbf{x}\|_1$ and $\text{TV}(\mathbf{x})$, respectively. An algorithm is proposed in [149] that uses forward-backward splitting twice; there are two forward steps and two backward steps, involving both (4.15) and (4.18).

The prox-linear iterations can be coupled with various techniques and achieve significant acceleration. Some of these techniques are reviewed in Section 4.10 below.

### 4.5.3   Convergence Rates

If the objective function $F(\mathbf{x})$ is strongly convex, namely,

$$\alpha F(\mathbf{x}) + (1 - \alpha)F(\mathbf{y}) \geq F(\alpha \mathbf{x} + (1 - \alpha)\mathbf{y}) + \frac{\nu}{2}\alpha(1 - \alpha)\|\mathbf{x} - \mathbf{y}\|_2^2$$

holds for any $\alpha \in [0, 1]$, then it is easy to show that prox-linear algorithm with proper stepsizes $\delta_k$ converges at a rate $O(1/c^k)$ where $0 < c < 1$ and $k$ is the number of iterations. However, problems such as (4.2b) are general not strongly convex, especially when $\mathbf{A}$ is a fat matrix, namely, the number of rows of $\mathbf{A}$ is less than that of its columns. With only convexity, the rate reduces to $O(1/k)$. For problems (4.1b) with $r(\mathbf{x}) = \|\mathbf{x}\|_1$ and smooth $\bar{h}(\mathbf{x})$ and (4.2b), linear convergence in the final stage of the algorithm can be obtained [151, 152], and [152] also shows that after a finite number of iterations, roughly speaking, $\mathbf{x}^k$ must have the optimal support supp$(\mathbf{x}_{\text{opt}})$. Nesterov's method [153] for minimizing smooth convex functions can be extended to the prox-linear algorithm in different ways for different settings [154, 155, 156, 157]; they have a better global rate at $O(1/k^2)$ while only marginally increasing the computing cost per iteration. These are provable rates; on practical problems, researchers report that the advantage of $O(1/k^2)$-complexity methods is more apparent in the initial phase, and when the iterations reduce to a small set of coordinates, they are not necessarily faster and can even be slower.

## 4.6   Dual Algorithms

While the algorithms in Section 4.5 solve the penalty model (4.2b), the algorithms in this section can generate a sequence of points that converges to a solution to the equality constrained model (4.2a). Interestingly, some of the points in this sequence, though not solving (4.2b) or (4.2c), have "better quality" in terms of sparsity and data fitness than the solutions of (4.2b) or (4.2c).

Coupled with splitting techniques, dual algorithms in this section can solve problems (4.2b) or (4.2c) as well.

### 4.6.1   Dual Formulations

For a constrained optimization problem like (4.2a), duality gives arise to a companion problem. Between the original (primal) and the companion (dual) problems, there exists a weak, and sometimes also a strong, duality relation, which makes solving one of the problems somewhat equivalent to solving the other. This brings new choices of efficient algorithms to solve problem (4.1a). We leave the dual-formulation derivation of the general form (4.1a) to textbooks such as [120] and focus on sparse optimization examples.

**Example 6** [Dual of basis pursuit]. For $\ell_1$ minimization problem (4.2a), its dual problem is

$$\max_{\mathbf{y}}\{\mathbf{b}^T\mathbf{y} : \|\mathbf{A}^T\mathbf{y}\|_\infty \leq 1\}. \tag{4.39}$$

To derive (4.39), we let the objective value of a minimization (maximization) problem be $+\infty$ $(-\infty)$ at infeasible points[2] and establish

$$\min_{\mathbf{x}}\{\|\mathbf{x}\|_1 : \mathbf{A}\mathbf{x} = \mathbf{b}\} = \min_{\mathbf{x}}\max_{\mathbf{y}} \|\mathbf{x}\|_1 - \mathbf{y}^T(\mathbf{A}\mathbf{x} - \mathbf{b})$$

$$= \max_{\mathbf{y}}\min_{\mathbf{x}} \|\mathbf{x}\|_1 - \mathbf{y}^T(\mathbf{A}\mathbf{x} - \mathbf{b})$$

$$= \max_{\mathbf{y}}\{\mathbf{b}^T\mathbf{y} : \|\mathbf{A}^T\mathbf{y}\|_\infty \leq 1\},$$

where the first equality follows from that $\mathbf{A}\mathbf{x} = \mathbf{b}$ if and only if $\max_\mathbf{y} -\mathbf{y}^T(\mathbf{A}\mathbf{x} - \mathbf{b})$ is finite, the second equality holds for the mini-max principle, and the last equality follows from the fact that $\min_\mathbf{x}\{\|\mathbf{x}\|_1 - \mathbf{y}^T\mathbf{A}\mathbf{x}\}$ is finite if and only if $\|\mathbf{A}^T\mathbf{y}\|_\infty \leq 1$.

**Example 7** [Group sparsity]. The dual of (4.5) is

$$\max_{\mathbf{y}}\{\mathbf{b}^T\mathbf{y} : \|\mathbf{A}_{G_s}^T\mathbf{y}\|_2 \leq w_i, \ s = 1, 2, \ldots, S\}, \tag{4.40}$$

which can be derived in a way similar to Example 6, or see [158].

**Example 8** [Nuclear norm]. In (4.3e), we rewrite $\mathcal{A}(\mathbf{X}) = [\mathbf{A}_1 \bullet \mathbf{X}, \mathbf{A}_2 \bullet \mathbf{X}, \ldots, \mathbf{A}_M \bullet \mathbf{X}]$, where $\mathbf{Y} \bullet \mathbf{X} = \sum_{m=1}^M \sum_{n=1}^N y_{mn}x_{mn}$. Thus, $\mathcal{A}(\mathbf{X}) = \mathbf{b}$ can be written as $\mathbf{A}_m \bullet \mathbf{X} = b_m$, for $m = 1, 2, \ldots, M$. The dual of (4.3e) is obtained through

$$\min_{\mathbf{X}}\{\|\mathbf{X}\|_* : \mathcal{A}(\mathbf{X}) = \mathbf{b}\} = \min_{\mathbf{X}}\max_{\mathbf{y}} \|\mathbf{X}\|_* - \mathbf{y}^T(\mathcal{A}(\mathbf{X}) - \mathbf{b})$$

$$= \max_{\mathbf{y}}\min_{\mathbf{X}} \|\mathbf{X}\|_* - (\sum_{m=1}^M y_m\mathbf{A}_m) \bullet \mathbf{X} + \mathbf{b}^T\mathbf{y}$$

$$\overset{(*)}{=} \max_{\mathbf{y}}\{\mathbf{b}^T\mathbf{y} : \|\sum_{m=1}^M y_m\mathbf{A}_m\|_2 \leq 1\},$$

where $\|\mathbf{X}\|_2 = \sigma_1(\mathbf{X})$ equals the maximum singular value of $\mathbf{X}$ (i.e., the operator norm of $\mathbf{X}$) and $(*)$ follows from the fact that $\|\cdot\|_*$ and $\|\cdot\|_2$ are dual to each other [159].

## 4.7      Bregman Method

There are different versions of Bregman algorithms: original Bregman, linearized Bregman, and split Bregman. This subsection describes the original version. The linearized

---

[2] Points that do not satisfy all the constraints.

version is given in Section 4.7.2 below, and the split version, also known as the ADMM method, has been discussed in the previous chapter.

The Bregman method is based on successively minimizing the Bregman distance (also called the Bregman divergence).

DEFINITION 1 ([160]). *The Bregman distance induced by convex function $r(\cdot)$ is defined as*

$$D_r(\mathbf{x}, \mathbf{y}; \mathbf{p}) = r(\mathbf{x}) - r(\mathbf{y}) - \langle \mathbf{p}, \mathbf{x} - \mathbf{y} \rangle, \quad \text{where } \mathbf{p} \in \partial r(\mathbf{y}). \tag{4.41}$$

Since $\partial r(\mathbf{y})$ may include multiple subgradients, different choices of $\mathbf{p} \in \partial r(\mathbf{y})$ lead to different Bregman distances, so we include $\mathbf{p}$ in $D_r(\mathbf{x}, \mathbf{y}; \mathbf{p})$. When $r$ is differentiable or the choice of $\mathbf{p}$ is clear from the context, we drop $\mathbf{p}$ and write $D_r(\mathbf{x}, \mathbf{y})$. Because we have $D_r(\mathbf{x}, \mathbf{y}; \mathbf{p}) \neq D_r(\mathbf{y}, \mathbf{x}; \mathbf{p})$ in general, the Bregman distance is not a distance in the mathematical sense. However, it measures the closeness between $\mathbf{x}$ and $\mathbf{y}$ in some sense. We have $D_r(\mathbf{x}, \mathbf{y}; \mathbf{p}) \geq 0$ uniformly (due to the convexity of $r$), and fixing $\mathbf{x}$ and $\mathbf{y}$, we have $D_r(\mathbf{x}, \mathbf{y}; \mathbf{p}) \geq D_r(\mathbf{w}, \mathbf{y}; \mathbf{p})$ for any point $\mathbf{w}$ picked on the line segment $[\mathbf{x}, \mathbf{y}]$. In addition, if $r$ is strictly convex, then $D_r(\mathbf{x}, \mathbf{y}; \mathbf{p}) = 0$ if and only if $\mathbf{x} = \mathbf{y}$. Figure 4.2 illustrates the Bregman distance.

The original Bregman algorithm iteration has two steps

$$\mathbf{x}^{k+1} = \arg\min_{\mathbf{x}} D_r(\mathbf{x}, \mathbf{x}^k; \mathbf{p}^k) + f(\mathbf{x}), \tag{4.42a}$$

$$\mathbf{p}^{k+1} = \mathbf{p}^k - \nabla f(\mathbf{x}^{k+1}), \tag{4.42b}$$

starting at $k = 0$ and $(\mathbf{x}^0, \mathbf{p}^0) = (\mathbf{0}, \mathbf{0})$. Step (4.42b) generates $\mathbf{p}^{k+1} \in \partial r(\mathbf{x}^{k+1})$ following the first-order optimality condition of (4.42a):

$$\mathbf{0} \in \partial r(\mathbf{x}^{k+1}) - \mathbf{p}^k + \nabla f(\mathbf{x}^{k+1}), \tag{4.43}$$

so the Bregman distance $D_r(\mathbf{x}, \mathbf{x}^{k+1}; \mathbf{p}^{k+1})$ becomes well defined.

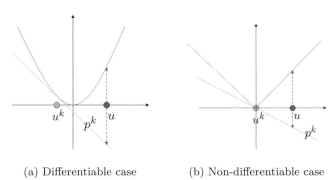

(a) Differentiable case          (b) Non-differentiable case

**Figure 4.2** Bregman distance (the length of the dashed line) between $u$ and $u^k$.

For $f(\mathbf{x}) = (\delta/2)\|\mathbf{Ax} - \mathbf{b}\|_2^2$, an interesting alternative form of the Bregman algorithm is the iteration of "adding back the residual":

$$\mathbf{b}^{k+1} = \mathbf{b} + (\mathbf{b}^k - \mathbf{Ax}^k), \tag{4.44a}$$

$$\mathbf{x}^{k+1} = \arg\min_{\mathbf{x}} r(\mathbf{x}) + \frac{\delta}{2}\|\mathbf{Ax} - \mathbf{b}^{k+1}\|_2^2, \tag{4.44b}$$

starting from $k = 0$ and $(\mathbf{x}^0, \mathbf{b}^0) = (\mathbf{0}, \mathbf{0})$.

LEMMA 2. *For $f(\mathbf{x}) = \frac{\delta}{2}\|\mathbf{Ax} - \mathbf{b}\|_2^2$, iterations (4.42) and (4.44) yield the same sequence $\{\mathbf{x}^k\}$. In particular, $\mathbf{p}^k = -\delta\mathbf{A}^T(\mathbf{Ax}^k - \mathbf{b}^k)$ for all $k$.*

*Proof.* Since $\mathbf{p}^0 = 0$ and $\mathbf{b}^1 = \mathbf{b}$, both iterations yield the same $\mathbf{x}^1$ and (4.42b) gives $\mathbf{p}^1 = -\delta\mathbf{A}^T(\mathbf{Ax}^1 - \mathbf{b}^1)$. We prove by induction. Suppose both iterations yield the same $\mathbf{x}^k$ and $\mathbf{p}^k = -\delta\mathbf{A}^T(\mathbf{Ax}^k - \mathbf{b}^k)$. From (4.42a) and (4.44a), it follows:

$$\begin{aligned}
\mathbf{x}^{k+1} &= \arg\min_{\mathbf{x}} r(\mathbf{x}) - \langle \mathbf{p}^k, \mathbf{x} \rangle + \frac{\delta}{2}\|\mathbf{Ax} - \mathbf{b}\|_2^2 \\
&= \arg\min_{\mathbf{x}} r(\mathbf{x}) + \frac{\delta}{2}\left\|\mathbf{Ax} - \left(\mathbf{b} + (\mathbf{b}^k - \mathbf{Ax}^k)\right)\right\|_2^2 \\
&= \arg\min_{\mathbf{x}} r(\mathbf{x}) + \frac{\delta}{2}\|\mathbf{Ax} - \mathbf{b}^{k+1}\|_2^2,
\end{aligned}$$

which is (4.44b). Hence, both iterations yield the same $\mathbf{x}^{k+1}$. Due to (4.42b) and (4.44b), we obtain

$$\begin{aligned}
\mathbf{p}^{k+1} &= \mathbf{p}^k - \delta\mathbf{A}^T(\mathbf{Ax}^{k+1} - \mathbf{b}) \\
&= -\delta\mathbf{A}^T(\mathbf{Ax}^k - \mathbf{b}^k) - \delta\mathbf{A}^T(\mathbf{Ax}^{k+1} - \mathbf{b}) \\
&= -\delta\mathbf{A}^T\left(\mathbf{Ax}^{k+1} - \left(\mathbf{b} + (\mathbf{b}^k - \mathbf{Ax}^k)\right)\right) \\
&= -\delta\mathbf{A}^T(\mathbf{Ax}^{k+1} - \mathbf{b}^{k+1}).
\end{aligned}$$

The claim is proved.                                              □

If subproblems (4.42a) and (4.44b) are solved *inexactly*, the two iterations no longer yield the same sequence! In light of Lemma 2, the identity $\mathbf{p}^k = -\delta\mathbf{A}^T(\mathbf{Ax}^k - \mathbf{b}^k)$ will fail to hold. The "adding-back" iteration (4.44) is stabler and, when $r$ is a piece-wise linear function (also referred to as a polyhedral function) and the subproblems are solved with sufficient (but not necessarily high) accuracies, such iteration enjoys the property of *error forgetting*, namely, the errors in $\mathbf{x}^k$ do not accumulate, and furthermore the errors at iterations $k$ and $k + 1$ may cancel each other, allowing $\mathbf{x}^k$ to converge to a solution to model (4.1a) up to the machine precision; see [161] for examples and analysis.

## 4.7.1    Bregman Iterations and Denoising

It is known that the Bregman iteration converges to a solution of the underlying equality-constrained problem (4.1a), which is a model for noiseless measurements $\mathbf{b}$. When there is noise in **b**, properly stopping the iterations before convergence may yield clean solutions with better fitting than the exact solutions to problems (4.1b) and (4.1c), even with their best parameters.

We illustrate this through an example and quoting a theorem. In this example, we generated a signal $\mathbf{x}^o$ of $n = 250$ dimensions with $k = 20$ nonzero components that were i.i.d. samples of the standard Gaussian distribution. $\mathbf{x}^o$ is depicted by the block dot • in Figure 4.3. Measurement matrix **A** had $m = 100$ rows and $n = 250$ columns with its entries sampled i.i.d. from the standard Gaussian distribution. Measurements were $\mathbf{b} = \mathbf{A}\mathbf{x}^o + \mathbf{n}$, where noise **n** followed the independent Gaussian noise of standard deviation 0.1. We compared

1. Model (4.2c) with $\sigma = \|\mathbf{n}\|_2$ in Figure 4.3(a).
2. Bregman iteration (4.44) with $r(\cdot) = \|\cdot\|_1$ and $\delta = 1/50$ at iterations 1, 3, and 5, in Figures 4.3(b), 4.3(c), and 4.3(d), respectively.

Due to the noise **n** in the measurements, neither approach gave exact reconstruction of $\mathbf{x}^0$. The BPDN solution had a large number of false nonzero components and slightly

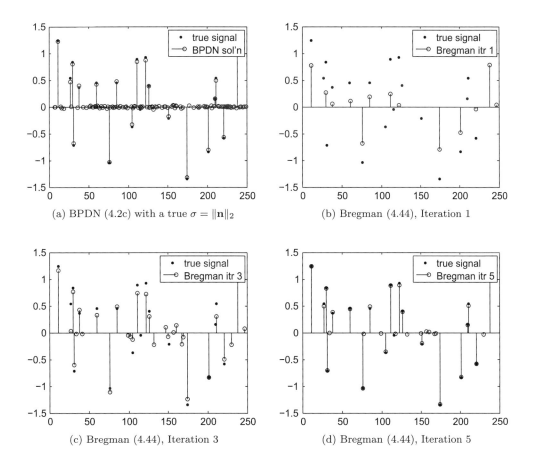

(a) BPDN (4.2c) with a true $\sigma = \|\mathbf{n}\|_2$

(b) Bregman (4.44), Iteration 1

(c) Bregman (4.44), Iteration 3

(d) Bregman (4.44), Iteration 5

**Figure 4.3** BPDN vs. Bregman iteration.

mismatched with the nonzero components of $\mathbf{x}^o$. The Bregman solutions had significantly fewer false nonzero components. Iterations 1 and 3 missed certain nonzero
components of $\mathbf{x}^o$. Iteration 5 had a much better solution that better matched with $\mathbf{x}^0$,
achieving a relative 2-norm error of 3.50%, than the solution of BPDN (relative 2-norm
error 6.19%).

In general, as long as noise is not too large, this phenomenon arises with other dual
algorithms such as the linearzied Bregman and alternating direction methods below,
and it also does with the logistic loss function. When the measurements contain
a moderate amount of noise, instead of solving models (4.1b) and (4.1c), solving
model (4.1a) using a dual algorithm and appropriately stopping the algorithm before
convergence can yield a better solution; more examples and discussions can be found
in [162, 163]. We adapt a result from [162], Theorem 3.5, to our problem:

THEOREM 1. *Assume that from the original signal* $\mathbf{x}^o$, *we obtain measurements* $\mathbf{b} =$
$\mathbf{A}\mathbf{x}^o + \mathbf{n}$, *where* $\mathbf{n}$ *is an arbitrary noise obeying* $\frac{1}{2}\|\mathbf{n}\|_2^2 \leq \sigma^2$. *Let* $r$ *be* $\ell_1$-norm. *Let* $\mathbf{x}^k$
*and* $\mathbf{p}^k$ *be generated by the Bregman iteration. Then, as long as* $\frac{1}{2}\|\mathbf{A}\mathbf{x}^k - \mathbf{b}\|_2^2 \leq \sigma^2$, *the*
*Bregman distance between* $\mathbf{x}^o$ *and* $\mathbf{x}^k$ *is monotonically non-increasing in* $k$, *namely,*

$$D_r(\mathbf{x}^o, \mathbf{x}^k; \mathbf{p}^k) \leq D_r(\mathbf{x}^o, \mathbf{x}^{k-1}; \mathbf{p}^{k-1}).$$

## 4.7.2    Linearized Bregman and Augmented Models

The linearized Bregman method is motivated by the challenge that step (4.42a) of Bregman iteration has no closed-form solution and needs another algorithm to solve. As the
name suggests, linearized Bregman iteration applies linearization to $f(\mathbf{x})$ in (4.42a) (with
an extra proximal term). The iteration is

$$\mathbf{x}^{k+1} \leftarrow \arg\min_{\mathbf{x}} D_r(\mathbf{x}, \mathbf{x}^k; \mathbf{p}^k) + \langle \nabla f(\mathbf{x}^k), \mathbf{x} \rangle + \frac{1}{2\alpha}\|\mathbf{x} - \mathbf{x}^k\|_2^2, \quad (4.45\text{a})$$

$$\mathbf{p}^{k+1} \leftarrow \mathbf{p}^k - \nabla f(\mathbf{x}^k) - \frac{1}{\alpha}(\mathbf{x}^{k+1} - \mathbf{x}^k). \quad (4.45\text{b})$$

Step (4.45b) generates a valid subgradient $\mathbf{p}^{k+1}$ following the first-order optimality
condition of (4.45a). If we introduce the *augmented* function of $r$

$$\bar{r}(\mathbf{x}) := r(\mathbf{x}) + \frac{1}{2\alpha}\|\mathbf{x}\|_2^2, \quad (4.46)$$

we can write

$$D_{\bar{r}}(\mathbf{x}, \mathbf{x}^k; \bar{\mathbf{p}}^k) = D_r(\mathbf{x}, \mathbf{x}^k; \mathbf{p}^k) + \frac{1}{2\alpha}\|\mathbf{x} - \mathbf{x}^k\|_2^2$$

for $\bar{\mathbf{p}}^k = \mathbf{p}^k + (1/\alpha)\mathbf{x}^k$ using the additivity property of the Bregman distance. Therefore,
iteration (4.45) can be equivalently written as

$$\mathbf{x}^{k+1} \leftarrow \arg\min_{\mathbf{x}} D_{\bar{r}}(\mathbf{x}, \mathbf{x}^k; \bar{\mathbf{p}}^k) + \langle \nabla f(\mathbf{x}^k), \mathbf{x} \rangle, \quad (4.47\text{a})$$

$$\bar{\mathbf{p}}^{k+1} \leftarrow \bar{\mathbf{p}}^k - \nabla f(\mathbf{x}^k), \quad (4.47\text{b})$$

where in (4.47a) the Bregman distance term and the term of $\langle \nabla f(\mathbf{x}^k), \mathbf{x} \rangle$ together remind us of the name *linearized Bregman*.

### 4.7.3 Handling Complex Data and Variables

Let $r(\cdot)$, $i(\cdot)$, and $|\cdot|$ denote the operators that return the real part, the imaginary part, and the magnitude of the input. Many researchers define the $\ell_1$-norm and $\ell_2$-norm of a complex-valued vector $\mathbf{x} \in \mathbb{C}^n$ as

$$\|\mathbf{x}\|_1 := \sum_{i=1}^{n} \left( r^2(x_i) + i^2(x_i) \right)^{1/2} = \|\mathbf{a}\|_1,$$

$$\|\mathbf{x}\|_2 := \left( \sum_{i=1}^{n} r^2(x_i) + i^2(x_i) \right)^{1/2} = \|\mathbf{a}\|_2,$$

respectively, where $\mathbf{a} = [|x_1|, \ldots, |x_n|]^T \in \mathbb{R}^n$. They reduce to the standard $\ell_1$-norm and $\ell_2$-norm if $\mathbf{x}$ is real-valued. In general, we define the $\ell_p$-norm of $\mathbf{x}$ by $\|\mathbf{a}\|_p$, which is convex for $1 \leq p \leq \infty$.

The *dual norm* of $\|\mathbf{x}\|_p$ is $\|\mathbf{y}\|_d = \max\{r(\mathbf{y}^*\mathbf{x}) : \|\mathbf{x}\|_p \leq 1\}$, where $\mathbf{y}^*$ denotes the conjugate transpose of $\mathbf{y}$.

We are interested in developing the dual problems of

$$\min_{\mathbf{x}}\{\|\mathbf{x}\|_p : \mathbf{Ax} = \mathbf{b}\}, \tag{4.48a}$$

$$\min_{\mathbf{x}} \|\mathbf{x}\|_p + \frac{\mu}{2}\|\mathbf{Ax} - \mathbf{b}\|_2^2, \tag{4.48b}$$

$$\min_{\mathbf{x}}\{\|\mathbf{x}\|_p : \|\mathbf{Ax} - \mathbf{b}\|_2 \leq \sigma\}. \tag{4.48c}$$

The constraints $\mathbf{Ax} = \mathbf{b}$ have their real and imaginary parts, namely, $r(\mathbf{Ax}) = r(\mathbf{b})$ and $i(\mathbf{Ax}) = i(\mathbf{b})$. To each of them, we shall assign a real-valued dual vector. If we assign real-valued vectors $\mathbf{p}$ and $\mathbf{q}$ to the real and imaginary parts, respectively, and introduce complex-valued vector $\mathbf{y} = \mathbf{p} + \mathbf{q}i$, then we obtain

$$\mathbf{p}^T(r(\mathbf{Ax}) - r(\mathbf{b})) + \mathbf{q}^T(i(\mathbf{Ax}) - i(\mathbf{b})) = r(\mathbf{y}^*(\mathbf{Ax} - \mathbf{b})).$$

Hence, we can obtain the dual of (4.48a) via

$$\min_{\mathbf{x}}\{\|\mathbf{x}\|_p : \mathbf{Ax} = \mathbf{b}\} = \min_{\mathbf{x}} \max_{\mathbf{y}} \|\mathbf{x}\|_p - r(\mathbf{y}^*(\mathbf{Ax} - \mathbf{b})) \tag{4.49a}$$

$$= \max_{\mathbf{y}} \min_{\mathbf{x}} \|\mathbf{x}\|_p - r(\mathbf{y}^*(\mathbf{Ax} - \mathbf{b})) \tag{4.49b}$$

$$= \max_{\mathbf{y}}\{r(\mathbf{b}^*\mathbf{y}) : \|\mathbf{A}^*\mathbf{y}\|_d \leq 1\}, \tag{4.49c}$$

where (4.49a) follows from the fact that $\max_{\mathbf{y}}\{-r(\mathbf{y}^*(\mathbf{Ax} - \mathbf{b}))\}$ stays bounded if and only if $\mathbf{Ax} = \mathbf{b}$, (4.49b) follows from the mini-max theorem, and (4.49c) follows from the dual relation between $\|\cdot\|_p$ and $\|\cdot\|_d$, as well as

$$\min_{\mathbf{x}} \|\mathbf{x}\|_p - r(\mathbf{y}^*(\mathbf{Ax})) = \begin{cases} 0, & \|\mathbf{A}^*\mathbf{y}\|_d \leq 1, \\ -\infty, & \text{otherwise.} \end{cases} \tag{4.50}$$

Leaving the details to the reader, the dual problem of (4.48b) is

$$\max_{\mathbf{y}}\{r(\mathbf{b}^*\mathbf{y}) - \frac{1}{2\mu}\|\mathbf{y}\|_2^2 : \|\mathbf{A}^*\mathbf{y}\|_d \le 1\}.$$

One way to dualize problem (4.48c) is

$$\min_{\mathbf{x}}\{\|\mathbf{x}\|_p : \|\mathbf{A}\mathbf{x} - \mathbf{b}\|_2 \le \sigma\} = \min_{\mathbf{x},\mathbf{z}}\{\|\mathbf{x}\|_p : \mathbf{A}\mathbf{x} + \mathbf{z} = \mathbf{b}, \|\mathbf{z}\|_2 \le \sigma\} \quad (4.51a)$$

$$= \min_{\mathbf{x},\mathbf{z}} \max_{\mathbf{y}}\{\|\mathbf{x}\|_p - r(\mathbf{y}^*(\mathbf{A}\mathbf{x} + \mathbf{z} - \mathbf{b})) : \|\mathbf{z}\|_2 \le \sigma\} \quad (4.51b)$$

$$= \max_{\mathbf{y}} \min_{\mathbf{x},\mathbf{z}}\{\|\mathbf{x}\|_p - r(\mathbf{y}^*(\mathbf{A}\mathbf{x} + \mathbf{z} - \mathbf{b})) : \|\mathbf{z}\|_2 \le \sigma\}. \quad (4.51c)$$

The inner problem of (4.51c), given a fixed $\mathbf{y}$, is separable in the unknowns $\mathbf{x}$ and $\mathbf{z}$. The optimal $\mathbf{z}$ is given by

$$\mathbf{z} = \begin{cases} \sigma\mathbf{y}/\|\mathbf{y}\|_2, & \mathbf{y} \ne \mathbf{0}, \\ \mathbf{0}, & \text{otherwise.} \end{cases}$$

In either case, $-r(\mathbf{y}^*\mathbf{z}) = -\sigma\|\mathbf{y}\|_2$. Also with (4.50), (4.51c) is equivalent to

$$\max_{\mathbf{y}}\{r(\mathbf{b}^*\mathbf{y}) - \sigma\|\mathbf{y}\|_2 : \|\mathbf{A}^*\mathbf{y}\|_d \le 1\}.$$

## 4.8     ADMM for Sparse Optimization

In this section, we discuss the application of ADMM for sparse optimization. The ADMM is useful in this setting because it endows the ability of "divide and conquer" – by appropriately introducing new variables and equality constraints, ADMM is capable of dealing with the nonsmooth functions and/or difficult constraints one at a time; see the following example:

**Example 9.**   Consider the following generic convex problem

$$\min_{\mathbf{x}} p(\mathbf{x}) + q(\mathbf{A}\mathbf{x}).$$

By "splitting" the variable $\mathbf{x}$ to two variables $\mathbf{x}$ and $\mathbf{y}$, we obtain the following equivalent problem

$$\min_{\mathbf{x},\mathbf{y}}\{p(\mathbf{x}) + q(\mathbf{y}) : \mathbf{A}\mathbf{x} - \mathbf{y} = \mathbf{0}\},$$

where the objective has separable terms $p(\mathbf{x})$ and $q(\mathbf{y})$.

**Example 10** [Basis pursuit denoising].   The penalty formulation (4.2b) can be written equivalently as

$$\min_{\mathbf{x}} \left\{ \|\mathbf{x}\|_1 + \frac{\mu}{2}\|\mathbf{y}\|_2^2 : \mathbf{A}\mathbf{x} + \mathbf{y} = \mathbf{b} \right\}. \quad (4.52)$$

With augmented Lagrangian

$$L(\mathbf{x}, \mathbf{y}; \lambda) = \|\mathbf{x}\|_1 + \frac{\mu}{2}\|\mathbf{y}\|_2^2 - \lambda^T(\mathbf{A}\mathbf{x} + \mathbf{y} - \mathbf{b}) + \frac{c}{2}\|\mathbf{A}\mathbf{x} + \mathbf{y} - \mathbf{b}\|_2^2,$$

applying the inexact ADMM iteration (3.51) to (4.52) yields the iteration

$$\mathbf{y}^{k+1} = \frac{c}{\mu + c}(\lambda^k/c - (\mathbf{A}\mathbf{x}^k - \mathbf{b})), \tag{4.53a}$$

$$\mathbf{x}^{k+1} = \text{shrink}(\mathbf{x}^k - \tau_k \nabla_{\mathbf{x}} L(\mathbf{x}^k, \mathbf{y}^{k+1}; \lambda^k), \tau_k), \tag{4.53b}$$

$$\lambda^{k+1} = \lambda^k - \gamma c(\mathbf{A}\mathbf{x}^{k+1} + \mathbf{y}^{k+1} - \mathbf{b}), \tag{4.53c}$$

The basis pursuit formulation (4.2a) corresponds to (4.52) with $\mu = \infty$ and thus $\mathbf{y}^k \equiv \mathbf{0}$ in (4.53), which reduces to the inexact augmented Lagrangian iteration.

**Example 11** [Basis pursuit denoising]. In the reformulation (4.52), $\mathbf{x}$ is associated with both $\|\cdot\|_1$ and $\mathbf{A}$; hence, the $\mathbf{x}$-subproblem is in the form of $\min_{\mathbf{x}} \|\mathbf{x}\|_1 + \frac{c}{2}\|\mathbf{A}\mathbf{x} - (\cdots)\|_2^2$. A different splitting can separate $\|\cdot\|_1$ and $\mathbf{A}$ as follows:

$$\min_{\mathbf{x}} \left\{ \|\mathbf{x}\|_1 + \frac{\mu}{2}\|\mathbf{A}\mathbf{y} - \mathbf{b}\|_2^2 : \mathbf{x} - \mathbf{y} = \mathbf{0} \right\}. \tag{4.54}$$

With augmented Lagrangian

$$L(\mathbf{x}, \mathbf{y}; \lambda) = \|\mathbf{x}\|_1 + \frac{\mu}{2}\|\mathbf{A}\mathbf{y} - \mathbf{b}\|_2^2 - \lambda^T(\mathbf{x} - \mathbf{y}) + \frac{c}{2}\|\mathbf{x} - \mathbf{y}\|_2^2,$$

applying the inexact ADMM to (4.54) yields the iteration

$$\mathbf{x}^{k+1} = \text{shrink}(\mathbf{y}^k + \lambda^k/c, c^{-1}), \tag{4.55a}$$

$$\mathbf{y}^{k+1} = \mathbf{y}^k - \tau_k \nabla_{\mathbf{y}} L(\mathbf{x}^{k+1}, \mathbf{y}^k; \lambda^k), \tag{4.55b}$$

$$\lambda^{k+1} = \lambda^k - \gamma c(\mathbf{A}\mathbf{x}^{k+1} + \mathbf{y}^{k+1} - \mathbf{b}), \tag{4.55c}$$

where (4.55a) is an exact solve and (4.55b) is a gradient step.

One advantage of separate $\|\cdot\|_1$ and $\mathbf{A}$ in two subproblems is that if $\mathbf{A}\mathbf{A}^T = I$, one can replace (4.55b) with an exact minimization step at a relatively small cost. Specifically, minimizing $L(\mathbf{x}^{k+1}, \mathbf{y}; \lambda^k)$ over $\mathbf{y}$ amounts to solving the equation system $(\mu\mathbf{A}^T\mathbf{A} + cI)\mathbf{y} = \mu\mathbf{A}^T\mathbf{b} - \lambda^k + c\mathbf{x}^{k+1}$, and one can apply the Sherman–Morrison–Woodbury formula (or the fact that $\mathbf{A}\mathbf{A}^T$ is an idempotent matrix) to derive $(\mu\mathbf{A}^T\mathbf{A} + cI)^{-1} = c^{-1}(I - ((\mu c)^{-1} + c^{-2})\mathbf{A}^T\mathbf{A})$.

Alternatively, one can pre-compute $(\mu\mathbf{A}^T\mathbf{A} + cI)^{-1}$ or the Cholesky factorization of $(\mu\mathbf{A}^T\mathbf{A} + cI)$ before starting the ADMM iterations. Then, at each iteration, instead of (4.55b), one can exactly compute

$$\mathbf{y}^{k+1} = \arg\min_{\mathbf{y}} L(\mathbf{x}^{k+1}, \mathbf{y}; \lambda^k)$$

at a low cost.

**Example 12** [$\ell_2$-constrained basis pursuit denoising]. Next, to apply inexact ADMM to (4.2c), we begin with rewriting (4.2c) as

$$\min_{\mathbf{x}} \{ \|\mathbf{x}\|_1 : \mathbf{A}\mathbf{x} + \mathbf{y} = \mathbf{b}, \|\mathbf{y}\|_2 \leq \sigma \}. \tag{4.56}$$

Treating the constraint $\|\mathbf{y}\|_2 \leq \sigma$ as an indicator function $\chi_{\|\mathbf{y}\|_2 \leq \sigma}$ in the objective, we obtain the Lagrangian

$$L(\mathbf{x}, \mathbf{y}; \lambda) = \|\mathbf{x}\|_1 + \chi_{\|\mathbf{y}\|_2 \leq \sigma} - \lambda^T(\mathbf{A}\mathbf{x} + \mathbf{y} - \mathbf{b}) + \frac{c}{2}\|\mathbf{A}\mathbf{x} + \mathbf{y} - \mathbf{b}\|_2^2$$

and the inexact ADMM iteration

$$\mathbf{y}^{k+1} = \mathrm{Proj}_{\mathbf{B}_\sigma}(\lambda^k/c - (\mathbf{A}\mathbf{y}^k - \mathbf{b})), \tag{4.57a}$$

$$\mathbf{x}^{k+1} = \mathrm{shrink}(\mathbf{x}^k - \tau_k \nabla_{\mathbf{x}} L(\mathbf{x}^k, \mathbf{y}^{k+1}; \lambda^k), \tau_k), \tag{4.57b}$$

$$\lambda^{k+1} = \lambda^k - \gamma c(\mathbf{A}\mathbf{x}^{k+1} + \mathbf{y}^{k+1} - \mathbf{b}), \tag{4.57c}$$

where $\mathrm{Proj}_{\mathbf{B}_\sigma}$ stands for Euclidean projection to the $\sigma$-ball $\mathbf{B}_\sigma = \{\mathbf{y} : \|\mathbf{y}\|_2 \leq \sigma\}$.

---

The exact and inexact ADMM can also be applied to the dual problems of (4.2a)–(4.2c), as well as the primal and dual problems of the augmented versions of various models. It can have a large number of different forms.

---

**Example 13** [Robust PCA]. Our next example is more complicated since it has two regularization terms. *Robust PCA* [164] aims to decompose a data matrix $\mathbf{M}$ into the sum of a low-rank matrix $\mathbf{L}$ and a sparse matrix $\mathbf{S}$, that is, $\mathbf{M} = \mathbf{L} + \mathbf{S}$, which has many applications in data mining, video processing, face recognition, and so on. The robust PCA model solves the problem

$$\min_{\mathbf{L},\mathbf{S}} \|\mathbf{L}\|_* + \alpha\|\mathbf{S}\|_1 + \frac{\mu}{2}\|\mathcal{A}(\mathbf{L}) + \mathcal{B}(\mathbf{S}) - \mathbf{b}\|_2^2, \tag{4.58}$$

where $\mathcal{A}$ and $\mathcal{B}$ are matrix linear operators, and $\alpha$ and $\mu$ are two weight parameters. Problem (4.58) is equivalent to

$$\min_{\mathbf{x},\mathbf{L},\mathbf{S}} \left\{ \|\mathbf{L}\|_* + \alpha\|\mathbf{S}\|_1 + \frac{\mu}{2}\|\mathbf{x}\|_2^2 : \mathbf{x} + \mathcal{A}(\mathbf{L}) + \mathcal{B}(\mathbf{S}) = \mathbf{b} \right\}. \tag{4.59}$$

Introduce

$$\mathbf{y} := (\mathbf{L}, \mathbf{S})$$

and linear operator $\mathcal{C}(\mathbf{y}) := \mathcal{A}(\mathbf{L}) + \mathcal{B}(\mathbf{S})$ and $h(\mathbf{y}) = \|\mathbf{L}\|_* + \alpha\|\mathbf{S}\|_1$. Then problem (4.59) can be rewritten as

$$\min_{\mathbf{x},\mathbf{y}} \left\{ \frac{\mu}{2}\|\mathbf{x}\|_2^2 + h(\mathbf{y}) : \mathbf{x} + \mathcal{C}(\mathbf{y}) = \mathbf{b} \right\}, \tag{4.60}$$

to which one can apply the inexact ADMM iterations in which the $\mathbf{y}$-subproblem decouples $\mathbf{L}$ and $\mathbf{S}$ with the linear proximal step. Specifically,

$$\mathbf{x}^{k+1} = \arg\min_{\mathbf{x}} \frac{\mu}{2}\|\mathbf{x}\|_2^2 + \frac{c}{2}\|\mathbf{x} + \mathcal{C}(\mathbf{y}^k) - \mathbf{b} - \lambda^k/c\|_2^2, \tag{4.61a}$$

$$\mathbf{L}^{k+1} = \arg\min_{\mathbf{L}} \|\mathbf{L}\|_* + c\langle \mathcal{A}^*(\mathbf{x}^{k+1} + \mathcal{C}(\mathbf{y}^k) - \mathbf{b} - \lambda^k/c), \mathbf{L}\rangle + \frac{1}{2t}\|\mathbf{L} - \mathbf{L}^k\|_2,$$

$$\tag{4.61b}$$

$$\mathbf{S}^{k+1} = \arg\min_{\mathbf{S}} \alpha\|\mathbf{S}\|_1 + c\langle \mathcal{B}^*(\mathbf{x}^{k+1} + \mathcal{C}(\mathbf{y}^k) - \mathbf{b} - \lambda^k/c), \mathbf{S}\rangle + \frac{1}{2t}\|\mathbf{S} - \mathbf{S}^k\|_2,$$

$$\text{(4.61c)}$$

$$\lambda^{k+1} = \lambda^k - \gamma c(\mathbf{x}^{k+1} + \mathcal{C}(\mathbf{y}^{k+1}) - \mathbf{b}). \tag{4.61d}$$

**Example 14** [$\ell_1$-$\ell_1$ model]. This example considers the $\ell_1$-$\ell_1$ model

$$\min_{\mathbf{x}} \|\mathbf{x}\|_1 + \mu\|\mathbf{A}\mathbf{x} - \mathbf{b}\|_1,$$

which can be written equivalently as

$$\min_{\mathbf{x}} \{\|\mathbf{x}\|_1 + \mu\|\mathbf{y}\|_1 : \mathbf{A}\mathbf{x} + \mathbf{y} = \mathbf{b}\}.$$

The augmented Lagrangian is

$$L(\mathbf{x}, \mathbf{y}; \lambda) = \|\mathbf{x}\|_1 + \mu\|\mathbf{y}\|_1 - \lambda^T(\mathbf{A}\mathbf{x} + \mathbf{y} - \mathbf{b}) + \frac{c}{2}\|\mathbf{A}\mathbf{x} + \mathbf{y} - \mathbf{b}\|_2^2,$$

to which one can apply the inexact ADMM iteration

$$\mathbf{y}^{k+1} = \text{shrink}(\lambda^k/c - (\mathbf{A}\mathbf{x}^k - \mathbf{b}), \mu/c), \tag{4.62a}$$

$$\mathbf{x}^{k+1} = \text{shrink}(\mathbf{x}^k - \tau_k \nabla_{\mathbf{x}} L(\mathbf{x}^k, \mathbf{y}^{k+1}; \lambda^k), \tau_k), \tag{4.62b}$$

$$\lambda^{k+1} = \lambda^k - \gamma c(\mathbf{A}\mathbf{x}^{k+1} + \mathbf{y}^{k+1} - \mathbf{b}). \tag{4.62c}$$

**Example 15** [Group basis pursuit]. Group basis pursuit is problem (4.5), for which primal and dual ADMM algorithms are presented in [158]. Problem (4.5) is equivalent to

$$\min_{\mathbf{x},\mathbf{z}} \{\|\mathbf{z}\|_{\mathcal{G},2,1} : \mathbf{A}\mathbf{x} = \mathbf{b}, \mathbf{x} = \mathbf{z}\}, \tag{4.63}$$

whose augmented Lagrangian is

$$L(\mathbf{x}, \mathbf{z}; \lambda_1, \lambda_2) = \|\mathbf{z}\|_{\mathcal{G},2,1} - \lambda_1^T(\mathbf{z} - \mathbf{x}) + \frac{c_1}{2}\|\mathbf{z} - \mathbf{x}\|_2^2 - \lambda_2^T(\mathbf{A}\mathbf{x} - \mathbf{b}) + \frac{c_2}{2}\|\mathbf{A}\mathbf{x} - \mathbf{b}\|_2^2.$$

The exact ADMM subproblems are

$$\mathbf{x}^{k+1} = (c_1\mathbf{I} + c_2\mathbf{A}^T\mathbf{A})^{-1}(c_1\mathbf{z}^k - \lambda_1 + c_2\mathbf{A}^T\mathbf{b} + \mathbf{A}^T\lambda_2), \tag{4.64a}$$

$$\mathbf{z}_{\mathcal{G}_s}^{k+1} = (4.18) \text{ with } \mathbf{x} \leftarrow (\mathbf{x}_{\mathcal{G}_s}^{k+1} + c_1^{-1}(\lambda_1)_{\mathcal{G}_s}) \text{ and } \tau \leftarrow (w_s/c_1), \ s = 1,\ldots,S \tag{4.64b}$$

$$\lambda^{k+1} = \lambda^k - \gamma c(\mathbf{A}\mathbf{x}^{k+1} + \mathbf{y}^{k+1} - \mathbf{b}). \tag{4.64c}$$

If it is easy to invert $(\mathbf{I} + \mathbf{A}\mathbf{A}^T)$ then one can consider the Sherman–Morrison–Woodbury formula to reduce the cost of (4.64a); otherwise, one can consider an inexact ADMM step such as the gradient descent type, or cache the inverse or the Cholesky factorization of $(\mathbf{I} + \mathbf{A}\mathbf{A}^T)$.

The group LASSO problem also has an ADMM algorithm, which is not very different from (4.64); see [158].

**Example 16** [Overlapping group basis pursuit]. In the last example or problem (4.5), the objective is defined with non-overlapping groups, namely, $\mathcal{G}_i \cap \mathcal{G}_j = \emptyset, \forall i \neq j$. If the groups *overlap*, then only an easy modification is needed. For $s = 1, \ldots, S$, define $\mathbf{z}_s \in \mathbb{R}^{|\mathcal{G}_s|}$ and introduce constraints $\mathbf{x}_{\mathcal{G}_s} = \mathbf{z}_s$. This set of constraints over all $s$ can be written compactly as $\mathbf{Gx} = \mathbf{z}$, where $\mathbf{z} = [\mathbf{z}_1, \ldots, \mathbf{z}_S]$ and $\mathbf{G}$ is a certain matrix, and it replaces constraints $\mathbf{x} = \mathbf{z}$ in problem (4.63). Then, one can just follow the previous example and derive the modified ADMM iterations. Note that since $\mathbf{z}_s$, $s = 1, \ldots, S$, do not share variables, the $\mathbf{z}$-subproblem still decomposes to $S$ $\mathbf{z}_s$-subproblems in parallel.

## 4.9   Homotopy Algorithms and Parametric Quadratic Programming

For model (4.2b), there is a method to compute its solutions corresponding to all values of $\mu > 0$ since, assuming the uniqueness of solution $\mathbf{x}^*$ for each $\mu$, the solution path $\mathbf{x}^*(\mu)$ is continuous and piece-wise linear in $\mu$. To see this, let us fix $\mu$ and study the optimality condition of (4.2b) satisfied by $\mathbf{x}^*$:

$$0 \in \partial \|\mathbf{x}^*\|_1 + \mu \mathbf{A}^T (\mathbf{A}\mathbf{x}^* - \mathbf{b}), \tag{4.65}$$

where the subdifferential of $\ell_1$ is given by

$$\partial \|\mathbf{x}\|_1 = \partial |x_1| \times \ldots \times \partial |x_n| \subset \mathbb{R}^n, \quad \partial |x_i| = \begin{cases} \{1\}, & x_i > 0, \\ [-1, 1], & x_i = 0, \\ \{-1\}, & x_i < 0. \end{cases}$$

Since $\partial \|\mathbf{x}\|_1$ is component-wise separable, the condition reduces to

$$0 \in \partial |x_i^*| + \mu \mathbf{a}_i^T (\mathbf{A}\mathbf{x}^* - \mathbf{b}), \quad i = 1, \ldots, n,$$

or equivalently,

$$\mu \mathbf{a}_i^T (\mathbf{A}\mathbf{x}^* - \mathbf{b}) \in -\partial |x_i^*|, \quad i = 1, \ldots, n.$$

By definition, we know $\mu \mathbf{a}_i^T (\mathbf{A}\mathbf{x}^* - \mathbf{b}) \in [-1, 1]$ and

$$x_i^* \begin{cases} \geq 0, & \text{if } \mu \mathbf{a}_i^T (\mathbf{A}\mathbf{x}^* - \mathbf{b}) = -1, \\ = 0, & \text{if } \mu \mathbf{a}_i^T (\mathbf{A}\mathbf{x}^* - \mathbf{b}) \in (-1, 1), \\ \leq 0, & \text{if } \mu \mathbf{a}_i^T (\mathbf{A}\mathbf{x}^* - \mathbf{b}) = 1. \end{cases}$$

This offers a way to rewrite the optimality conditions. If we let

$$\mathcal{S}_+ := \{i : \mu \mathbf{a}_i^T (\mathbf{A}\mathbf{x}^* - \mathbf{b}) = -1\},$$
$$\mathcal{S}_0 := \{i : \mu \mathbf{a}_i^T (\mathbf{A}\mathbf{x}^* - \mathbf{b}) \in (-1, 1)\},$$
$$\mathcal{S}_- := \{i : \mu \mathbf{a}_i^T (\mathbf{A}\mathbf{x}^* - \mathbf{b}) = +1\},$$
$$\mathcal{S} := \mathcal{S}_+ \cup \mathcal{S}_-,$$

then we have $\text{supp}(\mathbf{x}^*) \subseteq \mathcal{S}$ and can rewrite (4.65) equivalently as

$$\mathbf{1} + \mu \mathbf{A}_{\mathcal{S}_+}^T (\mathbf{A}_{\mathcal{S}} \mathbf{x}_{\mathcal{S}}^* - \mathbf{b}) = \mathbf{0}, \tag{4.66a}$$

$$-\mathbf{1} + \mu \mathbf{A}_{\mathcal{S}_-}^T (\mathbf{A}_{\mathcal{S}} \mathbf{x}_{\mathcal{S}}^* - \mathbf{b}) = \mathbf{0}, \tag{4.66b}$$

$$\mu \mathbf{a}_i^T (\mathbf{A}_{\mathcal{S}} \mathbf{x}_{\mathcal{S}}^* - \mathbf{b}) \in (-1, 1), \quad \forall i \notin \mathcal{S}, \tag{4.66c}$$

where $\mathbf{A}_{\mathcal{T}}$ denotes the submatrix of $\mathbf{A}$ formed by the columns of $\mathbf{A}$ in $\mathcal{T}$ and $\mathbf{a}_i$ is the $i$th column of $\mathbf{A}$. Note that the composition is *not* defined by the sign of $x_i^*$ but by the values of $\mu \mathbf{a}_i^T (\mathbf{Ax}^* - \mathbf{b})$. It is *possible* to have $x_i^* = 0$ for some $i \in (\mathcal{S}_+ \cup \mathcal{S}_-)$. However, if $i \in \mathcal{S}_0$, then $x_i^* = 0$. Combining (4.66a) and (4.66b), we obtain

$$(\mathbf{A}_{\mathcal{S}}^T \mathbf{A}_{\mathcal{S}}) \mathbf{x}_{\mathcal{S}}^* = \mathbf{A}_{\mathcal{S}}^T \mathbf{b} + \mu^{-1} \begin{bmatrix} -\mathbf{1} \\ \mathbf{1} \end{bmatrix}, \tag{4.67}$$

which uniquely determines $\mathbf{x}_{\mathcal{S}}^*$ if $(\mathbf{A}_{\mathcal{S}}^T \mathbf{A}_{\mathcal{S}})$ is non-singular. The solution $\mathbf{x}_{\mathcal{S}}^*$ is piece-wise linear in $\mu^{-1}$. If $(\mathbf{A}_{\mathcal{S}}^T \mathbf{A}_{\mathcal{S}})$ is singular, $\mathbf{A}_{\mathcal{S}} \mathbf{x}_{\mathcal{S}}^*$ is unique and piece-wise linear in $\mu^{-1}$; we leave the proof to the interested reader. Condition (4.66c) joins (4.66a) and (4.66b) in determining $\mathcal{S}$ for each given $\mu > 0$ so that there exists $\mathbf{x}_{\mathcal{S}}^*$ satisfying (4.66) and consistent with $\mathcal{S}$; (4.66c) must be satisfied but it does not directly determine the values of $\mathbf{x}_{\mathcal{S}}^*$.

For each $\mu > 0$, both $\mathbf{Ax}^* - \mathbf{b}$ and $\mathcal{S}$ are unique. This is because any point on the line segment between two solutions of a convex program is also a solution. If there are two distinct solutions, they must give the same value of $\|\mathbf{x}\|_1 + \frac{\mu}{2} \|\mathbf{Ax} - \mathbf{b}\|_2^2$. Over the line segment between the two points, $\|\mathbf{x}\|_1$ is piece-wise linear, but $\frac{\mu}{2} \|\mathbf{Ax} - \mathbf{b}\|_2^2$ cannot be piece-wise linear unless $\mathbf{Ax} - \mathbf{b}$ is constant over the line segment. So, not only is $\mathbf{Ax}^* - \mathbf{b}$ unique for each $\mu$, so is $\|\mathbf{x}^*\|_1$. $\mathcal{S}$ is defined by $\mathbf{A}^T (\mathbf{Ax}^* - \mathbf{b})$, so it is also unique. So far, we have shown that $\mu$ uniquely determines $\mathcal{S}$ and $\mathbf{x}_{\mathcal{S}}^*$ (or $\mathbf{A}_{\mathcal{S}} \mathbf{x}_{\mathcal{S}}^*$ if $(\mathbf{A}_{\mathcal{S}}^T \mathbf{A}_{\mathcal{S}})$ is singular), the latter of which is linearly in $\mu^{-1}$.

On the other hand, each $\mathcal{S}$ is associated with a (possibly empty) *interval* of $\mu$, which can be verified by plugging the solution of (4.67) into (4.66c). Therefore, $\mathbf{x}^*$ (or $\mathbf{Ax}^*$ if $(\mathbf{A}_{\mathcal{S}}^T \mathbf{A}_{\mathcal{S}})$ is singular) is piece-wise linear in $\mu^{-1}$. Not only is it piece-wise linear, the entire solution path is continuous in $\mu$. If at $\mu$ the solution path is discontinuous, consider the left line segment $\{\mathbf{Ax}^*(\mu - \epsilon) : 0 < \epsilon < \epsilon^0\}$ and the right one $\{\mathbf{Ax}^*(\mu + \epsilon) : 0 < \epsilon < \epsilon^0\}$, where $\epsilon^0 > 0$ is sufficiently small. Since the objective is continuous, taking $\epsilon \downarrow 0$, we get $\mathbf{Ax}^*(\mu_-)$ and $\mathbf{Ax}^*(\mu_+)$ that are different yet both optimal at $\mu$, which contradicts the uniqueness of $\mathbf{Ax}^*$ at $\mu$! To summarize, we have argued that $\mathbf{x}^*$, if it is unique for all $\mu$, and $\mathbf{Ax}^*$ are continuous and piece-wise linear in $\mu^{-1}$.

All the discussions above hint at a $\mu$-homotopy method, starting from $\mu = 0$ with the corresponding solution $\mathbf{x}^* = \mathbf{0}$ and increasing $\mu$ to compute the piece-wise linear solution path [165]. This is also known as parametric quadratic programming [166]. From $\mu$ equals 0 through the minimum value such that $\mu \mathbf{a}_i^T \mathbf{b} \in (-1, 1)$ for all $i$, $\mathcal{S} = \emptyset$ and $\mathbf{x}^* = \mathbf{0}$ satisfy the optimality condition (4.66). If one keeps $\mathcal{S} = \emptyset$ but further increases $\mu$, then either (4.66c) is always satisfied (possible but not likely in practice) or (4.66c) is eventually violated for some $i$. Right at the violating point of $\mu$, add that $i$ into $\mathcal{S}$ and compute $\mathbf{x}_{\mathcal{S}}^*$ from (4.67), which is in a $\mu$-parametric form as

$$\mathbf{x}_{\mathcal{S}}^* = (\mathbf{A}_{\mathcal{S}}^T \mathbf{A}_{\mathcal{S}})^{-1} \mathbf{A}_{\mathcal{S}}^T \mathbf{b} + \mu^{-1} (\mathbf{A}_{\mathcal{S}}^T \mathbf{A}_{\mathcal{S}})^{-1} \begin{bmatrix} -\mathbf{1} \\ \mathbf{1} \end{bmatrix}, \tag{4.68}$$

or

$$\mathbf{A}_{\mathcal{S}}\mathbf{x}_{\mathcal{S}}^* = \mathbf{A}_{\mathcal{S}}(\mathbf{A}_{\mathcal{S}}^T\mathbf{A}_{\mathcal{S}})^\dagger\mathbf{A}_{\mathcal{S}}^T\mathbf{b} + \mu^{-1}\mathbf{A}_{\mathcal{S}}(\mathbf{A}_{\mathcal{S}}^T\mathbf{A}_{\mathcal{S}})^\dagger \begin{bmatrix} -1 \\ 1 \end{bmatrix}, \qquad (4.69)$$

if $\mathbf{A}_{\mathcal{S}}^T\mathbf{A}_{\mathcal{S}}$ is singular. Note at the moment, $\mathcal{S}$ has a single entry, so one of $\mathcal{S}_+$ and $\mathcal{S}_-$ is empty and the other has a single entry. Therefore, all quantities in the above equation are scalars, and in particular, $\begin{bmatrix} -1 \\ 1 \end{bmatrix}$ reduces to either 1 or $-1$. Now one further increases $\mu$, until either (4.66c) is violated for another $i$ at a point of $\mu$ (if it exists, this violating point can be easily computed by plugging (4.68) and (4.69) into (4.66c)) or sign($\mathbf{x}_i^*$) violates $\mathcal{S}$. In the former case, one again adds that $i$ to $\mathcal{S}$, and in the latter case, one drops the violating $i$ from $\mathcal{S}$. This process of augmenting and shrinking $\mathcal{S}$ is continued until $\mu$ reaches a target value or $\mu$ can increase to $\infty$ without any violations. This is a finite process since there is a finite number of possibilities of $\mathcal{S}$. Although we have omitted the computation detail, this is roughly the algorithm in [166, 165].

The homotopy algorithm is extremely efficient when $\mathcal{S}$ only needs to be updated a small number of times before reaching the target $\mu$ since the computing cost for each linear piece is low when $\mathcal{S}$ is small. Therefore, the homotopy algorithm suits a rather small value of $\mu$ and problems with highly sparse solutions. For larger $\mu$ though one cannot make much saving on computing (4.68); checking the condition (4.66c) is time consuming. Fortunately, the check can be skipped on certain sets of $i$. This is referred to as the elimination of coordinates. A safe elimination rule is given in [167], and a more efficient heuristic is presented in [168].

## 4.10    Continuation, Varying Stepsizes, and Line Search

The convergence of the prox-linear iterations for (4.1b) depends critically on the parameter $\mu$ and the stepsize $\delta_k$. This section discusses how these parameters should be chosen in practice.

The discussion begins with the choice of $\mu$. A smaller $\mu$ gives more emphasis to the regularization function $r(\mathbf{x})$ and often leads to more structured, or sparser, solutions. Such solutions are simpler, and for technical reasons that we skip here, are quicker to obtain. In fact, a large $\mu$ can lead to prox-linear iterations that are extremely slow.

Continuation on $\mu$ is a simple technique used in several codes such as FPC [152] and SPARSA [169] to accelerate the convergence of prox-linear iterations. To apply this technique, one does not use the $\mu$ given in (4.1b) (which we call the target $\mu$) to start the iterations; instead, use a small $\bar{\mu}$ and gradually increase $\bar{\mu}$ over the iterations until it reaches the target $\mu$. This lets the iteration produce highly structured (low $r(\mathbf{x})$) yet less fitted (large $f(\mathbf{A}\mathbf{x} - \mathbf{b})$) solutions first, corresponding to smaller $\bar{\mu}$ values, and use them to "warm start" the iterations for producing solutions with more balanced structure and fitting, corresponding to smaller $\bar{\mu}$ values. Continuation does not have rigorous theories to give the optimal sequence of $\bar{\mu}$ and how many iterations are needed for each $\bar{\mu}$. It is more regarded as a very effective heuristic. One can geometrically increase $\bar{\mu}$ and, since warm starts do not require high accuracies, one just needs a moderate number of iterations for each $\bar{\mu}$ except after it reaches the target $\mu$ given in (4.1b).

It turns out even simple continuation strategy leads to much faster convergence. In [152], initial $\bar{\mu}$ is chosen simply to have a value large enough to avoid a zero solution for (4.1b) with $\mu = \bar{\mu}$. For $r(\cdot) = \|\cdot\|_2$ and $f(\cdot) = \frac{1}{2}\|\cdot\|_2^2$, the choice is $\|\mathbf{A}^T\mathbf{b}\|_\infty^{-1}$. During the iterations, once $\mathbf{x}^k$ and $\mathbf{x}^{k+1}$ become very close or the $\nabla f(\mathbf{x}^k)$ has a small size, $\bar{\mu}$ is increased by multiplying with a constant factor.

Line search is widely used in nonlinear optimization to speed up objective descent and, in some cases, to guarantee convergence. It allows one to vary the stepsize $\delta_k$ in a systematic way. Let

$$\psi(\mathbf{x}) = r(\mathbf{x}) + \mu f(\mathbf{x}).$$

One approach [170] replaces (4.26) by the following modified Armijo–Wolfe line search iteration:

$$\text{trial point:}\quad \bar{\mathbf{x}}^{k+1} = \arg\min_{\mathbf{x}}\ r(\mathbf{x}) + \mu f(\mathbf{x}^k) + \langle \mu \nabla f(\mathbf{x}), \mathbf{x} - \mathbf{x}^k \rangle$$

$$+ \frac{1}{2\delta}\|\mathbf{x} - \mathbf{x}^k\|_2^2, \tag{4.70a}$$

$$\text{search direction:}\quad \mathbf{d} = \bar{\mathbf{x}}^{k+1} - \mathbf{x}^k, \tag{4.70b}$$

$$\text{decay rate:}\quad \Delta = \mu \nabla f(\mathbf{k}^k)^T \mathbf{d} + r(\bar{\mathbf{x}}^{k+1}) - r(\mathbf{x}^k), \tag{4.70c}$$

$$\text{backtracking:}\quad h^* = \arg\min\{h \in \mathbb{Z}_+ : \psi(\mathbf{x}^k + \alpha\rho^h\mathbf{d}) \le \psi(\mathbf{x}^k) + \eta\alpha\rho^h\Delta\}, \tag{4.70d}$$

$$\text{new point:}\quad \mathbf{x}^{k+1} = \mathbf{x}^k + \alpha\rho^{h^*}\mathbf{d}. \tag{4.70e}$$

There is no universally best values for $\alpha$, $\eta$, and $\rho$, but one can start with $\alpha = 1$, $\rho = 0.85$, and $\eta = 0.001$. The idea is to exploit the trial point $\bar{\mathbf{x}}^{k+1}$ by making a large step along the direction $\mathbf{d}$ while guaranteeing that the new point $\mathbf{x}^{k+1}$ gives a sufficient descent in the sense of (4.70d). See [171] for a nice explanation of this "sufficient descent" condition. The main computing cost of line search is "backtracking" when the condition (4.70d) is not satisfied by $h = 0$. For one to benefit from line search, the cost of backtracking must be significantly smaller than that of the trial point computation. See [170] on how to efficiently compute $h^*$.

## 4.11  Greedy Algorithms

### 4.11.1  Greedy Pursuit Algorithms

Strictly speaking, the greedy algorithms for sparse optimization are not optimization algorithms since they are not driven by an overall objective. Instead, they build a solution – or the support of the solution – part by part, yet in many cases, the solution is indeed the sparsest solution.

A basic greedy algorithm is called the orthogonal matching pursuit (OMP) [172]. With an initial point $\mathbf{x}^0$ and empty initial support $\mathcal{S}^0$, starting at $k = 1$, it iterates (where $\mathbf{A}_i$ is the $i$th column of $\mathbf{A}$)

$$\mathbf{r}^k = \mathbf{b} - \mathbf{A}\mathbf{x}^{k-1}, \tag{4.71a}$$

$$\mathcal{S}^k = \mathcal{S}^{k-1} \cup \arg\min_i \{\|\mathbf{A}_i\alpha - \mathbf{r}^k\|_2 : i \notin \mathcal{S}^{k-1}, \alpha \in \mathbb{R}\}, \tag{4.71b}$$

$$\mathbf{x}^k = \min_{\mathbf{x}}\{\|\mathbf{A}\mathbf{x} - \mathbf{b}\|_2 : \mathrm{supp}(\mathbf{x}) \subseteq \mathcal{S}^k\}, \tag{4.71c}$$

until $\|\mathbf{r}^k\|_2 \leq \epsilon$ is satisfied. Step (4.71a) computes the measurement residual, step (4.71b) adds the columns of $\mathbf{A}$ that best explains the residual to the running support, and step (4.71c) updates the solution to one that best explains the measurements while confined to the running support.

In (4.71b), for each $i$, the minimizer is $\alpha = \mathbf{A}_i^T\mathbf{r}^k/\|\mathbf{A}_i\|_2$. Hence, the selected $i$ has the smallest $\|(\mathbf{A}_i^T\mathbf{r}^k/\|\mathbf{A}_i\|_2)\mathbf{A}_i - \mathbf{r}^k\|_2$ over all $i \in \mathcal{S}^{k-1}$. Step (4.71c) solves a least-squares problem, restricting $\mathbf{x}$ to the support $\mathcal{S}^k$. There are numerical skills to speed up these steps but we do not overview them here.

Many further developments to OMP lead to algorithms with improved solution quality and/or speed such as stage-wise OMP (StOMP) [173], regularized OMP [174], subspace pursuit [175], CoSaMP [176], as well as HTP [177].

Subspace pursuit and CoSaMP have better performance than the other greedy algorithms. Because they are similar, we describe only CoSaMP. There are more steps in each iteration to predict and refine the intermediate solution and support. With an initial point $\mathbf{x}^0$ and an estimate sparsity level $s$, starting at $k = 1$, CoSaMP iterates

$$\mathbf{r}^k \leftarrow \mathbf{b} - \mathbf{A}\mathbf{x}^{k-1} \qquad \text{(residual)} \tag{4.72a}$$

$$\mathbf{a} \leftarrow \mathbf{A}^*\mathbf{r}^k \qquad \text{(correlation)} \tag{4.72b}$$

$$\mathcal{T} \leftarrow \mathrm{supp}(\mathbf{x}^{k-1}) \cup \mathrm{supp}(\mathbf{a}_{2s}) \qquad \text{(merge supports)} \tag{4.72c}$$

$$\mathbf{c} \leftarrow \min_{\mathbf{x}}\{\|\mathbf{A}\mathbf{x} - \mathbf{b}\|_2 : \mathrm{supp}(\mathbf{x}) \subseteq \mathcal{T}\} \qquad \text{(least-squares)} \tag{4.72d}$$

$$\mathbf{x}^k \leftarrow \mathbf{c}_s, \qquad \text{(pruning)} \tag{4.72e}$$

until $\|\mathbf{r}^k\|_2 \leq \epsilon$ is satisfied, where $\mathbf{a}_{2s} = \arg\min\{\|\mathbf{x} - \mathbf{a}\|_2 : \|\mathbf{x}\|_0 \leq 2s\}$ is the best $2s$-approximate of $\mathbf{a}$ and similarly, $\mathbf{c}_s$ is the best $s$-approximate of $\mathbf{c}$. Over the iterations, $\mathrm{supp}(\mathbf{x}^k)$ are updated but kept to contain no more than $s$ components. Hence, $\mathcal{T}$ has no more than $3s$ components. Unlike OMP, CoSaMP's support is updated batch by batch.

## 4.11.2    Iterative Support Detection

In the previous subsection, greedy algorithms find a solution by iteratively growing (in OMP) or updating (in subspace pursuit and CoSaMP) the solution support. To add or update the support, greedy algorithms measure the column-residual correlation $\mathbf{A}^*\mathbf{r}^k$ or normalized correlation $\min\{\|\mathbf{A}_i\alpha - \mathbf{r}^k\|_2 : \alpha \in \mathbb{R}\}$ for each $i \notin \mathcal{S}^{k-1}$. Given the support, a least-squares restricted to the support estimates the solution value. In this section, we overview an algorithm that is based on a similar support-value alternating iteration but it uses different ways.

The iterative support detection (ISD) framework is the iteration with initial $\mathcal{T}$ set to $\{1, 2, \ldots, n\}$:

$$\mathbf{x}^k \leftarrow \arg\min\{\|\mathbf{x}_\mathcal{T}\|_1 : \mathbf{A}\mathbf{x} = \mathbf{b}\}, \tag{4.73a}$$

$$\mathcal{T} \leftarrow \{|x_i^k| < \rho^k\}, \tag{4.73b}$$

while $\|\mathbf{x}^k\|_0 > m$ and $|\mathcal{T}| \geq n - m$, where $\{\rho^k\}$ is a monotonically decreasing sequence of scalars and $\|\mathbf{x}_\mathcal{T}\|_1 = \sum_{i \in \mathcal{T}} |x_i|$. The iterative update of the support $\mathcal{T}$ is not like any greedy algorithms in the sense that it is typically growing yet is not always monotonic in $k$.

The iteration (4.73) is motivated by the fact that if the entire correct support – $\text{supp}(\mathbf{x}^o)$ – is out of $\mathcal{T}$, then (4.73a) will return the correct recovery, and a larger $\mathcal{T}^C \cap \text{supp}(\mathbf{x}^o)$ tends to let (4.73a) give a better estimate of $\mathbf{x}^o$.

Since at $k = 1$ step (4.73a) is equivalent to (4.2a). If it does not return a faithful reconstruction, ISD with $k \geq 2$ addresses failed reconstructions of (4.2a) by hopefully discovering a partial correct support from $\mathbf{x}^k$. Analysis in [178] gives conditions on $\mathcal{T}$ that will guarantee the recovery of $\mathbf{x}^o$.

There are two existing rules for choosing $\rho^k$. The simple one uses

$$\rho^k = \beta^{-k} \|\mathbf{x}^k\|_\infty, \tag{4.74}$$

where $\beta > 1$ can be chosen as, for example, 3. Since $\|\mathbf{x}^k\|_\infty$ is quite stable in $k$, $\rho^k$ will become sufficiently small to make $|\mathcal{T}| < n - m$ in a small number of iterations. Hence, the number of iterations of (4.73) is typically small.

A less simple rule for $\rho^k$ is based on the "first significant jump," which is effective for $\mathbf{x}^o$ with nonzero entries following a fast-decaying distribution. The idea is that with such $\mathbf{x}^o$, $\mathbf{x}^k$ tends to be fast-decaying as well, yet the false nonzero entries in $\mathbf{x}^k$ are often clustered and form the tail of $\mathbf{x}^k$. See Figure 2 of [178] for an illustration. When the components of $\mathbf{x}^k$ are sorted in the increasing order like $|x_{[1]}^k| \leq |x_{[2]}^k| \leq \cdots \leq |x_{[n]}^k|$, the tail and head are often separated by the first significant jump, which is the smallest $i$ such that

$$|x_{[i+1]}^k| - |x_{[i]}^k| > \beta^{-k}. \tag{4.75}$$

Both rules were tested empirically, and the results show that ISD and nonconvex approaches have roughly equal recovery quality. Yet, it is more versatile to incorporate model sparsity into the selection of $\mathcal{T}$.

## 4.11.3    Hard Thresholding

Hard thresholding refers to the operator that keeps the $K$ largest entries of a vector in magnitude for a given $K$. We write it as $\text{hardthr}(\mathbf{x})$. For vector $\mathbf{x}$ and the index set $\mathcal{K} = \{K \text{ largest entries of } \mathbf{x} \text{ in magnitude}\}$, $\text{hardthr}(\mathbf{x})$ obeys

$$(\text{hardthr}(\mathbf{x}))_i = \begin{cases} x_i, & i \in \mathcal{K}, \\ 0, & \text{otherwise.} \end{cases} \tag{4.76}$$

Unlike soft thresholding, the values of the $K$ largest entries are not changed. The basic iterative hard-thresholding iteration [179] is

$$\mathbf{x}^{k+1} \leftarrow \text{hardthr}\left(\mathbf{x}^k + \mathbf{A}^T(\mathbf{b} - \mathbf{A}\mathbf{x}^k)\right), \tag{4.77}$$

which differs from (4.33) only by the use of the hard, as opposed to the soft, thresholding, ignoring the difference in the step lengths. The iteration (4.77) can be used to solve the problem

$$\min_{\mathbf{x}} \left\{ \|\mathbf{A}\mathbf{x} - \mathbf{b}\|_2^2 : \|\mathbf{x}\|_0 \leq K \right\}, \tag{4.78}$$

with the guarantee to converge to one of its local minima provided that $\|\mathbf{A}\|_2 \leq 1$. Nevertheless, under the RIP assumption of $\mathbf{A}$, $\delta_{3K}(\mathbf{A}) \leq 32^{-1/2}$, this local minimum $\mathbf{x}^*$ has bounded distance to the original unknown signal $\mathbf{x}_0$. Specifically, $\|\mathbf{x}^* - \mathbf{x}_0\|_2 \leq \|\mathbf{x}_0 - \text{hardthr}(\mathbf{x}_0)\|_2 + \|\mathbf{x}_0 - \text{hardthr}(\mathbf{x}_0)\|_1/\sqrt{k} + \|\mathbf{n}\|_2$, where $\mathbf{n} = \mathbf{b} - \mathbf{A}\mathbf{x}_0$.

The iteration (4.77) is not very competitive compared to $\ell_1$ minimization, greedy algorithms, especially to nonconvex $\ell_q$ and iterative support detection algorithms. However, the normalized iterative hard-thresholding iteration

$$\mathbf{x}^{k+1} \leftarrow \text{hardthr}\left(\mathbf{x}^k + \mu^k \mathbf{A}^T(\mathbf{b} - \mathbf{A}\mathbf{x}^k)\right), \tag{4.79}$$

where $\mu^k$ is a changing stepsize, has much better performance. The rule of choosing $\mu^k$ is detailed in [180], and it depends whether the resulting $\text{supp}(\mathbf{x}^{k+1})$ equals $\text{supp}(\mathbf{x}^k)$ or not. Under this rule, (4.79) also converges to one of the local minima of (4.78) and obeys the bounded distance to $\mathbf{x}_0$ under a certain RIP assumption on $\mathbf{A}$. On the other hand, the recovery performance of (4.79) is significantly stronger than that of (4.77).

The work [177] proposes hard-thresholding pursuit (HTP) as a combination of the (original and normalized) IHT with the idea of matching pursuit. In (4.77), instead of applying the hard-thresholding operator to return $\mathbf{x}^{k+1}$, HTP performs

$$\mathcal{T} \leftarrow \{s \text{ largest entries of } \left(\mathbf{x}^k + \mathbf{A}^T(\mathbf{b} - \mathbf{A}\mathbf{x}^k)\right)\}, \tag{4.80a}$$

$$\mathbf{x}^{k+1} \leftarrow \underset{\{\text{supp}(\mathbf{x}) \in \mathcal{T}\}}{\arg \min} \ \|\mathbf{b} - \mathbf{A}\mathbf{x}\|_2. \tag{4.80b}$$

Reference [177] also applies the same change to (4.79) and introduces a fast version of (4.79). Exact and stable recoveries are guaranteed if $\mathbf{A}$ satisfies the RIP with $\delta_{3K}(\mathbf{A}) \leq 3^{-1/2}$.

## 4.12    Algorithms for Low-Rank Matrices

When the recovery of a low-rank matrix $\mathbf{X}^o \in \mathbb{R}^{n_1 \times n_2}$ is based on minimizing the nuclear norm $\|\cdot\|_*$ and solving a convex optimization problem, most convex optimization algorithms above for $\ell_1$ minimization can be extended for nuclear norm minimization. They include, but are not limited to, the singular value thresholding (SVT) algorithm [125] based on the linearized Bregman algorithm [181, 182], fixed-point continuation code

FPCA [126] extending FPC [152], the code APGL [183] extending [156], and the code [184] based on the alternating direction method [144]. Refer to Section 4.4.1 for the shrinkage operation for the nuclear norm. There are also algorithms with no vector predecessors.

Suppose that the rank of the underlying matrix is given as $s \ll \min\{n_1, n_2\}$ either exactly or as the current overestimate of the true rank, the underlying matrix can be factorized as $\mathbf{X} = \mathbf{USV}^*$ or $\mathbf{X} = \mathbf{PQ}$, where $\mathbf{U}$ and $\mathbf{P}$ have $s$ columns, $\mathbf{V}^*$ and $\mathbf{Q}$ have $s$ rows, and $\mathbf{S}$ is an $s \times s$ diagonal matrix. Instead of $\mathbf{X}$, if one solves for $(\mathbf{U}, \mathbf{S}, \mathbf{V})$ or $(\mathbf{P}, \mathbf{Q})$, then $\mathbf{X}$ can be recovered from these variables and has its rank no more than $s$. Hence, these factorizations and change of variable replace the role of minimizing the nuclear norm $\|\mathbf{X}\|_*$ and give rise to a low-rank matrix. Solvers based on these factorizations include OptSpace [185], LMaFit [186], and an SDPLR-based algorithm [159]. OptSpace [185] and LMaFit [186] are based on minimizing $\|\mathcal{P}_\Omega(\mathbf{USV}^*) - \mathbf{b}\|_2$ and $\|\mathcal{P}_\Omega(\mathbf{PQ}) - \mathbf{b}\|_2$ for recovering a low-rank matrix from a subset of its entries $\mathbf{b} = \mathbf{X}^o_\Omega$, where $\Omega$ is the set of entries observed, and the algorithm in [159] solves $\min\{\frac{1}{2}(\|\mathbf{P}\|_F^2 + \|\mathbf{Q}\|_F^2) : \mathcal{A}(\mathbf{PQ}) = \mathbf{b}\}$, which is a nonconvex problem whose global solution $(\mathbf{P}^*, \mathbf{Q}^*)$ generates $\mathbf{X}^* = \mathbf{P}^*\mathbf{Q}^*$ that is the solution to $\min\{\|\mathbf{X}\|_* : \mathcal{A}(\mathbf{X}) = \mathbf{b}\}$. The common advantage of these algorithms is the few unknown variables, a point easy to see by comparing $n_1 \times n_2$ to $n_1 \times s + s \times n_2$. On the other hand, these methods require judiciously selecting $s$, which can be known, approximately known, or completely unknown based on applications. (Arguably, the nuclear norm convex models in the form of (4.1b) and (4.1c) also have parameters that determine the rank of their solutions. However, their parameters are affected by the noise in $\mathbf{b}$ and how close $\mathbf{X}^o$ is from its best low-rank approximation, while $s$ in the above algorithms is determined by the unknown solution and is critical to the success of those algorithms.) To determine $s$, a dynamic update rule for $s$ is included in LMaFit [186], which is reportedly very effective.

## 4.13 How to Choose an Algorithm

This chapter has gone through several algorithms and various techniques for sparse optimization, and they are illustrated in Fig. 4.4.

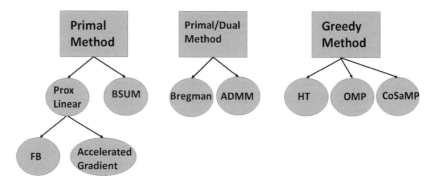

**Figure 4.4** Relations of different types of algorithms discussed in this section.

Due to our limited knowledge and the fast-growing nature of this active field of research, many efficient algorithms are not covered. But already, to recover sparse vectors and low-rank matrices, one has multiple choices. In the previous sections, when each approach is introduced or discussed, we have mentioned their strengths (also weaknesses for some). Here, we attempt to give a more direct comparison and provide some guidance on selecting algorithms. Before we start, we would like to stress that the judgment on different algorithms may have become out of date due to new developments after this chapter is written. After all, sparse optimization is an amazingly common field where a very wide spectrum of optimization, as well as non-optimization, techniques can play active roles by providing new approaches or improving the existing algorithms. Another reason is that, as this part is being written, novel sparse optimization applications are being discovered each day! An algorithm that is less competitive today might have unique features that make it one of the best choices tomorrow under a new setting. We have witnessed such shifts happening on, for example, the ADM algorithms – it was introduced a long time ago, then it lost its favor in nonlinear optimization for a while, but it has recently re-emerged and led to a large number of successful applications. Furthermore, although numerical experiments are a good means to assess the efficiencies of algorithms, it is very difficult to be comprehensive and nearly impossible to reveal the potentials of different algorithmic approaches. The IHT approach in its original form (with a fixed thresholding parameter), as an example, has very poor performance, yet in just three years, works [180, 177] have released its power and made it competitive. The same has happened on many other approaches covered in this chapter to different extents. These make it very tricky to compare different approaches and make judgments. The reader should be aware of our limitations and is advised to check the literature for latest developments.

**Greedy vs. optimization algorithms.** First, both approaches enjoy exact and stable recovery guarantees, or in other words, they can be trusted for recovering sparse solutions given sufficient measurements. Second, due to the nature of greedy algorithms that progressively build (and correct) the solution support, their performance on signals with entries following a *faster* decaying distribution is better – namely fewer measurements are required when the decay is *faster*. On the other hand, $\ell_1$ minimization tends to have more consistent recovery performance – the solution quality tends to be less affected by the decay speed. That said, the speed and accuracies specific $\ell_1$ algorithms are still affected by the decay speed. Some algorithms prefer faster decays, and the other prefer slower decays. Notably, the algorithm in Section 4.11.2 integrates $\ell_1$ minimization with greedy support selection, so it has the advantages of the two. Thirdly, the two approaches are extended in different ways, which to a large extent determine which one is more suitable for a specific application. The greedy approach can be extended to model-base CS [187], in which the underlying signals are not only sparse but their supports are restricted to certain models (e.g., the tree structure). Some of such models are difficult to express as optimization models, not to mention as convex optimization models. On the other hand, it is difficult to extend greedy approaches to handle general objective or energy functions, when they are present along with the sparse objective in a model. Sparse optimization naturally accepts objective functions and constraints of many kinds, especially if they are convex. Algorithms based on prox-linear, variable splitting, and block coordinate descent are very flexible, so they are among the best choices for engineering models with "side" objectives and constraints. Overall, the two approaches are very complementary to each other. When a problem can be solved by either one, one should examine the decay speed. But after all, they are both quite easy to implement and try out.

**Very large-scale sparse optimization.** To solve very large-scale optimization problems, we are facing memory and computing bottlenecks. Many existing algorithms will fail to run if the problem data, such as the matrix $\mathbf{A}$, or the solution, or both does not fit in the available memory. There are a few approaches to address this issue. First of all, if options are available, try to avoid general data and use data that have computation-friendly structures. In CS, examples of structured sensing matrices $\mathbf{A}$ include the discrete Fourier ensemble, a partial Toeplitz or circulant matrix, and so on, which allow fast computation of $\mathbf{Ax}$ and $\mathbf{A}^T\mathbf{y}$ (or complex-valued $\mathbf{A}^*\mathbf{y}$) (or even $(\mathbf{AA}^T)^{-1}$) without storing the entire matrix in memory. When an algorithm is based on such matrix-vector operations, it can take advantages of faster computation and be *matrix-free*. After all, if an algorithm must perform large-scale dense matrix-vector multiplications or solve dense linear systems, these bottlenecks will seriously hamper its ability to solve huge problems.

When the data is given and fast operations are not available, a viable approach is distributed computation. This approach divides the original problem in a series of smaller subproblems, each involving just a subset of data, or solution, or both. The smaller subproblems are solved, typically in parallel, on distributed computing nodes. There is a trade-off between the acceleration due to parallel computing and the overheads of synchronization and splitting.

Another very effective approach, which is popular among the community of large-scale machine learning, is based on stochastic approximation. It takes advantage of the fact that in many problems, from a small (random) set of data (even a single point), one can generate a gradient (or Hessian) that approximates the actual one, whose computation would require the entire batch of data. Between using just one point and the entire batch of data, there lies an optimal trade-off among memory, speed, and accuracy. Stochastic approximation methods use a very small amount of (random chosen) data to guide its iterations and are usually very quick at decreasing the objective until getting close to the solution, where further improvement would require more accurate function evaluations, gradients (and Hessians). If higher solution accuracies are needed, one can try the stochastic approximation methods with dynamic sample sizes.

Solution structure also provides us with opportunities to address the memory and computing bottlenecks. When the solution is a highly sparse vector, algorithms such as BCD with the Gauss–Southwell update rule in Section 3.3, homotopy algorithms in Section 4.9, and some dual algorithms can keep all the intermediate solutions sparse, which effectively reduces the data/solution storage and access. When the solution is a low-rank matrix (or tensor) that has a very large size, algorithms such as those in Section 4.12 that store and update the factors, which have much smaller sizes, instead of the matrix (or tensor) have significant advantages.

# 5    Sublinear Algorithms

## 5.1    Introduction

Sublinear algorithms were initially developed in the theoretical computer science community. The sublinear algorithm can be seen as one further classification of the approximation algorithm. It studies the classic trade-off between algorithm processing time and algorithm output quality.

In a conventional approximation algorithm, the algorithm can output an approximate result that deviates from the optimal result (the deviation is bounded), so that the algorithm processing time can become faster. One hidden implication of the approximation algorithm is that the approximate result is 100% guaranteed within the bound. In a sublinear algorithm, such an implication is relaxed. More specifically, a sublinear algorithm outputs an approximate result that deviates from the optimal result (the deviation is bounded) for a (usually) majority of the time. As a concrete example, a sublinear algorithm usually says that the output of the algorithm differs from the optimal solution by at most 0.1 (the bound) at least 95% of the time (the confidence).

From the theoretical research point of view, sublinear algorithms become a new category of approximation algorithms. From the practical point of view, sublinear algorithms provide two controlling parameters for the user in making trade-offs, while approximation algorithms have only one controlling parameter.

Sublinear algorithms are developed based on random and probabilistic techniques. Note, however, that the guarantee of a sublinear algorithm is on the individual outputs of this algorithm. In this sense, the sublinear algorithm differs from stochastic techniques, which analyze the mean and variance of a system in a steady state. For example, a typical queuing theory result is that the expected waiting time is 100 seconds.

In the theoretical computer sciences in the past few years, there have been many studies on sublinear algorithms. Sublinear algorithms have been developed for many classic computer science problems, such as finding the frequently element, finding distinct elements, and others; and for graph problems, such as finding the minimum spanning tree, and others; and for geometry problems, such as finding the intersection of two polygons, and others. Sublinear algorithms can be broadly classified into sublinear time algorithms, sublinear space algorithms, and sublinear communication algorithms, where the amount of time, storage space, or communications needed is $o(N)$ with $N$ as the input size.

Sublinear algorithms are a good match of big data analytics. For a big data application, decisions can be drawn by only looking at a subset of the data. In particular, sublinear algorithms are suitable for situations where the total amount of data is so massive that even linear processing time is not affordable. Sublinear algorithms are also suitable for situations, where some initial investigations need to be made before looking into the full data set. In many situations, the data are massive but it is not known whether the value of the data is big or not. As such, sublinear algorithms can serve to give an initial "peek" of the data before a more in-depth analysis is carried out. For example, in bioinformatics, we need to test whether certain DNA sequences are periodic. Sublinear algorithms, when appropriately designed to test periodicity in data sequences, can be applied to rule out data that are not useful for further investigation.

## 5.1.1    Organization

In this chapter, we first study the theoretical foundation of sublinear algorithms. We present the foundation of approximation and randomization and we present the history of the development of sublinear algorithms in the theoretical research line. As said, sublinear algorithms can be considered as one branch of approximation algorithms with confidence guarantees. A sublinear algorithm says that the accuracy of the algorithm output will not deviate from an error bound and there is high confidence that the error bound will be satisfied. More rigidly, a sublinear algorithm is commonly written as $(1+\epsilon, \delta)$-approximation in a mathematical form. Here $\epsilon$ is commonly called an *accuracy parameter* and $\delta$ is commonly called a *confidence parameter*. This accuracy parameter is the same to the approximate factor in approximation algorithms. This confidence parameter is the key trade-off where the complexity of a sublinear algorithm can reduce to a sublinear size. We will rigidly define these parameters in this chapter.

Then we present some inequalities, such as Chernoff inequality, Hoeffding inequality, and so on, which are commonly used to derive the bounds for sublinear algorithms. We further present the classification of sublinear algorithms, namely sublinear algorithms in time, sublinear algorithms in space, and sublinear algorithms in communication.

We present four examples in this chapter to illustrate the techniques that are commonly used to derive the bounds in the design of sublinear algorithms. The first example shows a direct application of Hoeffding inequality to derive a simple bound of a sublinear algorithm. The second example is a classic computer science problem of finding distinct elements. We use this example to show how to derive a standard, yet more complex bound for a sublinear algorithm. In addition, we study a *median trick* in boosting the confidence parameter $\delta$, which is widely used in sublinear algorithm designs. The third example is also a classic computer science problem of computing the number of connected components. In the third example, we study how we can split the error bound into two parts. The overall connected component problem can be solved by first dividing the problem into two subproblems and deriving the bounds of each of the subproblems. Finally the error bound of the overall problem is a merge of the two error bounds. Such divide and merge technique is widely used in sublinear algorithm designs. The fourth example is a two-cat problem. Its solution is a sublinear algorithm that does not belong to the standard form of $(\epsilon, \delta)$ approximation. We use this two-cat problem to further broaden the view on the sublinear algorithm designs.

## 5.2    Foundations

### 5.2.1    Approximation and Randomization

We start by considering algorithms. An algorithm is a step-by-step calculating procedure for solving a problem and outputting a result. In common sense, an algorithm tries to output an optimal result. When evaluating an algorithm, an important metric is its complexity. There are different complexity classes. Two most important classes are P and NP. The problems in P are those that can be solved in polynomial times and the problems in NP are those that must be solved in super-polynomial times. Using today's computing architecture, polynomial time algorithms are considered feasible for their running times, yet super-polynomial time algorithms are considered infeasible for their running times.

To handle the problems in NP, a development from theoretical computer science is to introduce a trade-off where we sacrifice the optimality of the output result in order to reduce the algorithm complexity. More specifically, we do not need to achieve the exact optimal solution; yet it is acceptable if we know that the output is close to the optimal solution. This is called *approximation*. Approximation can be rigidly defined. We show one example on a $(1 + \epsilon)$-approximation.

Let $Y$ be a problem space and $f(Y)$ be the procedure to output a result. We call an algorithm a $(1 + \epsilon)$-approximation if this algorithm returns $\hat{f}(Y)$ instead of the optimal solution $f^*(Y)$, and

$$|\hat{f}(Y) - f^*(Y)| \leq \epsilon f^*(Y)$$

We would like to make two comments. First, there can be other approximation criteria beyond $(1+\epsilon)$-approximation. Second, approximation, though introduced mostly for NP problems, is not restricted to NP problems. One can design an approximation algorithm for the problems in P to further reduce the complexity of the algorithm as well.

A hidden assumption of approximation is that an approximation algorithm requests that its output is always (i.e., 100%) within an $\epsilon$ factor of the optimal solution. A further development from theoretical computer sciences is to introduce another trade-off between optimality and algorithm complexity; that is, it is acceptable that the algorithm output is close to the optimal most of the times. For example, 95% of time, the output result is within the error bound to the optimal result. Such probabilistic nature requires an introduction of *randomization*. We call an algorithm a $(1+\epsilon, \delta)$-approximation if this algorithm returns $\hat{f}(Y)$ instead of the optimal solution $f^*(Y)$, and

$$Pr[|\hat{f}(Y) - f^*(Y)| \leq \epsilon f^*(Y)] \geq 1 - \delta$$

Here $\epsilon$ is usually called an *accuracy parameter* (error bound) and $\delta$ is usually called a *confidence parameter*.

**Discussion:** We have seen two steps in theoretical computer sciences in trading-off optimality and complexity. Such trade-off does not immediately lead to an algorithm that is sublinear to its input (i.e., $(1 + \epsilon, \delta)$-approximation is not necessarily sublinear). Nevertheless, these two advancements provide better categorization on algorithms. In particular, the second advancement in randomization makes a sublinear algorithm possible. As discussed in the introduction, processing the full data may not be tolerable in the big data era. As a matter of fact, practitioners have already designed many schemes using only partial data. These designs may be ad hoc in nature and may not have rigid proofs in their quality. On the one hand, sublinear algorithm designs may provide insights in application designs. On the other hand, from a quality control's point of view, the $(1+\epsilon, \delta)$-approximation brings to the practitioners a rigid theoretical evaluation benchmark when evaluating their designs.

### 5.2.2   Inequalities and Bounds

One may recall that the above formulas are similar to those inequalities in probability theory. The difference is that the above formulas and bounds are used on algorithms, yet in probability theory, the formulas and bounds are used on variables.

In reality, many developments of sublinear algorithms heavily apply probability inequalities. Therefore, we state a few mostly used inequalities here and we will use examples to show how they will be applied to sublinear algorithm development.

**Markov inequality**: For a non-negative random variable $X$, and any $a > 0$, we have

$$Pr[X \geq a] \leq \frac{E[X]}{a}$$

Markov inequality is a loose bound. The good thing is that Markov inequality requires no assumptions on the random variable $X$.

**Chernoff inequality**: For independent random Bernoulli variables $X_i$, let $X = \sum X_i$. For any $\Delta$, we have

$$Pr[X \leq (1 - \Delta)E[X]] \leq e^{-\frac{E[X]\Delta^2}{2}}$$

Chernoff bound is tighter. Note, however, that it requires the random variables to be independent.

**Discussion:** From probability theory, the intuition of Chernoff inequality is very simple. It says that the probability of the value of a random variable deviating from its expectation decreases very fast. From the sublinear algorithm point of view, the insight is that if we develop an algorithm and run this algorithm many times upon different subsets of randomly chosen partial data, the probability that the output of the algorithm deviating from the optimal solution decreases very fast. This is also called a *median trick*. We will see more on how to materialize this insight using examples throughout this book.

Chernoff inequality has many variations. Practitioners may often encounter a problem of computing $Pr[X \leq k]$ where $k$ is a parameter of real-world importance. Especially, one may want to link $k$ with $\delta$. For example, given that the expectation of $X$ is known, how can the $k$ be determined so that the probability $Pr[X \leq k]$ is at least $1 - \delta$. Such linkage between $k$ and $\delta$ can be derived from Chernoff inequality as follows:

$$Pr[X \leq k] = Pr[X \leq \frac{k}{E[X]}E[X]]$$

Let $1 - \Delta = \frac{k}{E[X]}$ and with Chernoff inequality we have:

$$Pr[X \leq k] \leq e^{-\frac{E[X](1-\frac{k}{E[X]})^2}{2}}$$

Then, to link $\delta$ and $k$, we have

$$Pr[X \leq k] \leq e^{-\frac{E[X](1-\frac{k}{E[X]})^2}{2}} \leq 1 - \delta$$

Note that the last inequality provides a connection between $k$ and $\delta$.

**Chebyshev inequality**: For any $X$ with $E[X] = \mu$ and $Var[X] = \sigma^2$, and for any $a > 0$,

$$Pr[|X - \mu| \geq a\sigma] \leq \frac{1}{a^2}$$

**Hoeffding inequality**: Assume we have $k$ random identical and independent variables $X_i$, for any $\epsilon$, we have

$$Pr[|X - E[X]| \geq \epsilon] \leq e^{-2\epsilon^2 k}$$

Hoeffding inequality is commonly used to bound the deviation from the mean.

### 5.2.3    Classification of Sublinear Algorithms

The most common classification of sublinear algorithms is to see whether a sublinear algorithm uses $o(N)$ in space or $o(N)$ in time or $o(N)$ in communication, where $N$ is the input size. Respectively, they are called sublinear algorithms in time, sublinear algorithms in space, or sublinear algorithms in communication.

Sublinear algorithms in time mean that one needs to make decisions yet it is impossible for one to look at all data; note that it takes a linear amount of time to look at all data. The result of the algorithm is using $o(N)$ time, where $N$ is the input size. Sublinear algorithms in space mean that one can look at all data because the data is coming in a streaming fashion. In other words, the data comes in an online fashion and it is possible to read each piece of data as time progresses. Yet the challenge is that it is impossible to store all these data in storage because the data is too big. The result of the algorithm is using $o(N)$ space, where $N$ is the storage space. Such category is also commonly called data stream algorithms. Sublinear algorithms in communication mean that the data is too big to be stored in a single machine and one needs to make decisions through collaboration between machines. It is only possible to use $o(N)$ communications, where $N$ is the total number of communications.

There are algorithms that do not fall into the $((1 + \epsilon), \delta)$-approximation category. A typical example is when there needs of a balance between the resources such as storage, communications, time, and so on. Therefore, algorithms can be developed where the contribution of each type of resource is sublinear; and they collectively achieve the task. One example of such kind can be found from a sensor data collection application in [188]. In this example, a data collection task is achieved with a sublinear sacrifice of storage and a sublinear sacrifice of communication.

In this chapter, we will present a few examples. The first one is a simple example on estimating percentage. We show how the bound of a sublinear algorithm can be derived using inequalities. This is a sublinear algorithm in time. Then we discuss a classic sublinear algorithm to find distinct elements. The idea is to see how we can go beyond simple sampling and quantify an idea and develop quantitative bounds. In this example, we also show the median trick, a classic trick in managing $\delta$. This is a sublinear algorithm in space. Then we discuss another classic sublinear algorithm to compute the number of connected components in a big graph. We show how to divide the problem into two subproblems, bound each of the two subproblems, and merge the results. This is a sublinear algorithm in time. Finally, we discuss a two-cat problem, where its intuition is applied in [188]. We show in this problem how to divide two resources so as to collectively achieve an overall task.

## 5.3 Examples

### 5.3.1 Estimating the User Percentage: The Very First Example

We start from a simple example. Assume that there is a group of people who can be classified into different categories. One category is the housewife. What we want to know is the percentage of this group who are housewives, but the group is too big to examine every person. A simple way is to sample a subset of people and see how many of these people belong to the housewife group. This is where the question arise: how many samples are enough?

Assume that the percentage who are housewives in this group of people is $\alpha$. We do not know $\alpha$ in advance. Let $\epsilon$ be the error allowed to deviate from $\alpha$ and $\delta$ be a confidence interval. For example, if $\alpha = 70\%$, $\epsilon = 0.05$, and $\delta = 0.05$, it means that we can output a result where we have a 95% confidence/probability that this result falls in the range of 65%–75%. The following theorem states the number of samples $k$ we need and its relationship with $\epsilon, \delta$.

THEOREM 2. *Given $\epsilon, \delta$, to guarantee that we have a probability of $1 - \delta$ success that the percentage (e.g., of housewife) will not diviate from $\alpha$ for more than $\epsilon$, the number of users we need to sample must be at least $-\frac{\log \delta}{2\epsilon^2}$.*

We first conduct some analysis. Let $N$ be the total number of users and let $m$ be the number of users we sample. Let $Y_i$ be an indicator random variable where

$$Y_i = \begin{cases} 1, & \text{housewife} \\ 0, & \text{otherwise} \end{cases}$$

We assume that $Y_i$ are independent (i.e., Alice belongs to the housewife group is independent of whether Mary belongs to housewife or not).

Let $Y = \sum_{i=1}^{N} Y_i$. By definition, we have $\alpha = \frac{1}{N}E[Y]$. Since $Y_i$ are all independent, $E[Y_i] = \alpha$. Let $X = \sum_{i=1}^{m} Y_i$. Let $\overline{X} = \frac{1}{m}X$. The next lemma says that the expectation $\overline{X}$ of the sampled set is the same as the expectation of the whole set.

LEMMA 3. $E[\overline{X}] = \alpha$.

*Proof.* $E[\overline{X}] = \frac{1}{m}E[\sum_{1}^{m} Y_i] = \frac{1}{m} \times m\alpha = \alpha$.   □

We next proof Theorem 2.

*Proof.*

$$Pr[(\overline{X} - \alpha) > \epsilon] = Pr[(\overline{X} - E[\overline{X}]) > \epsilon] \leq e^{-2\epsilon^2 m}$$

The last inequality is derived by Hoeffding inequality. To make sure that $e^{-2\epsilon^2 m} < \delta$, we need to have $m > -\frac{\log \delta}{2\epsilon^2}$.   □

**Discussion**: Sampling is not a new idea. Many practitioners naturally use sampling techniques to solve their problems. Usually, practitioners discuss the expected values, which ends up with a statistical estimation. In this example, the key idea is to transform a statistical estimation of the expected value into a bound.

## 5.3.2     Finding Distinct Elements

We now study a classic problem by using sublinear algorithms. We want to count the total number of distinct elements in a data stream. For example, suppose that we have a data stream $S = \{1, 2, 3, 1, 2, 3, 1, 2\}$. Clearly, the total number of distinct elements in $S$ is 3.

We look for an algorithm that is sublinear in space. This means that at any single point of time, only a subset of elements can be stored in the memory. The algorithm will go over one pass of the data stream. Our algorithm will only store $O(\log N)$ data, where $N$ is the total number of elements.

### The Initial Algorithm

Let the number of distinct elements in $S$ be $F$. Let $w = \log N$. Assume we have a hash function $h(\cdot)$, which can uniformly hash an element $k$ into $[0, 2^w - 1]$. Let $r(\cdot)$ be a function that calculates the trailing 0's (counting from the right) in the binary representation of $h(\cdot)$. Let $R = \max r(\cdot)$.

We explain these notations through examples. Consider the above stream $S$. A hash function can be $h(k) = 3k + 1 \mod 8$. Then $S$ is transformed into 4, 7, 1, 4, 7, 1, 4, 7. The $r(h(k))$ is then 2, 0, 0, 2, 0, 0, 2, 0. Hence, $R = 2$.

The algorithm is shown in Algorithm 3. We only need to store $R$, and clearly, $R$ can be stored in $w = O(\log N)$ bits.

Still using our example of $S$ where $R = 2$, the output result is $\hat{F} = 2^2 = 4$. This is an approximate to the true result $F = 3$.

This algorithm is not a direct application of sampling. The basic idea is as follows. The first step is to map the elements uniformly in the range of $[0, 2^w - 1]$. This avoids

---

**Algorithm 3** Finding Distinct Elements

---

**Input:** $S$. **Output:** $\hat{F}$.

  1: **for** Read each element $k$ in $S$; **do**
  2:    Calculate $r(h(k))$;
  3:    Update $R$;
  4:    Discard $k$;
  5: **end for**
  6: Output $\hat{F} = 2^R$;

---

the problem that some elements are clustered in a small range. The second step is to convert each of the mapping results into the number of zeros starting counting from the right (counting from the left has a similar effect). Intuitively, if the number of distinct elements is big, there is a greater probability that such hashing hits a number with more zeros starting counting from the right.

Next we analyze this algorithm. The next theorem shows that the approximate $\hat{F}$ is neither too big (overestimate), nor too small (underestimate) as compared to $F$.

THEOREM 3. *For any integer $c > 2$, the probability that $\frac{1}{c} \leq \frac{\hat{F}}{F} \leq c$ is at least $1 - \frac{2}{c}$.*

We need a set of lemma before finally proving this theorem. First, next lemma states that the probability that we will hit a $r(h(k))$ with a large number of trailing 0s is exponentially decreasing.

LEMMA 4. *For any integer $j \in [0, w]$, $Pr[r(h(k)) \geq j] = \frac{1}{2^j}$.*

*Proof.* Intrinsically, we are looking for $\underbrace{1 \ldots 1}_{w-j} \underbrace{0 \ldots 0}_{j}$. Since the hashing makes the elements of $h(k)$ uniformly distributed in $[0, 2^w]$, we have $Pr[r(h(k)) \geq j] = \frac{1}{2^j}$.    □

Now we consider that the approximate $\hat{F}$ is an overestimation or an underestimation, respectively.

We start from bounding that $\hat{F}$ is an overestimation. More specifically, given a constant $c$, we do not want $\hat{F}$ to be greater than $cF$.

Let $Z_j$ be the number of distinct items in the stream $S$ for which $r(f(k)) \geq j$. We are interested in the smallest $j$ such that $2^j > cF$. If we do not want an overestimation, this $Z_j$ should not be big, because if $Z_j = 1$, our output will be at least $2^j$. Next lemma states that this is indeed true. In other words, the probability that $Z_j \geq 1$ can be bounded.

LEMMA 5. *$Pr[Z_j \geq 1] \leq \frac{1}{c}$.*

*Proof.* Clearly, $Z_j$ is an indicator variable such that

$$Z_j = \begin{cases} 1, & \text{if } r(f(k)) \geq j \\ 0, & \text{otherwise} \end{cases}$$

Thus,

$$E[Z_j] = F \times Pr[r(h(k)) \geq j] = \frac{F}{2^j}$$

by Markov inequality, we have

$$Pr[Z_j \geq 1] \leq E[Z_j]/1$$

Therefore,

$$Pr[Z_j] \geq 1 \leq E[Z_j]/1 = \frac{F}{2^j} \leq \frac{1}{c}$$

and this completes the proof.    □

We now see that the approximate $\hat{F}$ is an underestimation. More specifically, given a constant $c$, we do not want $\hat{F}$ that is less than $\frac{F}{c}$.

Again, let $Z_l$ be the number of distinct items in the stream $S$ for which $r(f(k)) \geq l$. We are interested in the smallest $l$ such that $2^l < \frac{F}{c}$. If we do not want an underestimation, this $Z_l$ should be at least 1, because if $Z_l = 0$, our output will be less than $2^l$. Next lemma states that this is indeed true. In other words, the probability that $Z_l = 0$ can be bounded.

LEMMA 6.  $Pr[Z_l = 0] \leq \frac{1}{c}$.

*Proof.* Clearly, and again, $Z_l$ is an indicator variable such that

$$Z_l = \begin{cases} 1, & \text{if } r(f(k)) \geq l \\ 0, & \text{otherwise} \end{cases}$$

Thus,

$$E[Z_l] = F \times Pr[r(h(k)) \geq l] = \frac{F}{2^l}$$

and

$$Var[Z_l] = F\frac{1}{2^l}(1 - \frac{1}{2^l})$$

$$Pr[Z_l = 0] = Pr[Z_l - E[Z_l]] \geq E[Z_l] \text{ (assigning } Z_l = 0 \text{ in the right-hand side)}$$

$$\leq \frac{Var[Z_l]}{E[Z_l]^2} \text{ (by Chebyshev inequality)}$$

$$< \frac{E[Z_l]}{E[Z_l]^2} \text{ (see } E[Z_l] \text{ and } Var[Z_l] \text{ developed above)}$$

$$= \frac{1}{E[Z_l]} = \frac{2^l}{F} < \frac{1}{c}.$$

and this completes the proof.    □

By Lemmas 5 and 6, we will not overestimate and underestimate a combined probability of more than $\frac{2}{c}$. We have thus proved Theorem 3.

## Median Trick in Boosting Confidence

Algorithm 3 can output an approximate $\hat{F}$ of the true $F$ with a constant probability. In reality, we may want the probability to be arbitrarily close to 1. A common trick to do this (i.e., boost the success probability) is called the median trick.

The algorithm is shown in Algorithm 4.

The next theorem states that $t$ can be as small as $\log \frac{1}{\delta}$. Thus, the total storage required is $O(\log \frac{1}{\delta} \log N)$.

---

**Algorithm 4** Final Algorithm

---
1: Run $t$ copies of Algorithm 3 using mutually independent random has functions;
2: Output the median of the $t$ answers;

---

THEOREM 4. *There is a $t = O(\log \frac{1}{\delta})$ ensuring that $\frac{F}{c} \leq F \leq cF$ happens with probability at least $1 - \delta$.*

*Proof.* Define $x_i = 0$ if $\frac{\hat{F}}{F}$ is in $[\frac{1}{c}, c]$ or 1 otherwise. Let $X = \sum_{i=1}^{t} x_i$.

Note that we can associate $x_i$ with each copy of the algorithm running in parallel and $X$ indicates the total number of failure. Because we will output the median, we fail only if more than half of the parallel-running algorithms fail. In other words, $X > \frac{t}{2}$. Our objective is to find a $t$ that this happens with very small probability $\delta$.

From another angle, we want $X < \frac{t}{2}$ to be $1 - \delta$.

$$Pr[X \leq \frac{t}{2}] \leq e^{-\frac{E[X](1-\frac{t/2}{E[X]})^2}{2}} \leq (1 - \delta)$$

We know that $E[x_i] = \frac{2}{c}$ from Theorem 3. Thus $E[X] = t\frac{2}{c}$.

To solve this inequality, we have

$$t \geq c(1 - \frac{c}{4})^2 \log \delta$$

and this completes the proof. □

**Discussion**: The bound in this example is not tight. We use $c$ instead of $\epsilon$ as $c$ is an integer constant. There are other bounds for finding distinct elements. Nevertheless, our goal is to show some core development methods for sublinear algorithms. Most notably, we see how to develop bounds given some key insights. Again this is related to the fact that the probability that deviates from the expectation can be bounded and the variance can be bounded. The median trick is a common trick to further boost the probability. In addition, one may find that sublinear algorithms are very simple in implementation, yet the analysis is usually complex.

### 5.3.3  Computing the Number of Connected Components

We now study a sublinear time problem. It is a classical problem where we want to compute the number of connected components in a big graph (see an example of a graph with six connected component in Figure 5.1). More specifically, assume we have

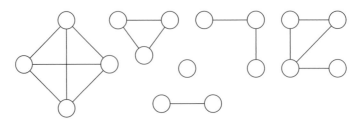

**Figure 5.1** A graph with six connected components.

a graph $G = (V, E)$, where $V$ is the set of vertices and $E$ is the set of edges. Let $n = |V|$ be the number of vertices. The reader can verify that to obtain an exact answer, the time complexity is $O(n^2)$.

Assume we are given an approximation ratio $\epsilon$. We now study a sublinear algorithm, where, with a probability greater than $\frac{2}{3}$, the time complexity is $O(\frac{1}{\epsilon^4})$. More specifically, assume that the number of connected component is $Q$, we would like to output $\hat{Q}$, the approximate number of connected components of the graph, where $|\hat{Q} - Q| \leq \epsilon$ with probability greater than $\frac{2}{3}$.

The core idea of the algorithm is to first sample some vertices $u_i$ from $V$. Starting from these vertices, expand to search their neighbors (e.g., by breadth first search (BFS)) for a few rounds. The stopping condition for BFS is either the BFS reaches all the nodes in the component of $u_i$, or a threshold (discuss later) is reached. The stopping condition is a balance between the following two: (1) a threshold is reached and not all vertices are visited by the BFS expansion. This means that this component is big and has a lot of vertices. From another point of view, we can also conclude that this component does not contribute to the counting of $Q$, because the graph will not have many very big components; and (2) the component is small enough and all nodes are reach. These components are the major contribution to the counting of $Q$. Because these components are small, for each of these components, even if we run all the vertices, (e.g., by BFS) it will not penalize too much to our time complexity.

This idea is developed into the algorithm and analyzed as follows. Let $u_i$ be an arbitrary vertex, and $n_u$ be the number of vertices in $u$'s connected component. A fact that can be easily seen is that for each component $\mathcal{A}$, $\sum_{u \in \mathcal{A}} \frac{1}{n_u} = 1$ and $\sum_{u \in V} \frac{1}{n_u} = Q$. Similar to many sublinear algorithms, we will analyze the expectation and variance of $\frac{1}{n_u}$. Let $\hat{n} = \min\{n_u, \frac{2}{\epsilon}\}$. Here the idea of $\hat{n}_u$ is to balance either we hit a small component, which will not penalize our running time, $\hat{n}_u = n_u$ or we hit a big component, so we set a stopping condition threshold $\frac{2}{\epsilon}$.

Our algorithm is shown in Algorithm 5. The computational complexity of Algorithm 5 is $O(\frac{1}{\epsilon^4})$. The algorithm runs for $\frac{1}{\epsilon^2}$ iterations. In each iteration, there is a BFS computation for $O(\frac{1}{\epsilon})$ vertices, resulting in $O(\frac{1}{\epsilon^2})$. Clearly, better BFS implementation can improve the complexity. There are better sublinear algorithms as well. We focus on the ideas in this chapter. For more advances on connected components, see [189].

---

**Algorithm 5** Connected Component

---

1: $i = 0$;

2: $s = \frac{1}{\epsilon^2}$;

3: **while** $i \leq s$ **do**

4:     randomly pick a vertex $u_i$;

5:     starting from $u_i$, run BFS until we hit $\frac{2}{\epsilon}$ new vertices, count the number of vertices $n_{u_i}$;

6:     $\hat{n}_i = \min\{n_{u_i}, \frac{2}{\epsilon}\}$;

7:     $i++$;

8: **end while**

9: $\hat{Q} = n \times (\sum_{i=0}^{s} \frac{1}{\hat{n_{u_i}}})/s$;

10: output $\hat{Q}$;

---

The algorithm will introduce two errors. First, bounding errors of a single vertex $u$. Here the error is that for each connected component, we have a stopping condition. Second, bounding errors of sampling $u$. Here the error is that we will not sample all vertices in the graph. We will see that we split the total error $\epsilon n$ into two $\frac{\epsilon n}{2}$, and bound the two errors respectively.

**Bounding Errors from BFS for a vertex** $u$: We now analyze the property of $\hat{n}_u$. We show that $|\sum_{u \in V} \frac{1}{\hat{n}} - \sum_{u \in V} \frac{1}{n}| \leq \frac{\epsilon n}{2}$. This is how the threshold of $\frac{\epsilon}{2}$ was derived.

LEMMA 7. *The $\hat{n}_u$ computed by Algorithm 5 satisfies $|\frac{1}{\hat{n}_u} - \frac{1}{n_u}| \leq \frac{\epsilon}{2}$.*

*Proof.* There are two parts:

(1) if $u$'s component has less than $\frac{2}{\epsilon}$ nodes, $\hat{n}_u = n_u$
(2) if $u$'s component has more than $\frac{2}{\epsilon}$ nodes, $\hat{n}_u = \frac{2}{\epsilon}$. Then, $\frac{1}{\hat{n}_u} - \frac{1}{n_u} < \frac{1}{\hat{n}_u} = \frac{\epsilon}{2}$.

Overall, $|\frac{1}{\hat{n}_u} - \frac{1}{n_u}| \leq \frac{\epsilon}{2}$.     □

This lemma quantifies our intuition where even we ignore the big connected components, the accuracy loss is bounded by $\frac{\epsilon}{2}$. Therefore, the total estimation error is bounded by the following lemma

LEMMA 8. *The $\hat{n}_u$ computed by Algorithm 5 satisfies $|\sum_{u \in V} \frac{1}{\hat{n}_u} - \sum_{u \in V} \frac{1}{n_u}| \leq \frac{\epsilon n}{2}$*

*Proof.* $|\sum_{u \in V} \frac{1}{\hat{n}_u} - \sum_{u \in V} \frac{1}{n_u}| \leq \sum_{u \in V} |\frac{1}{\hat{n}_u} - \frac{1}{n_u}| \leq \frac{\epsilon n}{2}$.     □

**Bounding Errors from sampling** $u$: We now look at the number of $u$ that we need to sample (i.e., how we derived $s = \frac{1}{\epsilon^2}$).

This follows similar techniques described in the previous two examples. Let $Y_1, Y_2, Y_3, \ldots, Y_s$ be independently distributed random variables for vertices $u_1, u_2, u_3, \ldots, u_s$, where $Y_i = \frac{1}{n_{u_i}}$ for vertex $u_i$. Let $Y = \sum_{i=1}^{s} Y_i$.

Recall that in Algorithm 5, we compute $Y = \sum_{i=1}^{s} \frac{1}{n_{u_i}}$, and output $Y$ with an extension factor of $\frac{n}{s}$ to approximate the total vertices (i.e., $\hat{Q} = \frac{Y \times n}{s}$).

Clearly, we need to bound the error of $\sum_{u \in V} \frac{1}{n_u}$ of all vertices in the graph and its deviation from $\sum_{i=1}^{s} \frac{1}{n_{u_i}}$ of the subset of $s$ vertices.

For ease of presentation, let $Q' = \sum_{u \in V} \frac{1}{n_u}$. We need to analyze $|\hat{Q} - Q'|$ and how many $s$ is needed to give a bound to $|\hat{Q} - Q'|$.

LEMMA 9. *Let* $Q' = \sum_{u \in V} \frac{1}{n_u}$. *Let* $s = O(\frac{1}{\epsilon^2})$. *The output from Algorithm 5 is* $\hat{Q}$. *Then* $Pr[|Q' - \hat{Q}| \geq \frac{\epsilon n}{2}] \leq \frac{1}{3}$.

*Proof.* The output from Algorithm 5 is $\hat{Q} = \frac{(\sum_{i=0}^{s} \frac{1}{n_u}) \times n}{s}$. Since, $Y_i = \frac{1}{n_{u_i}}$ and $Y = \sum_{i=1}^{s} Y_i$, we have $Y = \frac{\hat{Q} \times s}{n}$ or $\hat{Q} = \frac{Y \times n}{s}$.

The linkage of $Q'$ and $\hat{Q}$ comes from the fact that the expectation of every $Y_i$ is associated with $Q'$ and $Y$ is associated with $\hat{Q}$. More specifically, $E[Y_i] = \frac{\sum_{u \in V} \frac{1}{n_u}}{n}$. Thus, $Q' = E[Y_i] \times n$. Since $E[Y] = \sum_{i=0}^{s} E[Y_i] = sE[Y_i]$, $Q' = \frac{E[Y] \times n}{s}$.

Thus, $Pr[|\hat{Q} - Q'| \geq \frac{\epsilon n}{2}] = Pr[|\frac{n}{s} Y - \frac{n}{s} E[Y]| \geq \frac{\epsilon n}{2}] = Pr[|Y - E[Y]| \geq \frac{\epsilon s}{2}]$.

Through Hoeffding bound, we know that $Pr[|Y - E[Y]| \geq \frac{\epsilon s}{2}] \leq 2e^{-\frac{\epsilon^2 s}{2}}$. Since $s = O(\frac{1}{\epsilon^2})$, we have $Pr[|\hat{Q} - Q'| \geq \frac{\epsilon n}{2}] \leq \frac{1}{3}$. □

Now we can combine the two errors and prove the whole theorem as follows.

THEOREM 5. *Let Q be the number of connected components in a graph with n vertices. Algorithm 5 Connected Component runs in time* $O(\frac{1}{\epsilon^4})$ *and with probability at least* $\frac{2}{3}$ *outputs* $\hat{Q}$ *where* $|\hat{Q} - Q| \leq \epsilon n$.

*Proof.* We have $|Q' - Q| \leq \frac{\epsilon n}{2}$. We also have $Pr[|\hat{Q} - Q'| \geq \frac{\epsilon n}{2}] \leq \frac{1}{3}$. Then, we have $|\hat{Q} - Q| \leq |\hat{Q} - Q'| + |Q' - Q| \leq \frac{\epsilon n}{2} + \frac{\epsilon n}{2} = \epsilon n$ with probability at least $\frac{2}{3}$. □

**Discussion:** This is a sublinear time algorithm. The graph is stored and we only look at a small number of nodes on the graph, resulting in a sublinear time complexity. In this problem, we also see that the algorithm complexity is $O(\frac{1}{\epsilon^4})$ which is independent of $n$. The intuition is that the approximation bound has $n$ (i.e., $\epsilon n$). This is meaningful when we only need to know a fraction of the entire components. For example, if we want an error of 1% or less, we can set epsilon = 0.01. In practice, this is useful when both the number of vertices in the graph, and the number of connected components are big. Of course, if the number of connected components is small (e.g., 2) we will not make much error whatever our output is.

We also see some additional design skills for sublinear algorithms. We have seen dividing errors, and again, quantifying samples, etc. The important thing is how to analyze an insight. The insight is that by doing BFS, we need a balance between reaching out for more vertices and stop at some point of time to keep the time complexity small.

### 5.3.4 The Two-Cat Problem

We now study one problem that does not fall in the form of $(1 + \epsilon, \delta)$ approximation. Yet, the problem can be solved in a sublinear amount of resources. The problem is as follows.

**The Two-Cat Problem**: Consider a tall skyscraper building and you do not know the total number of floors of this building. Assume you have cats. The cats have the following properties: when you throw a cat out of window from floor $1, 2, \cdots, n$, the cat will survive; and when you throw a cat out of window from floor $n + 1, n + 2, \cdots, \infty$, the cat will die. The question is to develop an efficient method to determine $n$ given that you have *two* cats.

We first look at when there is only one cat. It is easy to see that we have to use this cat to test floor one by one from $1, 2, \cdots, n$. This will take a total of $n$ tests, which is linear. This $n$ is also a lower bound.

Now we see the case that we have two cats. Before we give out the final solution, we first analyze the algorithm of an exponential increase algorithm. The algorithm is shown in Algorithm 6.

---
**Algorithm 6** Exponential Increase Algorithm
---
1: $i = -1$
2: **for** the first cat is still alive **do**
3:    $i + +$;
4:    test floor $2^i$;
5: **end for** $k = 0$;
6: **for** the second cat is still alive **do**
7:    $k + +$;
8:    test floor $2^{i-1} + k$;
9: **end for**
10: Output $2^{i-1} + k$;

---

The first cat in this algorithm will be used to test floors of $1, 2, 4, 8, 16, \cdots$. Assume that the first cat dies when it is used to test floor $l$; then the second cat will be used to test floors $2^{(\log l - 1)} + 1, 2^{(\log l - 1)} + 2, \cdots, l - 1, l$. For example, assume that $n = 23$. Using the above exponential algorithm, the first cat survives when it is used to test floor 16 and dies when it is used to test floor 32. Then we use the second cat to test floor 16 to 32. We finally conclude that $n = 23$.

It is easy to see that this exponential algorithm also takes linear time. The first cat takes $O(\log n)$ time, and the second cat, where the primary complexity comes from, takes $O(\frac{n}{2})$. This leads to a linear complexity to the overall algorithm.

Now we present a sublinear algorithm in Algorithm 7. The first cat in this algorithm will be used to test floors of $1, 3, 6, 10, 15, 21, 28, \cdots$. Assume that the first cat dies when it is used to test floor $l$ at the $i$th round; then the second cat will be used to test floors $l - i + 1, l - i + 2, \cdots, l - 1, l$. For example, assume that $n = 23$. Using the above algorithm, the first cat survives when it is used to test floor 21 and dies when it is used

---

**Algorithm 7** The Two-Cat Algorithm

---

1: $i = -1, l = 1$
2: **for** the first cat is still alive **do**
3:     $i + +$;
4:     test floor $l = l + i$;
5: **end for**$l = l - i$;
6: **for** the second cat is still alive **do**
7:     $l + +$;
8:     test floor $l$;
9: **end for**
10: Output $l$;

---

to test floor 28. Then we use the second cat to test floor 21 to 28. We finally conclude that $n = 23$. Now we analyze this algorithm.

LEMMA 10. *The Two-Cat Algorithm is sublinear and it takes $O(\sqrt{n})$ steps.*

*Proof.* Sketch: Assume the first cat takes $x$ step and the second one takes $y$ steps.

For the first cat, the final floor $l$ it reaches is equal to $1 + 2 + 3 + 4 + \cdots$. More specifically, we have $l = \frac{(1+x)x}{2}$. Clearly, $l = O(n)$. Thus, $x = O(\sqrt{n})$.

For the second cat, we look at the total number of floors $l'$ before the first cat dies. This is $l' = \frac{x(x-1)}{2}$. The maximum number of floors this second cat should test is equal to $l - l'$, that is, $y = O(l - l')$. Therefore, $y = O(\frac{(1+x)x}{2} - \frac{x(x-1)}{2}) = O(x)$. Hence, $y = O(\sqrt{n})$.

Combining $x$ and $y$, we have the complexity of the algorithm $O(\sqrt{n})$.     $\square$

For this two-cat problem, $O(\sqrt{n})$ is also a lower bound (i.e., this algorithm is the fastest algorithm that we can gain). We omit the proof. One may be interested in investigating the case, if we have three cats. The result will be $O(\sqrt[3]{n})$.

**Discussion:** This algorithm has important indication. Intrinsically we may consider that the two cats are two pieces of resources. This problem shows that, to collectively achieve a certain task, we can divide the resources where each piece of the resource undertakes a sublinear overhead.

One work of Partial Network Coding (PNC) applies this idea [188]. In PNC, there are two pieces of resources, namely, communication and storage. To collectively achieve a task, either the total amount of communication needs to be linear or the total amount of storage needs to be linear. In [188], it is shown that we can divide the overhead into a $O(\sqrt{N})$ factor for the communication and a $O(\sqrt{N})$ factor to storage, so that each resource has a sublinear overhead.

## 5.4    Summary

In this chapter, we presented the foundation, some examples, and common algorithm development techniques for sublinear algorithms. We first presented how sublinear algorithms are developed from the theoretical point of view, in particular, its connections with approximation algorithms. Then we presented a set of commonly used inequalities that are important in developing approximation bounds. We started from a very simple example that directly applies Hoeffding inequality. Then we studied a classic sublinear space algorithm to find distinct elements in data streams. We presented some commonly used tricks such as using the median trick to boost the confidence of the sublinear algorithm. Then we studied a classic sublinear time algorithm to compute the number of connected components in a big graph. We showed a technique in dividing errors, analyzing a trade-off between time complexity and errors, and merging the errors. Finally, we presented the two-cat problem.

These examples reveal how sublinear algorithms are developed. From a theoretical point of view, the most important aspect in sublinear algorithm theory development is to develop techniques that can be widely applied to the algorithm design of different problems. These techniques are sharpened by continuously investigating the fundamental problems in computer science, such as finding distinct elements, computing the number of connected components, and others. A comprehensive survey in [190] has more sublinear algorithms in literature.

In some of the following chapters, we will study applications of sublinear algorithms. We will need to derive bounds given specific application scenarios. We can either apply existing results from the sublinear algorithms designed for the fundamental problems or apply the techniques developed from existing sublinear algorithms.

# 6    Tensor for Big Data

The concepts of tensor analysis arose from the work of Carl Friedrich Gauss in differential geometry, and the formulation was much influenced by the theory of algebraic forms and invariants developed during the middle of the nineteenth century. The word "tensor" itself was introduced in 1846 by William Rowan Hamilton to describe something different from what is now meant by a tensor. The contemporary usage was introduced by Woldemar Voigt in 1898. Tensor calculus was developed around 1890 by Gregorio Ricci-Curbastro in 1892. In the twentieth century, the subject came to be known as tensor analysis, and achieved broader acceptance with the introduction of Einstein's theory of general relativity, around 1915.

Tensors are geometric objects that describe linear relations between geometric vectors, scalars, and other tensors. Given a coordinate basis or fixed frame of reference, a tensor can be represented as an organized multidimensional array of numerical values. The order (also degree or rank) of a tensor is the dimensionality of the array needed to represent it, or equivalently, the number of indices needed to label a component of that array. Tensors are important in engineering because they provide a concise mathematical framework for formulating and solving physics problems in areas such as elasticity, fluid mechanics, and general relativity. Recently, tenors have been applied in many big data analyses [191, 192, 193, 194, 195].

In this chapter, first we study the basics of tensor concepts in Section 6.1. Then we investigate the tensor decomposition for tensor compression and tensor completion in Section 6.2. We change to another application for tensor voting in Section 6.3. Finally, some other applications are briefly mentioned in Section 6.4.

## 6.1    Tensor Basics

In this section, we first define some basics tensor definitions. Second, we illustrate rules of tensor operations. We follow the logic flow according to [196].

### 6.1.1    Tensor Definitions

Tensor can be generally regarded as three-way array, which looks like a "shoe box." The definition of a tensor is given as follows.

DEFINITION 2. ***Tensor Definition***: *For two vectors* $\mathbf{a}_{I \times 1}$ *and* $\mathbf{b}_{J \times 1}$, *define product* $\mathbf{a} \circ \mathbf{b}$ *is an* $I \times J$ *matrix with the* $(i, j)$*th element* $\mathbf{a}_i \mathbf{b}_j$, *that is,* $\mathbf{a} \circ \mathbf{b} = \mathbf{a} \mathbf{b}^T$. *For three vector,* $\mathbf{a} \circ \mathbf{b} \circ \mathbf{c}$ *is an* $I \times J \times K$ *rank-one three-way tensor with the* $(i, j, k)$ *element* $\mathbf{a}_i \mathbf{b}_j \mathbf{c}_k$.

Then the rank of a tensor is defined as

DEFINITION 3. ***Tensor Rank***: *The rank of a tensor is the smallest number of outer products needed to synthesize the tensor.*

One example of tensor is language as "subject-verb-object" which naturally leads to a 3-mode tensor as illustrated in Figure 6.1, where $\mathbf{a}$ is the subject set, $\mathbf{b}$ is the verb set, and $\mathbf{c}$ is the object set. Each rank-one factor corresponds to a concept such as "this book" "is" "the best."

Before we study the rank decomposition of a tensor, let's define two products first.

DEFINITION 4. ***Kronecker Product***: *For matrix* $\mathbf{A}$ *and matrix* $\mathbf{B}$, *the Kronecker Product is*

$$\mathbf{A} \otimes \mathbf{B} = \begin{bmatrix} \mathbf{B}\mathbf{A}(1,1), & \mathbf{B}\mathbf{A}(1,2), & \ldots \\ \mathbf{B}\mathbf{A}(2,1), & \mathbf{B}\mathbf{A}(2,2), & \ldots \\ \vdots & \vdots & \vdots \end{bmatrix}. \tag{6.1}$$

We can show $vec(\mathbf{ABC}) = (\mathbf{C}^T \otimes \mathbf{A})vec(\mathbf{B})$, where $vec$ is vectorization.

DEFINITION 5. ***Khatri-Rao (Column-wise Kronecker) Product***: *Given* $\mathbf{A}_{I \times F}$ *and* $\mathbf{B}_{J \times F}$, $\mathbf{A} \odot \mathbf{B}$ *is an* $JI \times F$ *matrix as:*

$$\mathbf{A} \odot \mathbf{B} = [\mathbf{A}(:, 1) \otimes \mathbf{B}(:, 1) \cdots \mathbf{A}(:, F) \otimes \mathbf{B}(:, F)]. \tag{6.2}$$

We can show if $\mathbf{D} = \mathrm{diag}(\mathbf{d})$, then $vec(\mathbf{ADC}) = (\mathbf{C}^T \odot \mathbf{A})\mathbf{d}$.

Next, we explain how to conduct rank decomposition for three-way tensor.

DEFINITION 6. ***Tensor***:

$$X = \sum_{f=1}^{F} \mathbf{a}_f \circ \mathbf{b}_f \circ \mathbf{c}_f. \tag{6.3}$$

*Scalar:*

$$X(i, j, k) = \sum_{f=1}^{F} a_{i,f} b_{j,f} c_{k,f}, \forall i \in \{1, \ldots, I\}, \forall j \in \{1, \ldots, J\}, \forall k \in \{1, \ldots, K\}. \tag{6.4}$$

*Slabs:*

$$\mathbf{X}_k = \mathbf{A}\mathbf{D}_k(\mathbf{C})\mathbf{B}^T, k = 1, \ldots, K. \tag{6.5}$$

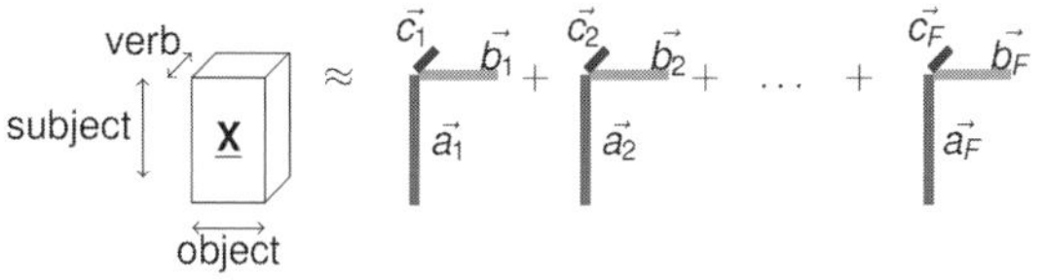

**Figure 6.1** Illustration of tensor for language.

*Matrix:*

$$\mathbf{X}^{(KJ \times I)} = (\mathbf{B} \odot \mathbf{C})\mathbf{A}^T. \tag{6.6}$$

*Tall Vector:*

$$\mathbf{x}^{(KJI)} = vec\left(\mathbf{X}^{(KJ \times I)}\right) = (\mathbf{A} \odot \mathbf{B} \odot \mathbf{C})\mathbf{1}_{F \times 1}. \tag{6.7}$$

Some applications need high-way tensors beyond 3. In those cases, the $N$-order tensors have the following rank decomposition.

DEFINITION 7. *Tensor:*

$$X = \sum_{f=1}^{F} \mathbf{a}_f^{(1)} \circ \cdots \circ \mathbf{a}_f^{(N)}. \tag{6.8}$$

*Scalar:*

$$X(i_1, \ldots, i_N) = \sum_{f=1}^{F} \prod_{n=1}^{N} a_{I_n,f}^{(n)}. \tag{6.9}$$

*Matrix:*

$$\mathbf{X}^{(I_1 I_2 \ldots I_{N-1} \times I_N)} = (\mathbf{A}^{(N-1)} \odot \mathbf{A}^{(N-2)} \odot \cdots \odot \mathbf{A}^{(1)})(\mathbf{A}^{(N)})^T. \tag{6.10}$$

*Tall Vector:*

$$\mathbf{x}^{(I_1 \ldots I_N)} = vec\left(\mathbf{X}^{(I_1 I_2 \ldots I_{N-1} \times I_N)}\right) = \left(\mathbf{A}^{(N)} \odot \mathbf{A}^{(N-1)} \odot \cdots \odot \mathbf{A}^{(1)}\right)\mathbf{1}_{F \times 1}. \tag{6.11}$$

## 6.1.2    Tensor Operations

There are a number of basic operations that may be conducted on tensors that again produce a tensor. The linear nature of tensor implies that two tensors of the same type may be added together, and that tensors may be multiplied by a scalar with results analogous to the scaling of a vector. On components, these operations are simply performed component for component. These operations do not change the type of the tensor, however there also exist operations that change the type of the tensors. Some basic tensor operations are defined as follows.

DEFINITION 8. *Addition: Two tensors of the same type can be added term by term. Tensor addition is commutative. Tensors of different ranks cannot be added.*

DEFINITION 9. *Multiplication by a Scalar: Each coordinate of a tensor can be multiplied by a scalar to have a new tensor of the same type. We can also show $(a + b)\mathbf{X} = a\mathbf{X} + b\mathbf{X}$, $a(\mathbf{X} + \mathbf{Y}) = a\mathbf{X} + a\mathbf{Y}$, and $(ab)\mathbf{X} = a(b\mathbf{X}) = b(a\mathbf{X})$.*

DEFINITION 10. *Tensor n-Mode Product: The n-mode matrix product of a tensor $\mathbf{X} \in \mathbb{R}^{I_1 \times I_2 \times \cdots \times I_N}$ with a matrix $\mathbf{U} \in \mathbb{R}^{J \times I_n}$ has the size of $I_1 \times \cdots \times I_{n-1} \times J \times I_{n+1} \times \cdots \times I_N$, and*

$$\mathbf{X} \otimes_n \mathbf{U} = \sum_{i_n=1}^{I_n} x_{i_1 i_2 \ldots i_N} u_{j i_n}. \tag{6.12}$$

*The n-mode vector product of a tensor* $\mathbf{X} \in \mathbb{R}^{I_1 \times I_2 \times \cdots \times I_N}$ *with a vector* $\mathbf{v} \in \mathbb{R}^{I_n}$ *has the size of* $I_1 \times \cdots \times I_{n-1} \times I_{n+1} \times \cdots \times I_N$, *and*

$$\mathbf{X} \otimes_n \mathbf{v} = \sum_{i_n=1}^{I_n} x_{i_1 i_2 \cdots I_N} v_{i_n}. \tag{6.13}$$

## 6.2    Tensor Decomposition

In this section, we first show two ways of singular value decomposition for tensors. Then we discuss how to compress the big data tensor. Third, we discuss the tensor completion problem, and finally we provide some tensor algorithm design literature survey.

### 6.2.1    Tucker3 and PARAFAC

For matrix decomposition, singular value decomposition (SVD) can factorize a real or complex matrix and has wide applications. However, there is single tensor SVD for a tensor equivalent to the matrix SVD. There are two two basic decompositions as follows, both of which are outer product decompositions but with very different structural properties.

- Tucker3, which is orthogonal without loss of generality but non-unique except for very special cases. Tucker3 is used for subspace estimation and tensor approximation (e.g., compression applications).
- CANonical DECOMPosition (CANDECOMP), also known as PARAllel FACtor (PARAFAC) analysis, or CANDECOMP-PARAFAC (CP) for short, which is non-orthogonal and unique under certain conditions. PARAFAC is used for latent parameter estimation for recovering the "hidden" rank-one factors.

**Tucker3**

In Figure 6.2, we show an $I \times J \times K$ three-way array $\mathbf{X}$ decomposed into a core tensor $\mathbf{G} \in \mathbb{R}^{P \times R \times Q}$ multiplied by a matrix along each mode. So it can be regarded as a form of high-order PCA, that is,

$$\mathbf{X} \approx \mathbf{G} \otimes_1 \mathbf{A} \otimes_2 \mathbf{B} \otimes_3 \mathbf{C} = \sum_{p=1}^{P} \sum_{q=1}^{Q} \sum_{r=1}^{R} g_{pqr} \mathbf{a}_p \circ \mathbf{b}_q \circ \mathbf{c}_r, \tag{6.14}$$

where $\mathbf{A} \in \mathbb{R}^{I \times P}$, $\mathbf{B} \in \mathbb{R}^{J \times Q}$, and $\mathbf{C} \in \mathbb{R}^{K \times R}$ are the factor matrices as the principal components in each mode. Element-wise, it is shown in Figure 6.2. Typically $P, Q, R$ are smaller than $I, J, K$, and so data compression for $\mathbf{X}$ is possible.

The Tucker model can be extended to $N$-way tensors as

$$\mathbf{X} \approx \mathbf{G} \otimes_1 \mathbf{A}^{(1)} \otimes_2 \mathbf{A}^{(2)} \cdots \otimes_N \mathbf{A}^{(N)} = \sum_{r_1=1}^{R_1} \sum_{r_2=1}^{R_2} \cdots \sum_{r_N=1}^{R_N} g_{r_1 r_2 \cdots r_N} \mathbf{a}_{r_1}^1 \circ \mathbf{a}_{r_2}^2 \cdots \mathbf{a}_{r_N}^N. \tag{6.15}$$

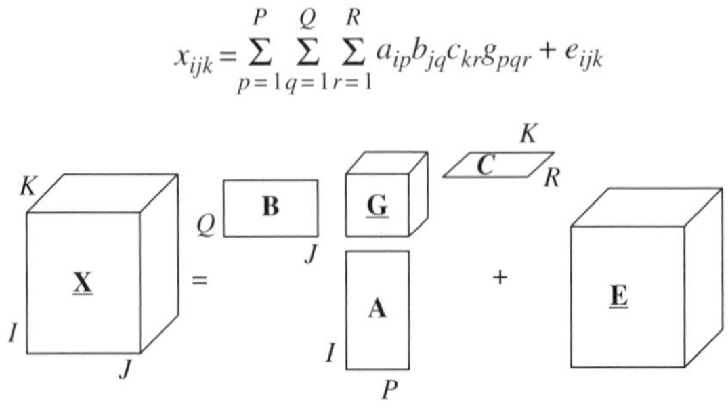

$$x_{ijk} = \sum_{p=1}^{P} \sum_{q=1}^{Q} \sum_{r=1}^{R} a_{ip} b_{jq} c_{kr} g_{pqr} + e_{ijk}$$

**Figure 6.2** Tucker3 model for tensor decomposition.

The $n$-rank of $\mathbf{X}$ of size $I_1 \times I_2 \times \cdots \times I_N$, $R_n = rank_n(\mathbf{X})$, is the column rank of $\mathbf{X}_{(n)}$, and then tensor $\mathbf{X}$ has rank of $rank - (R_1, R_2, \cdots, R_N)$.

To calculate the Tucker3 decomposition, the optimization can be written as

$$\min_{\mathbf{G}, \mathbf{A}^{(1)}, \cdots, \mathbf{A}^{(N)}} \| \mathbf{X} - \mathbf{G} \otimes_1 \mathbf{A}^{(1)} \otimes_2 \mathbf{A}^{(2)} \cdots \otimes_N \mathbf{A}^{(N)} \| \tag{6.16}$$

$$s.t. \ \mathbf{G} \in \mathbb{R}^{R_1 \times R_2 \times \cdots \times R_N}, \ \mathbf{A}^n \in \mathbb{R}^{I_n \times R_n}, \forall n.$$

As shown in [196], the above problem can be recast as a series of subproblems to solve for the $n$th component matrix as

$$\max_{\mathbf{A}^{(n)}} \| \mathbf{X} \otimes_1 \mathbf{A}^{(1)T} \otimes_2 \mathbf{A}^{(2)T} \cdots \otimes_N \mathbf{A}^{(N)T} \| \tag{6.17}$$

$$s.t. \ \mathbf{A}^n \in \mathbb{R}^{I_n \times R_n} \text{ and column-wise orthogonal.}$$

Some algorithms such as Alternating Least Squares (ALS) can be employed. The Tucker decompositions are not unique, but we can choose transformations that simplify the core structure so that most elements of core $\mathbf{G}$ are zero.

### PARAFAC

PARAFAC decomposition factorizes a tensor $\mathbf{X} \in \mathbb{R}^{I \times J \times K}$ into a sum of component rank-one tensors as shown in Figure 6.3, that is,

$$\mathbf{X} \approx \sum_{r=1}^{R} \mathbf{a}_r \circ \mathbf{b}_r \circ \mathbf{c}_r, \tag{6.18}$$

where $R$ is a positive integer and $\mathbf{a}_r \in \mathbb{R}^I$, $\mathbf{b}_r \in \mathbb{R}^J$, and $\mathbf{c}_r \in \mathbb{R}^K$, for $r = 1, \dots, R$. Element-wise, we have

$$x_{ijk} = \sum_{r=1}^{R} a_{ir} b_{jr} c_{kr}, \ i = 1, \dots, I, \ j = 1, \dots, J, \ k = 1, \dots, K. \tag{6.19}$$

The rank of the tensor is loosely bounded by $rank(\mathbf{X}) \leq \min\{IK, IK, JK\}$.

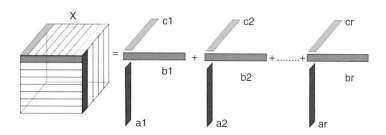

**Figure 6.3** PARAFAC model for tensor decomposition.

If we do normalization with weights in to a vector $\lambda \in \mathbf{R}^R$, for a general $N$th order tensor $\mathbf{X} \in \mathbb{R}^{I_1 \times I_2 \times \cdots \times I_N}$, the PARAFAC decomposition is

$$\mathbf{X} \approx \sum_{r=1}^{R} \lambda_r \mathbf{a}_r^{(1)} \circ \mathbf{a}_r^{(2)} \circ \cdots \circ a_r^{(N)}, \tag{6.20}$$

where $\mathbf{A}^{(n)} \in \mathbb{R}^{I_n \times R}$, $n = 1, \ldots, N$.

To calculate PARAFAC decomposition, the goal is to find $R$ components to best approximate the original tensor, that is,

$$\min_{\hat{\mathbf{X}}} \|\mathbf{X} - \hat{\mathbf{X}}\| \tag{6.21}$$

$$s.t. \ \hat{\mathbf{X}} = \sum_{r=1}^{R} \lambda_r \mathbf{a}_r \circ \mathbf{b}_r \circ \mathbf{c}_r.$$

One solution is ALS algorithm, which fixes $\mathbf{B}$ and $\mathbf{C}$ to solve $\mathbf{A}$, then fixes $\mathbf{A}$ and $\mathbf{C}$ to solve $\mathbf{B}$, and then fixes $\mathbf{A}$ and $\mathbf{B}$ to solve $\mathbf{C}$, and repeat until convergence. For each step, the optimal solution is

$$\hat{\mathbf{A}} = \mathbf{X}_{(1)}[(\mathbf{C} \odot \mathbf{B})^T]^\dagger. \tag{6.22}$$

The nice property of high-order tensors is that the PARAFAC decomposition is often unique.

**Other Decompositions**

There are some other tensor decompositions such as individual differences in scaling (INDSCAL) [197], parallel factors for cross products (PARAFAC2) [198], CANDE-COMP with linear constraints (CANDELINC) [199], decomposition into directional components (DEDICOM) [200], and PARAFAC & Tucker2 (PARATUCK2) [201].

### 6.2.2 Tensor Compression

Tensors can easily become gigantic, since the size is exponential as the number of dimensions. Moreover, the data sets typically have millions of items per mode, which cannot load in main memory and may have to reside in cloud storage. Typically, tensors are very sparse, and can store and process as (i, j, k, value) list, nonzero column indices for each row, runlength coding, etc. In this section, we study some theoretical results for tensor compression, followed by a powerful algorithm PArallel Randomly COMPressed Cubes (PARACOMP) and its MapReduce implementation.

In literature, one straightforward way to implement parallel algorithms for matrix algebra is to use data partitioning. In [202], each sub-block is independently decomposed, then low-rank factors of the big tensor are stitched together. The decomposition of each sub-block must be identifiable, which are stringent ID conditions. Each sub-block has different permutation and scaling of the factors, and all these must be reconciled before stitching. More recently, the work such as in [203] also relies on partitioning (so sub-block identification conditions are required), but is more robust, based on collaborative consensus-averaging and the matching of different permutations and scales is explicitly taken into account. However, the work requires considerable communication overhead across clusters and nodes (i.e., the parallel threads are not independent). One commonly used compression method for moderate-size tensors is to fit orthogonal Tucker3 model together with regress data onto fitted mode bases. Some implementation of such an idea can be found in COMFAC [204]. The method is lossless if exact mode bases used (such as CANDELINC), but Tucker3 fitting is itself cumbersome for big tensors (due to big matrix SVDs), and cannot compress below mode ranks without introducing errors.

Specifically for tensor compression, we want to compress $\mathbf{x} = \vec{(\underline{\mathbf{X}})}$ into $\mathbf{y} = \mathbf{S}\mathbf{x}$, where $\mathbf{S}$ is $d \times IJK$. Here we have $d \ll IJK$. Here we consider the special structured compression matrix with the following form

$$\mathbf{S} = \mathbf{U}^T \otimes \mathbf{V}^T \otimes \mathbf{W}^T. \tag{6.23}$$

The above equation corresponds to multiplying $\mathbf{X}$ from the $I$-mode with $\mathbf{U}^T$, from the $J$-mode with $\mathbf{V}^T$, and from the $K$-mode with $\mathbf{W}^T$, where $\mathbf{U}$ is $I \times L$, $\mathbf{V}$ is $J \times M$, and $\mathbf{W}$ is $K \times N$, with $L \le I$, $M \le J$, $N \le K$ and $LMN \le IJK$. The compression is illustrated in Figure 6.4.

Due to a property of the Kronecker product

$$\left(\mathbf{U}^T \otimes \mathbf{V}^T \otimes \mathbf{W}^T\right)(\mathbf{A} \odot \mathbf{B} \odot \mathbf{C}) = \left((\mathbf{U}^T\mathbf{A}) \odot (\mathbf{V}^T\mathbf{B}) \odot (\mathbf{W}^T\mathbf{C})\right), \tag{6.24}$$

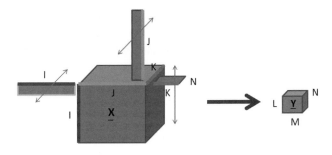

**Figure 6.4** Tensor compression [205].

we have the following

$$\mathbf{y} = \left( (\mathbf{U}^T \mathbf{A}) \odot (\mathbf{V}^T \mathbf{B}) \odot (\mathbf{W}^T \mathbf{C}) \right) \mathbf{1} = \left( \tilde{\mathbf{A}} \odot \tilde{\mathbf{B}} \odot \tilde{\mathbf{C}} \right) \mathbf{1}. \tag{6.25}$$

In other words, the compressed data follow a PARAFAC model of size $L \times M \times N$ and order $F$ parameterized by $(\tilde{\mathbf{A}}, \tilde{\mathbf{B}}, \tilde{\mathbf{C}})$, with $\tilde{\mathbf{A}} = \mathbf{U}^T \mathbf{A}$, $\tilde{\mathbf{B}} = \mathbf{V}^T \mathbf{B}$, and $\tilde{\mathbf{C}} = \mathbf{W}^T \mathbf{C}$.

According to [206], with random multiway compression, the following theoretical guarantee for tensor compression can be achieved.

THEOREM 6. *Assume that the columns of* $\mathbf{A}$, $\mathbf{B}$, *and* $\mathbf{C}$ *are sparse, and let* $n_a$, $n_b$, *and* $n_c$ *be the upper bounds on the number of nonzero elements per column of* $\mathbf{A}$, $\mathbf{B}$, $\mathbf{C}$, *respectively. Let the mode-compression matrices* $\mathbf{U}(I \times L, L \leq I)$, $\mathbf{V}(J \times M, M \leq J)$, *and* $\mathbf{W}(K \times N, N \leq K)$ *be randomly drawn from an absolutely continuous distribution with respect to the Lebesgue measure in* $\mathcal{R}^{IL}$, $\mathcal{R}^{JM}$, *and* $\mathcal{R}^{KN}$, *respectively. If the following conditions hold:*

$$r_A = r_B = r_C = F \tag{6.26}$$

$$L(L-1)M(M-1) \geq 2F(F-1), \ N \geq F, \tag{6.27}$$

$$L \geq 2n_a, \ M \geq 2n_b, \ N \geq 2n_c, \tag{6.28}$$

*then the original factor loadings* $\mathbf{A}$, $\mathbf{B}$, *and* $\mathbf{C}$ *are almost surely identifiable from the compressed data up to a common column permutation and scaling. The theorem can be extended to higher-way arrays.*

PARAFAC can be fitted in compressed space, and then the sparse $\mathbf{A}$, $\mathbf{B}$, and $\mathbf{C}$ can be recovered from the fitted compressed factors with complexity $O(LMNF + (I^{3.5} + J^{3.5} + K^{3.5})F)$. Using sparsity first and then fitting PARAFAC in raw space entails complexity $O(IJKF + (IJK)^{3.5})$. The difference between two approaches can be significant. Also note that the proposed approach does not require computations in the uncompressed data domain, which is important for big data that do not fit in memory for processing. Moreover, according to [207], under some conditions, PARAFAC has a unique decomposition almost surely.

Next, we explain a parallel implementation algorithm, PARACOMP [195], for tensor compression, as shown in Figure 6.5. The parallel architecture can conduct parallel processing of a set of randomly compressed, reduced-size "replicas" of the big tensor. Each replica is independently decomposed, and the results are joined via a master linear equation per tensor mode, which fits the MapReduce structure. Theorem shows that fully parallel computation of the big tensor decomposition is possible, which guarantees identifiability of the big tensor decomposition from the small tensor decompositions, without stringent additional constraints: if the big tensor is indeed of low rank and the system parameters are appropriately chosen, the rank-one factors of the big tensor can be recovered from the analysis of the reduced-size replicas. Moreover, the architecture affords memory, storage, and complexity gains of a high order, and sparsity can be further exploited to improve those gains.

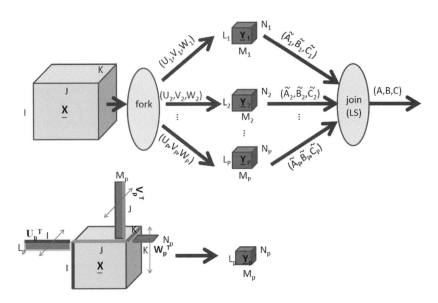

**Figure 6.5** PArallel Randomly COMPressed Cubes (PARACOMP) [195].

## 6.2.3    Tensor Completion

The purpose of tensor completion is to recover missing entries of low-rank tensors. This problem is known as the low-rank tensor completion problem. In the following, we provide several examples of how the tensor completion problem can be formulated. The solutions to those optimization problems can be referred to the other chapters.

**Convex:** The tensor completion problem can be written in the following convex optimization problem as an extension of matrix completion problem:

$$\min_{\mathbf{X}} \|\mathbf{X}\|_* \tag{6.29}$$

$$s.t. \ \mathbf{X}_\Omega = \mathbf{T}_\Omega,$$

where $\mathbf{X} \in \mathbb{R}^{I_1 \times \cdots \times I_N}$ and $\mathbf{T} \in \mathbb{R}^{I_1 \times \cdots \times I_N}$ are $n$-mode tensors with identical size in each mode, $\Omega$ is the set without missing data, and there typically $\mathbf{T}$ is the tensor that needs to recover the missing entries. $\| \cdot \|_*$ is the trace norm that approximate the rank, with one example definition as

$$\|\mathbf{X}\|_* = \sum_{i=1}^{n} \alpha_i \mathbf{X}_i, \tag{6.30}$$

where $\alpha_i$ is the weight factor. This approximation reduces the complexity of calculating the tensor rank.

**Tucker:** With the Tucker model for tensor factorization, the tensor completion problem can be written as

$$\min_{\mathbf{X},\mathbf{G},\mathbf{A}^{(1)},\cdots,\mathbf{A}^{(N)}} \|\mathbf{X} - \mathbf{G} \otimes_1 \mathbf{A}^{(1)} \otimes_2 \mathbf{A}^{(2)} \cdots \otimes_N \mathbf{A}^{(N)}\|_F \quad (6.31)$$

$$s.t.\ \mathbf{X}_\Omega = \mathbf{T}_\Omega,$$

where $\mathbf{G} \otimes_1 \mathbf{A}^{(1)} \otimes_2 \mathbf{A}^{(2)} \cdots \otimes_N \mathbf{A}^{(N)}$ is the Tucker3 model-based tensor factorization. Here $\| \cdot \|_F$ is the Frobenius norm of a tensor.

**PARAFAC:**

$$\min_{\mathbf{X},\mathbf{a}_r,\mathbf{b}_r,\mathbf{c}_r} \|\mathbf{X} - \sum_{r=1}^{R} \mathbf{a}_r \circ \mathbf{b}_r \circ \mathbf{c}_r\|_F \quad (6.32)$$

$$s.t.\ \mathbf{X}_\Omega = \mathbf{T}_\Omega,$$

with the PARAFAC model-based decomposition.

**SVD:** Another alternative is to conser the tensor as multiple matrices and make the unfolding matrix along each mode of the tensor to have low rank, that is,

$$\min_{\mathbf{X},\mathbf{M}_1,\cdots,\mathbf{M}_N} \sum_{i=1}^{N} \|\mathbf{X}_{(i)} - M_i\|_F \quad (6.33)$$

$$s.t.\ \mathbf{X}_\Omega = \mathbf{T}_\Omega,\ \mathrm{rank}(\mathbf{M}_i) \le r_i, \forall i,$$

where $\mathbf{M}_i \in \mathbb{R}^{I_i \times (\prod_{k \ne i} I_k)}$.

In literature, there are many tutorials and applications using tensor completion. In [208], an algorithm is proposed to estimate missing values in tensors of visual datal, where the values can be missing due to problems in the acquisition process, or because the user manually identified unwanted outliers. In [209], an important sparse-vector approximation problem (compressed sensing) and the low-rank matrix recovery problem, using a convex relaxation technique, are proved to be a valuable solution strategy. In [210], completion of partially observed spectral images is studied, which can be naturally represented as third-order tensors and typically exhibit intraband correlations. In [211], a new algorithm performs Riemannian optimization techniques on the manifold of tensors of fixed multilinear rank. In [212], a new model to recover a low-rank tensor is proposed by simultaneously performing low-rank matrix factorizations to the all-mode matricizations of the underlying tensor. In [213], a convex optimization approach is investigated to tensor completion by directly minimizing a tensor nuclear norm and prove that this leads to an improved sample size requirement.

## 6.3    Tensor Voting

The tensor voting algorithm in artificial vision allows us to infer information concerning the structures geometrically described by a partial data set. It is inspired by the principles of Gestalt regarding animal visual system: the individual continually receives partial information, incorrect or corrupted information from the noise, but has the ability to distinguish noise, in order to eliminate, and to reconstruct the missing information. Two basic perceptual Gestalt principles are: (1) Close to: the proximity between geometric elements (e.g., points) leads the perceptual system to group them in an attempt to describe a single geometric structure; (2) Continuation: the orientation of geometric structures, for which you have enough information, is diffused to the surrounding geometric entities. The tensor voting algorithm aims are to identify outliers to their removal, estimate the dimensionality of the geometric structures described by the data, estimate the orientation of the geometric structures, and allow accurate reconstruction of these structures.

Specifically in order to infer the hidden or missing structures, we need to model the structures mathematically first. Generally, the structure types in the form of 2D images can be classified into two categories: curves and regions. Curves are modeled as the structures that have a 1-$d$ normal space, which is referred to as a stick. Regions are modeled as the structures that have a 2D normal space, referred to as a ball. Normal space represents the structure types well, but it is required to know how salient the structures are in order to adequately model the structure. Hence, the parameter defined as saliency associated with each structure type is employed to indicate the size of structure. Both the normal space and saliency information are encoded in a tensor via specific calculation that will be described later. After developing the mathematical models, we can further explain the hints obtained from the Gestalt principles [214]: (1) a token "communicates" its structure information to its surrounding tokens in a certain way with respect to its normal space (i.e., the surrounding tokens under its influence are supposed to have the same kind of normal space); (2) in the real world, a token may contain a combination of information of both structure types. For instance, a token that actually belongs to a curve has a dominant saliency in the 1-$d$ normal space while it probably has minor saliency in the 2D normal space.

In this section, we illustrate the tensor voting framework. In subsection 6.3.1, we present the way to encode the normal space with tensor representation. In subsection 6.3.2, the fundamental stick tensor voting is addressed as the basis for inferring geometric structure via the encoded normal space. In subsection 6.3.3, the initialization procedure for tensor voting is introduced under the circumstance that no prior structure information is known. In subsection 6.3.4, the inference method based on tensor decomposition is explained. Finally, in subsection 6.3.5, several inference algorithms are briefly discussed.

## 6.3.1    Encode Normal Space with Tensor

In math, a normal space is an $N$-by-$N$ matrix for objects in $N$-dimension, denoted by $N_d$. Consider a $d$-dimension normal space in the $N$-dimension world, which is spanned by the first $d$ out of $N$ orthonormal basis vectors $\mathbf{e}_k, k = 1, 2, \ldots, N$, we have the normal space matrix expressed as

$$\mathbf{N}_d = \sum_{k=1}^{d} \mathbf{e}_k \mathbf{e}_k^T. \tag{6.34}$$

The projection of any vector $\mathbf{v}$ into this normal space is

$$\mathbf{v}_n = \mathbf{N}_d \mathbf{v}. \tag{6.35}$$

It can be easily proved that $\mathbf{v}_n$ is the projection of $\mathbf{v}$ into the normal space by showing that: (1) $\mathbf{v}_n$ can be linearly expressed by the basis vectors $\mathbf{e}_k, k = 1, 2, \ldots, N$; (2) dot product $\langle \mathbf{v}_n - \mathbf{v}, \mathbf{v}_n \rangle = 0$.

The exact tensor matrix $\mathbf{V}$ that encodes both normal spaces and their saliency is assumed to be known. From the previous discussion, we know that $\mathbf{V}$ is a symmetric, positive semidefinite $N$-by-$N$ matrix. Suppose its eigenvalues are ordered as $\lambda_1 \geq \ldots \geq \lambda_N \geq 0$ with corresponding eigenvectors $\mathbf{e}_1, \ldots, \mathbf{e}_N$. By the knowledge of linear algebra, we know that the eigenvectors of $\mathbf{V}$ are orthogonal with each other (for the eigenvectors that belong to 0 eigenvalue, we can generate those eigenvectors in a way that they meet this requirement). Then if we normalize these eigenvectors into unit vectors, we have a set of orthonormal basis $\hat{\mathbf{e}}_1, \ldots, \hat{\mathbf{e}}_N$. Furthermore, we have the following derivations as

$$\mathbf{V}\hat{\mathbf{e}}_d = \lambda_d \hat{\mathbf{e}}_d, \tag{6.36}$$

$$\mathbf{V}\hat{\mathbf{e}}_d \hat{\mathbf{e}}_d^T = \lambda_d \hat{\mathbf{e}}_d \hat{\mathbf{e}}_d^T, \tag{6.37}$$

$$\mathbf{V}\sum_{d=1}^{N} \hat{\mathbf{e}}_d \hat{\mathbf{e}}_d^T = \sum_{d=1}^{N} \lambda_d \hat{\mathbf{e}}_d \hat{\mathbf{e}}_d^T, \tag{6.38}$$

$$\mathbf{V} = \sum_{d=1}^{N} \lambda_d \hat{\mathbf{e}}_d \hat{\mathbf{e}}_d^T = \sum_{d=1}^{N-1} (\lambda_d - \lambda_{d+1}) \sum_{k=1}^{d} \hat{\mathbf{e}}_k \hat{\mathbf{e}}_k^T + \lambda_N \sum_{k=1}^{N} \hat{\mathbf{e}}_k \hat{\mathbf{e}}_k^T, \tag{6.39}$$

and

$$\mathbf{V} = \sum_{d=1}^{N-1} (\lambda_d - \lambda_{d+1}) N_d + \lambda_N N_N. \tag{6.40}$$

From (6.40), we define the saliency $s_d$ in the straightforward fashion: $s_d = \lambda_d - \lambda_{d+1}$, if $d < N$; $s_d = \lambda_N$, if $d = N$. Thus, substituting the saliency into (6.40), we have

$$\mathbf{V} = \sum_{d=1}^{N} s_d N_d. \tag{6.41}$$

## 6.3.2    Inferring Structure

To simplify the illustration procedure, we will assume that we already know the structure types and saliencies for now. We will discuss how to obtain this information later.

We start with the simplest case that is the fundamental unit stick vote when the normal space is 1-$d$. Consider a voter point $p$ (a token that passes its structure information to others) on a curve. Its normal is a known unit vector $\hat{\mathbf{v}}_n$. We want to know how it influences its neighboring votee point $x$ (a token that receives structure information from voters). Based on the previous discussion, we assume that $p$ and $x$ share the same structure type when we consider $p$ is influencing $x$. To approximate the path of the same structure type that passes through $p$ and $x$, we take the arc of the osculating circle centered at $o$ passing through $p$ and $x$ as the most likely smooth path [215]. Figure 6.6 shows the geometric relationship between $p$ and $x$. $\hat{\mathbf{v}}_t$ is the known tangent vector at $p$, and $\mathbf{v}$ is the vector from $p$ to $x$. $\hat{\mathbf{v}}_c$ is the normal vector at $x$ that we want to calculate. $\theta$ is the vote angle between $\mathbf{v}$ and $\hat{\mathbf{v}}_t$. We take the influence as 0 when $\theta$ is larger than $\pi/4$ for the reason that two points that have an angle larger than 90 degrees between their normals are least likely to influence each other. $a$ is the arc length between $p$ and $x$. $k$ is the curvature of the osculating circle, which is the reciprocal of the radius $R = ox$. To calculate $\theta$ and $\hat{\mathbf{v}}_c$, we have

$$\theta = \arcsin \mathbf{v}^T \hat{\mathbf{v}}_n \tag{6.42}$$

and

$$\hat{\mathbf{v}}_c = \hat{\mathbf{v}}_n \cos 2\theta - \hat{\mathbf{v}}_t \sin 2\theta. \tag{6.43}$$

We also add a decay profile to the tensor to model the decaying influence of the information going through the structure. Therefore, the complete unit stick vote (tensor) that encodes the normal space information received by the votee is

$$\mathbf{V}^p = DF(a,k,\sigma)\hat{\mathbf{v}}_c\hat{\mathbf{v}}_c^T. \tag{6.44}$$

$DF(a,k,\sigma)$ is the decay profile that takes $a$, $k$, and $\sigma$ as parameters. $\sigma$ is the free parameter set by the user to control the scale of voting. The decay profile can be given empirically or based on a traditional choice as

$$DF(a,k,\sigma) = e^{-(\frac{a^2+ck^2}{\sigma^2})}, \tag{6.45}$$

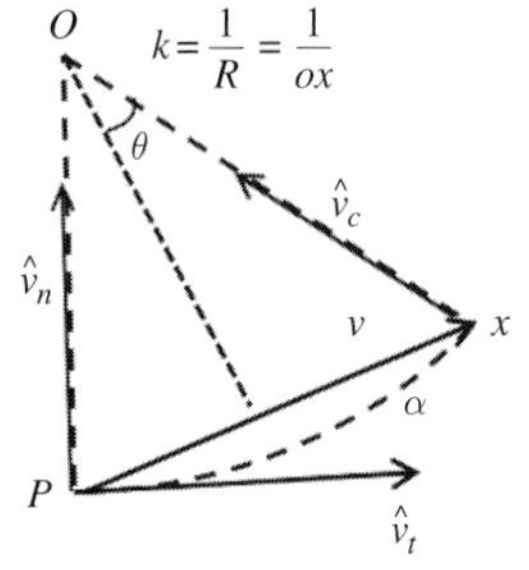

**Figure 6.6** Illustration of the fundamental stick vote.

where the parameters can be derived by the geometry as

$$c = \frac{-16log(0.1)(\sigma - 1)}{\pi^2},\tag{6.46}$$

$$a = \frac{\theta\|\mathbf{v}\|}{\sin\theta},\tag{6.47}$$

and

$$a = \frac{2\sin\theta}{\|\mathbf{v}\|}.\tag{6.48}$$

Usually, a voter's stick vote is not a unit vector. If so, the unit tensor expressed in (6.44) is multiplied with the corresponding saliency of the 1-$d$ normal space of the voter.

Likewise, when inferring structures for the 2D normal space, we attempt to find the same normal space at the votee. In that case, 2 basis vectors are required for the 2D normal space. The procedure is equivalent to: (1) find out the basis vectors that span voter's normal space; (2) vote or project these basis vectors, respectively, in the way the fundamental stick vote does to the votee; and (3) reconstruct the complete normal space information at the votee by combining the newly generated normal space information (i.e., adding the tensor matrices created by the stick vote, respectively).

To find out the basis vectors of the voter's normal space, Mordohai proposes a method [215] that projects the voter-to-votee vector into the voter's normal space and then computes the orthonormal basis vectors for the voter's normal space based on the projection and Gram-Schmidt orthogonalization procedure. This method significantly reduces the computation when calculating the basis vectors of the high-dimensional normal space.

Suppose the voter's normal space is known and encoded as the tensor matrix $\mathbf{N}_d^p$, where $p$ represents the voter point and $d$ represents the dimension of its normal space. For any fixed votee point $x$ that receives $p$'s vote, the voter-to-votee vector $\mathbf{v}$ is known. Then the projected vector is

$$\mathbf{v}_n = N_d^p\mathbf{v}.\tag{6.49}$$

Thus, the tangent vector $\mathbf{v}_n$ is computed by

$$\mathbf{v}_t = (I - N_d^p)\mathbf{v} = \mathbf{v} - \mathbf{v}_n,\tag{6.50}$$

where $I$ is the identity matrix of dimension $N$-by-$N$. Based on (6.50), these two vectors are then normalized as $\hat{\mathbf{v}}_n$ and $\hat{\mathbf{v}}_t$. The first constructed basis vector for the normal space is selected as $\hat{\mathbf{v}}_{n,1} = \hat{\mathbf{v}}_n$. Next, the Gram-Schmidt procedure is employed to construct the rest $d - 1$ orthonormal basis vectors $\hat{\mathbf{v}}_{n,i}$, $i = 2, 3, \ldots, d$. As a consequence, each $\hat{\mathbf{v}}_{n,i}$ is considered as the fundamental stick vote and voted to the votee as the voting procedure described above. Each stick vote results in a tensor matrix $\mathbf{V}_i^p$, $i = 1, 2, \ldots, d$ for the $d$-dimension normal space. Finally, these tensors are summed up into one matrix that represents the complete information for the $d$-dimension normal space at votee $x$. The proposed method of generating the basis set facilitates the computation because for $i = 2, 3, \ldots, d$, the basis vector $\hat{\mathbf{v}}_i$ is orthogonal to $\mathbf{v}$, which means the vote angle $\theta = 0$. Hence, (6.43) is simplified during the computation.

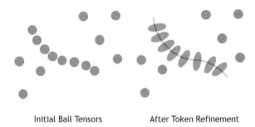

Initial Ball Tensors          After Token Refinement

**Figure 6.7** Illustration of the token refinement procedure: each point is initialized with a ball tensor; points nearby with each other form stick tensors while points far away from others remain ball tensors.

## 6.3.3    Token Refinement

So far, the discussions are based on the presumption that the normal space and saliencies information are known at a given voter site. Nevertheless, in most cases, it is impossible to obtain this kind of prior knowledge. Thus, the proper initialization called token refinement, which estimates the prior information, is needed.

Figure 6.7 illustrates the token refinement procedure in a 2D space. In the token refinement procedure, each input token is initialized with a unit ball tensor indicating neither direction preference nor prior saliency information. The input tokens are then considered one by one as the voter and voted to its neighboring input tokens. In the end, all the tensors received by each input token are summed up and stored as the known normal space and saliency information. If a cluster of tokens actually belong to the same curve in the real world, then the way they influence each other using their initial ball tensors will put major emphasis on the stick tensor, namely the 1-$d$ normal space in a 2D world. As can be seen in Figure 6.7, the tokens along the curve influence each other during token refinement, resulting in elongating their tensors to become stick tensors in the 2D space, while the tokens that sparsely spread out in the space receive little information from others, causing the existing ball tensors to remain.

## 6.3.4    Token Decomposition

After token refinement, the tensor voting procedure can be completed by the method described in Section 6.3.2. The result in the 2D image case is that each pixel is associated with a 2x2 matrix $\mathbf{T}$. The ultimate objective is to decide which structure type the candidate pixel should belong to. Hence, the tensor matrix $\mathbf{T}$ needs to be decomposed by (6.40) to extract the saliencies for the two structure types. In 2D case, it becomes

$$\mathbf{T} = \lambda_1 \hat{\mathbf{e}}_1 \hat{\mathbf{e}}_1^T = (\lambda_1 - \lambda_2)\hat{\mathbf{e}}_1 \hat{\mathbf{e}}_1^T + \lambda_2(\hat{\mathbf{e}}_1 \hat{\mathbf{e}}_1^T + \hat{\mathbf{e}}_2 \hat{\mathbf{e}}_2^T). \qquad (6.51)$$

If $\lambda_1 - \lambda_2 > \lambda_2 > 0$, the stick saliency is the dominant one, which indicates the certainty of one normal orientation. Therefore, the token is inferred as the part of a curve, with its estimated normal being $\hat{\mathbf{e}}_1$. If $\lambda_1 \approx \lambda_2 > 0$, the dominant component is the ball saliency, which means there is no preference of orientation. Thus, the token is estimated as the part of a region or a junction where two or more curves intersect with multiple orientations present simultaneously. Note that, if both the saliency values are very small, the candidate token is likely an outlier. This makes tensor voting capable of filtering noise.

## 6.3.5 Inference Algorithm

Based on the previous discussion, there are multiple choices with respect to how to implement the tensor voting technique. One feasible way to implement the tensor voting technique is to compute the so-called voting field for each token; this method is referred to as the per-voter scheme. After token refinement procedure is complete, the per-voter algorithm examines every input token as a voter and computes the set of votes it casts to all its neighbors. That set of votes is referred to as the voting field. The algorithm then integrates all the voting fields and performs tensor decomposition at each site. In [216], tensor voting is implemented based on the per-voter scheme. In addition, the algorithm is combined with the steerable filter theory [217] to rewrite the tensor voting operation as a linear combination of complex-valued convolutions, which significantly reduces the computation load.

In [218], a straightforward implementation method of tensor voting technique referred to as the per-votee scheme is proposed to infer the human mobility trace encoded in the location data with some recordings missing. The per-voter scheme calculates one vote from point to point at one time. In order to reduce the computation, we also implement tensor voting in a sparse sense. When examining one site as a votee, we only consider the influence it received from the neighboring pixels $\{C_{ee}\}$ within the radius of approximately $3\sigma$ as reported in [219]. Furthermore, we define the sparse voting region, gather all the tensors received by each votee only in that region and decompose the result tensor matrix $\mathbf{M}_{ee}$ to determine its actual structure type. Finally, we make the voting procedure iterative so that it is able to fill the large gaps.

## 6.4 Other Tensor Applications

Tensor factorizations have already had many applications in signal processing such as speech, audio, communications, radar, signal intelligence, and machine learning. For example, tensor factorization can be employed to blindly separate unknown mixtures of speech signals in reverberant environments [220]; unravel mixtures of code-division communication signals without knowledge of their spreading codes [221]; localize emitters in radar and communication applications [222]; detect cliques in social networks [223]; analyze fluorescence spectroscopy data [224]; and for additional machine learning applications [225].

Tensor computation has been developed to address various big data tasks in higher dimensions. In [226], a scalable and distributed version of the Tucker model, MR-T, is implemented using the Hadoop MapReduce framework. Authors in [227] propose a new constrained tensor factorization framework, building upon the Alternating Direction method of Multipliers (ADMM). Work in [228] permeates benefits from rank minimization to scalable imputation of missing data, via tracking low-dimensional subspaces and unraveling latent structure from incomplete streaming data. Work in [225] addresses the problems of computing decompositions of full tensors by using compressed sensing (CS) methods working on incomplete tensors (i.e., tensors with only a few known elements). In [229], regularized block multiconvex optimization is considered, where the feasible set and objective function are generally nonconvex but convex in each block of variables. In [230], an alternative convex relaxation is proposed on the Euclidean ball, and then a technique is described to solve the associated regularization problem, which builds upon the alternating direction method of multipliers. In [231], a recovery task is studied to find a low multilinear-rank tensor that fulfills some linear constraints in the general settings, which has many applications in computer vision and graphics. In [232, 233], sparse non-negative Tucker decomposition is considered to decompose a given tensor into the product of a core tensor and several factor matrices with sparsity and non-negativity constraints. In [234], to have an efficient low-rank tensor completion approach, a matrix factorization idea is introduced into the tensor nuclear norm model, and then a much smaller-scale matrix nuclear norm minimization problem is obtained. In [235], a class of structural regularization terms is introduced that extends nuclear norm and enables estimation of low-rank tensors through convex optimization problems. In [236], a novel core tensor trace norm minimization (CTNM) method is proposed for simultaneous tensor learning and decomposition, and has a much lower computational complexity. In [227], a parallel trace norm regularized tensor decomposition method is formulated as a convex optimization problem solved by parallel ADMM. In [237], low-rank matrix and tensor completion are studied with novel algorithms that employ adaptive sampling schemes to obtain strong performance guarantees. In [211], a new algorithm is proposed that performs Riemannian optimization techniques on the manifold of tensors of fixed multilinear rank. In [238], theories and computational methods are developed for overcomplete, non-orthogonal tensor decomposition using convex optimization. In [212], a new model is proposed to recover a low-rank tensor by simultaneously performing low-rank matrix factorizations to the all-mode matricizations of the underlying tensor. In [239], tailored optimization algorithms are proposed with global convergence guarantees for solving both the constrained and the Lagrangian formulations of the problem. In [240], an algorithm is presented for decomposing a symmetric tensor of dimension $n$ and order $d$ as a sum of rank-l symmetric tensors, extending the algorithm of Sylvester devised in 1886 for symmetric tensors of dimension 2. In [241], a tensor-based method considering the full spatial temporal information of traffic flow is proposed to fuse the traffic flow data from multiple detecting locations. In [242], for estimating missing values in tensors of visual data, the definition of the trace norm for tensors is defined with a working algorithm. In [243], an approach is presented for penalized tensor decomposition that estimates smoothly varying latent factors in multiway data. In [244], the general tensor PCA problem is reducible to its special case where the tensor in question is super-symmetric with an even degree. In [245], a tensor $n$-mode matrix unfolding truncated nuclear norm is proposed, which is extended from the matrix truncated nuclear norm, to the tensor completion problem.

Some tracking problems can be essentially transformed as the image-processing problems where the tensor voting technique has been widely utilized. Guy elaborated tensor voting theory with insightful analysis in his PhD thesis [219]. The theory is then ameliorated by other researchers in the past decade [246]. In tracking applications, tensor voting systematically infers hidden or incomplete structures (e.g., gaps and broken parts in the trace curve). Tensor voting is also referred to as perceptual grouping [247], and emphasizes the contribution of the Gestalt principles on which the theory is based. In brief, the Gestalt principles state that the presence of each input token (site, pixel, signal, etc.) implies a hypothesis that the structure passes through it. For example, considering the 2D imaging process of a chair, if one pixel has recorded the chair signal, it is highly likely that some of its neighboring pixels should have captured the same structure/object signals. In other words, it is human nature to configure simple elements into the perception of complex structures. In [248] and [249], the object signals observed by fixed cameras are represented using spatiotemporal features which facilitates the application of tensor voting theory. The advantage of applying tensor voting is that it brings several geometric properties including smooth continuous trajectories and bounding boxes with minimum registration error. Although there are extensive research efforts dedicated to tensor voting study in the image-processing field [250], little work involving the tensor voting theory has been done in the communication realm to the best of our knowledge. Moreover, very little research has dealt with the missing data problem in the tracking context.

Another topic in tracking problems is trajectory analysis [251], which has wide application scenarios including vehicle traffic management, vessel classification by satellites images, and so forth. In [252] a unifying framework is constructed to mine trajectory patterns of various temporal tightness. The proposed framework consists of two phases: initial pattern discovery and granularity adjustment. Authors in [253] employ the fractal analysis of a fin whale's trajectory tracked by the satellite. The implemented fractal analysis provides the scale-independent measurement to summarize interactions between an organism and its ecosystem and depends on the heterogeneity of the whale's environment and the whale's ability to perceive it. In [254], several real-world human mobility traces are employed to analyze network robustness in the time domain. Authors in [255] propose a novel integrated framework for multiple human trajectory detection, learning, and analysis in complicated environments. In [256], a new approach for abnormal loitering detection using trajectory analysis is described and Inverse Perspective Mapping (IPM) is presented to resolve distortion of trajectory direction.

# 7  Deep Learning and Applications

Deep learning (also known as deep structured learning, hierarchical learning, or deep machine learning) is a branch of machine learning based on a set of algorithms that attempt to model high-level abstractions in data by using a deep graph with multiple processing layers, composed of multiple linear and nonlinear transformations. Deep learning has been characterized as a class of machine learning algorithms with the following characteristics [257]:

- They use a cascade of many layers of nonlinear processing units for feature extraction and transformation. Each successive layer uses the output from the previous layer as input. The algorithms may be supervised or unsupervised and applications include pattern analysis (unsupervised) and classification (supervised).
- They are based on the (unsupervised) learning of multiple levels of features or representations of the data. Higher-level features are derived from lower-level features to form a hierarchical representation.
- They are part of the broader machine learning field of learning representations of data.
- They learn multiple levels of representations that correspond to different levels of abstraction; the levels form a hierarchy of concepts.

These definitions have in common: multiple layers of nonlinear processing units and the supervised or unsupervised learning of feature representations in each layer, with the layers forming a hierarchy from low-level to high-level features. Various deep learning architectures such as deep neural networks, convolutional deep neural networks, deep belief networks (DBN), and recurrent neural networks have been applied to fields like computer vision, automatic speech recognition, natural language processing, audio recognition, and bioinformatics where they have been shown to produce state-of-the-art results on various tasks.

In this chapter we start in Section 7.1 with an introduction, giving a brief history of this field, the relevant literature, and its applications. Then we study some basic concepts of deep learning such as convolutional neural networks, recurrent neural networks, backpropagation algorithm, restricted Boltzmann machines, and deep learning networks in Section 7.2. Then we illustrate three examples for Apache Spark implementation for mobile big data (MBD), user moving pattern extraction, and combination with nonparametric Bayesian learning, respectively, in Sections 7.3 through 7.5. Finally, we have summary in Section 7.6.

## 7.1 Introduction

### 7.1.1 History

One of first learning algorithms for supervised deep feedforward multilayer percep-trons was published in in 1965 in [258], and a deep network with eight layers trained is described by the group method of data handling algorithm in 1971 in [259]. For other deep learning working architectures such as artificial neural networks (ANN), the chal-lenge was how to train networks with multiple layers. The backpropagation algorithm was employed in a deep neural network in 1980 in [260]. In 1993, the neural history compressor [261] implemented as an unsupervised stack of recurrent neural networks (RNNs) solved a "Very Deep Learning" task [262] that required more than 1,000 sub-sequent layers in an RNN unfolded in time [263]. In 1995, it was possible to train a network containing six fully connected layers and several hundred hidden units using the wake-sleep algorithm [264]. Many factors contribute to the slow speed, one being the vanishing gradient problem analyzed in 1991 by [265, 266].

According to a survey [267], the expression "deep learning" was introduced to the Machine Learning community by [268] in 1986, and later to ANNs by [269] in 2000. In 2006, it was shown in [270] how many-layered feedforward neural network could be effectively pretrained one layer at a time, treating each layer in turn as an unsupervised restricted Boltzmann machine, then fine-tuning it using supervised backpropagation. Since its resurgence, deep learning has become part of many state-of-the-art systems in various disciplines, particularly computer vision [271, 272, 269, 273, 274, 275] and automatic speech recognition (ASR) [276, 277, 278, 279, 280].

The impact of deep learning in industry apparently began in the early 2000s, when CNNs already processed an estimated 10% to 20% of all the checks written in the United States in the early 2000s, according to [281]. The history of this significant development in deep learning has been described and analyzed in recent books and articles [282, 283, 284]. Advances in hardware have also been important in enabling the renewed interest in deep learning. In particular, powerful graphics processing units (GPUs) are well-suited for the kind of number crunching, matrix/vector math involved in machine learning [285, 286]. GPUs have been shown to speed up training algorithms by orders of magnitude, bringing running times of weeks back to days [287, 288].

### 7.1.2 Milestone Literature

In this subsection, we list several milestone papers listed by [289]. The report pub-lished in [290] was a breakthrough that used convolutional nets to almost halve the error rate for object recognition, and precipitated the rapid adoption of deep learning by the computer vision community. The paper in [276] from the major speech recognition laboratories, summarizing the breakthrough achieved with deep learning on the task of phonetic classification for automatic speech recognition, was the first major indus-trial application of deep learning. The paper in [291] showed state-of-the-art machine translation results with the architecture introduced in [289], with a recurrent network trained to read a sentence in one language, produce a semantic representation of its meaning, and generate a translation in another language. The paper in [292] showed that supervised training of very deep neural networks is much faster if the hidden layers are composed of ReLU. The paper in [293] introduced a novel and effective way of training very deep neural networks by pretraining one hidden layer at a time using the unsupervised learning procedure for restricted Boltzmann machines. The report in [294] demonstrated that the unsupervised pretraining method introduced in [293] significantly improves performance on test data and generalizes the method to other unsupervised representation-learning techniques, such as autoencoders. In [295] is the first paper on convolutional networks trained by backpropagation for the task of classifying low-resolution images of handwritten digits. The overview paper [296] on the principles of end-to-end training of modular systems such as deep neural networks using gradient-based optimization showed how neural networks (and in particular convolutional nets) can be combined with search or inference mechanisms to model complex outputs that are interdependent, such as sequences of characters associated with the content of a document. The paper in [297] introduced neural language models, which learn to convert a word symbol into a word vector or word embedding composed of learned semantic features in order to predict the next word in a sequence. The paper in [271] introduced LSTM recurrent networks, which have become a crucial ingredient in recent advances with recurrent networks because they are good at learning long-range dependencies.

## 7.1.3    Applications

### Speech Recognition

Speech recognition has been revolutionized by deep learning, especially by long short-term memory (LSTM), a recurrent neural network published by [271]. LSTM RNNs circumvent the vanishing gradient problem and can learn "Very Deep Learning" tasks [262] that involve speech events separated by thousands of discrete 10 ms time steps. In 2003, LSTM with forget gates [298] became competitive with traditional speech recognizers on certain tasks [299]. In 2007, LSTM trained by Connectionist Temporal Classification (CTC) [272] achieved excellent results in certain applications [300], although computers were much slower than today. In 2015, Google's large-scale speech recognition suddenly almost doubled its performance through CTC-trained LSTM, now available to all smartphone users [269]. All major commercial speech recognition systems (e.g., Microsoft Cortana, Xbox, Skype Translator, Amazon Alexa, Google Now, Apple Siri, Baidu and iFlyTek voice search, and a range of Nuance speech products, etc.) are all based on deep learning methods [282, 301, 302, 303]. Recent progress (and future directions) can be summarized into eight major areas [282, 304, 283]: scaling up/out and accelerate DNN training and decoding; sequence discriminative training of DNNs; feature processing by deep models with solid understanding of the underlying mechanisms; adaptation of DNNs and of related deep models; multitask and transfer learning by DNNs and related deep models; convolution neural networks and how to design them to best exploit domain knowledge of speech; recurrent neural network and its rich LSTM variants; and other types of deep models including tensor-based models and integrated deep generative/discriminative models.

**Image Recognition**

A common evaluation set for image classification is the MNIST database data set, composed of handwritten digits and includes 60,000 training examples and 10,000 test examples. The current best result on MNIST is an error rate of 0.23%, achieved in 2012 [305]. Until 2011, CNNs did not play a major role at computer vision conferences, but in June 2012, a paper in [305] showed how max-pooling CNNs on GPU can dramatically improve many vision benchmark records, sometimes with human-competitive or even superhuman performance. As in the ambitious moves from automatic speech recognition toward automatic speech translation and understanding, image classification has recently been extended to the more challenging task of automatic image captioning, in which deep learning (often as a combination of CNNs and LSTMs) is the essential underlying technology [306, 307, 308, 309, 310].

One major application of image recognition is the autonomous car (driverless car, self-driving car, robotic car, etc.), which is a vehicle capable of sensing its environment and navigating without human input [311, 312, 313, 314]. Autonomous vehicles detect surroundings using radar, lidar, GPS, odometry, and computer vision. Advanced control systems interpret sensory information to identify appropriate navigation paths, as well as obstacles and relevant signage [315, 316]. Autonomous cars have control systems that are capable of analyzing sensory data to distinguish between different cars on the road, which is very useful in planning a path to the desired destination [317]. Google Self-Driving Car is one in a range of autonomous cars, developed by Google X as part of its project to develop technology for mainly electric cars. The software installed in Google's cars is named Google Chauffeur [311]. Tesla also has its self-driving features in Model S and Model X.

**Natural Language Processing**

Neural networks have been used for implementing language models since the early 2000s [318, 297]. Recurrent neural networks, especially LSTM [271] are most appropriate for sequential data such as language. LSTM helped to improve machine translation [291] and Language [319, 320]. LSTM combined with CNNs also improved automatic image captioning [306] and a plethora of other applications [262]. Deep neural architectures have achieved state-of-the-art results in many natural language processing tasks such as constituency parsing [321], sentiment analysis [322], information retrieval [323, 324], spoken language understanding [325], machine translation [291, 326], contextual entity linking [327], and others [328].

**AlphaGo**

AlphaGo is a computer program developed by Google DeepMind to play the board game Go. In October 2015, it became the first Computer Go program to beat a professional human Go player without handicaps on a full-sized $19 \times 19$ board. In March 2016, it beat Lee Sedol in a five-game match, the first time a computer Go program has beaten a 9-dan professional without handicaps. AlphaGo's algorithm uses a Monte Carlo tree search to find its moves based on knowledge previously "learned" by machine learning, specifically by an ANN (a deep learning method) by extensive training, both from human and computer play.

### Others

There are some other applications of deep learning such as drug discovery and toxicology [329, 330, 331, 332, 333], customer relationship management [334], recommendation systems [335, 336], and bioinformatics [337].

## 7.2    Deep Learning Basics

As shown in Figure 7.1, deep learning algorithms are based on distributed representations, whose underlying assumption is that observed data are generated by the interactions of factors organized in layers. Deep learning adds the assumption that these layers of factors correspond to levels of abstraction or composition. Varying numbers of layers and layer sizes can be used to provide different amounts of abstraction. Deep learning exploits this idea of hierarchical explanatory factors where higher-level, more abstract concepts are learned from the lower-level ones. Deep learning helps disentangle these abstractions and pick out which features are useful for learning [338].

For supervised learning tasks, deep learning methods translate the data into compact intermediate representations akin to principal components, and derive layered structures which remove redundancy in representation [282]. Many deep learning algorithms are applied to unsupervised learning tasks. This is an important benefit because unlabeled data are usually more abundant than labeled data [338, 261, 339].

In this section, we first study the basic structure of each layer. Then we investigate the algorithms on training multiple layers. Finally, we discuss three types of deep learning networks and then compare them.

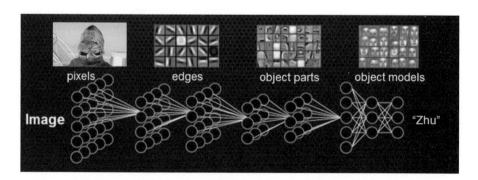

**Figure 7.1**  Deep learning structure.

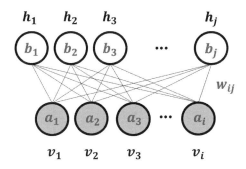

**Figure 7.2** Restricted Boltzmann machine.

## 7.2.1  Restricted Boltzmann Machine

A restricted Boltzmann machine (RBM) as shown in Figure 7.2 is a generative stochastic ANN that can learn a probability distribution over its set of inputs. RBMs are a variant of Boltzmann machines, with the restriction that their neurons must form a bipartite graph: a pair of nodes from each of the two groups of units (commonly referred to as the "visible" and "hidden" units, respectively) may have a symmetric connection between them; and there are no connections between nodes within a group.[1] This restriction allows for more efficient training algorithms than are available for the general class of Boltzmann machines, in particular the gradient-based contrastive divergence algorithm [340]. DBN can be formed by "stacking" RBMs and optionally fine-tuning the resulting deep network with gradient descent and backpropagation [339], which we will discuss in the next subsection.

The standard type of RBM has binary-valued (Boolean/Bernoulli) hidden and visible units, and consists of a matrix of weights $W = (w_{i,j})$ (size $m \times n$) associated with the connection between hidden unit $h_j$ and visible unit $v_i$, as well as bias weights (offsets) $a_i$ for the visible units and $b_j$ for the hidden units. Given these, the energy of a configuration (pair of boolean vectors) $(\mathbf{v}, \mathbf{h})$ is defined as

$$E(\mathbf{v}, \mathbf{h}) = -\sum_i a_i v_i - \sum_j b_j h_j - \sum_i \sum_j v_i w_{ij} h_j. \tag{7.1}$$

As in general Boltzmann machines, probability distributions over hidden and/or visible vectors are defined in terms of the energy function [341]:

$$P(\mathbf{v}, \mathbf{h}) = \frac{1}{Z} e^{-E(\mathbf{v}, \mathbf{h})} \tag{7.2}$$

where $Z$ is a partition function defined as the sum of $e^{-E(\mathbf{v}, \mathbf{h})}$ over all possible configurations (i.e., a normalizing constant to ensure the probability distribution sums to 1).

---

[1] By contrast, "unrestricted" Boltzmann machines may have connections between hidden units.

Similarly, the (marginal) probability of a visible (input) vector of booleans is the sum over all possible hidden-layer configurations:

$$P(\mathbf{v}) = \frac{1}{Z} \sum_h e^{-E(\mathbf{v},\mathbf{h})}. \tag{7.3}$$

Since the RBM has the shape of a bipartite graph, with no intralayer connections, the hidden unit activations are mutually independent given the visible unit activations and conversely, the visible unit activations are mutually independent given the hidden unit activations [340]. That is, for $m$ visible units and $n$ hidden units, the conditional probability of a configuration of the visible units $v$, given a configuration of the hidden units $h$, is

$$P(\mathbf{v}|\mathbf{h}) = \prod_{i=1}^{m} P(v_i|\mathbf{h}). \tag{7.4}$$

Conversely, the conditional probability of $\mathbf{h}$ given $\mathbf{v}$ is

$$P(\mathbf{h}|\mathbf{v}) = \prod_{j=1}^{n} P(h_j|\mathbf{v}). \tag{7.5}$$

The individual activation probabilities are given by

$$P(h_j = 1|\mathbf{v}) = \sigma \left( b_j + \sum_{i=1}^{m} w_{i,j} v_i \right), \tag{7.6}$$

$$P(v_i = 1|\mathbf{h}) = \sigma \left( a_i + \sum_{j=1}^{n} w_{i,j} h_j \right), \tag{7.7}$$

where $\sigma$ denotes the logistic sigmoid.

The visible units of RBM can be multinomial, although the hidden units are Bernoulli. In this case, the logistic function for visible units is replaced by

$$P(v_i^k = 1|\mathbf{h}) = \frac{\exp(a_i^k + \Sigma_j W_{ij}^k h_j)}{\Sigma_{k=1}^{K} \exp(a_i^k + \Sigma_j W_{ij}^k h_j)}, \tag{7.8}$$

where $K$ is the number of discrete values that the visible values have.

RBMs are trained to maximize the product of probabilities assigned to some training set $V$ (a matrix, each row of which is treated as a visible vector $v$),

$$\arg \max_{\mathbf{W}} \prod_{v \in V} P(v). \tag{7.9}$$

One algorithm often used to train RBMs (i.e., to optimize the weight vector $\mathbf{W}$) is the contrastive divergence (CD) algorithm [342, 343]. The algorithm performs Gibbs sampling and is used inside a gradient descent procedure (similar to the way back-propagation is used inside such a procedure when training feedforward neural nets) to compute weight update. The basic, single-step contrastive divergence (CD-1) procedure for a single sample can be summarized as follows:

1. Take a training sample $\mathbf{v}$, compute the probabilities of the hidden units, and sample a hidden activation vector $\mathbf{h}$ from this probability distribution.
2. Compute the outer product of $\mathbf{v}$ and $\mathbf{h}$ and call this the positive gradient.
3. From $\mathbf{h}$, sample a reconstruction $\mathbf{v}'$ of the visible units, then resample the hidden activations $\mathbf{h}'$ from this (Gibbs sampling step).
4. Compute the outer product of $\mathbf{v}'$ and $\mathbf{h}'$ and call this the negative gradient.
5. Let the weight update to $w_{ij}$ be the positive gradient minus the negative gradient, times some learning rate:

$$\Delta w_{ij} = \epsilon(\mathbf{vh}^\mathsf{T} - \mathbf{v}'\mathbf{h}'T). \tag{7.10}$$

### 7.2.2 Stochastic Gradient Descent and Backpropagation

A deep neural network (DNN) is an ANN with multiple hidden layers of units between the input and output layers [344, 262]. Similar to shallow ANNs, DNNs can model complex nonlinear relationships. DNN architectures (e.g., for object detection and parsing) generate compositional models where the object is expressed as a layered composition of image primitives [345]. The extra layers enable composition of features from lower layers, giving the potential of modeling complex data with fewer units than a similarly performing shallow network [344].

A DNN can be discriminatively trained with the standard backpropagation algorithm. The weight updates of backpropagation can be done via stochastic gradient descent using the following equation:

$$w_{ij}(t + 1) = w_{ij}(t) + \eta \frac{\partial C}{\partial w_{ij}}, \tag{7.11}$$

where $t$ is the iteration index, $\eta$ is the learning rate, and $C$ is the cost function. The choice of the cost function depends on factors such as the learning type (supervised, unsupervised, reinforcement, etc.) and the activation function.

For example, when performing supervised learning on a multiclass classification problem, common choices for the activation function and cost function are the softmax function and cross entropy function, respectively. The softmax function is defined as

$$p_j = \frac{\exp(x_j)}{\sum_k \exp(x_k)}, \tag{7.12}$$

where $p_j$ represents the class probability (output of the unit $j$), and $x_j$ and $x_k$ represent the total input to units $j$ and $k$ of the same level, respectively. Cross entropy is defined as

$$C = -\sum_j d_j \log(p_j), \tag{7.13}$$

where $d_j$ represents the target probability for output unit $j$ and $p_j$ is the probability output for $j$ after applying the activation function [276].

These can be used to output object bounding boxes in the form of a binary mask. They are also used for multiscale regression to increase localization precision. DNN-based regression can learn features that capture geometric information in addition to being a good classifier. They remove the limitation of designing a model which will capture parts and their relations explicitly. This helps learn a wide variety of objects. The model consists of multiple layers, each of which has a rectified linear unit for nonlinear transformation. Some layers are convolutional, while others are fully connected. Every convolutional layer has an additional max pooling. The network is trained to minimize $L_2$ error for predicting the mask ranging over the entire training set containing bounding boxes represented as masks.

For another concept, in the context of ANNs, the rectifier is an activation function defined as

$$f(x) = \max(0, x), \tag{7.14}$$

where $x$ is the input to a neuron. This is also known as a ramp function and is analogous to half-wave rectification in electrical engineering. The rectifier is the most popular activation function for deep neural networks. A unit employing the rectifier is also called a rectified linear unit (ReLU).

## 7.2.3    Deep Learning Networks

### Deep Belief Networks

A DBN is a probabilistic, generative model made up of multiple layers of hidden units. It can be considered a composition of simple learning modules that make up each layer [339]. A DBN can be used to generatively pretrain a DNN by using the learned DBN weights as the initial DNN weights. Backpropagation or other discriminative algorithms can then be applied for fine-tuning of these weights. This is particularly helpful when limited training data are available, because poorly initialized weights can significantly hinder the learned model's performance. These pretrained weights are in a region of the weight space that is closer to the optimal weights than are randomly chosen initial weights. This allows for both improved modeling and faster convergence of the fine-tuning phase [346].

A DBN can be efficiently trained in an unsupervised, layer-by-layer manner, where the layers are typically made of RBM. An RBM is an undirected, generative energy-based model with a "visible" input layer and a hidden layer, and connections between the layers but not within layers. The training method for RBMs proposed by Geoffrey Hinton for use with training 11 Product of Expert models is called contrastive divergence (CD) [343]. CD provides an approximation to the maximum likelihood method that would ideally be applied for learning the weights of the RBM [341, 347]. Although the approximation of CD to maximum likelihood is very crude (CD has been shown to not follow the gradient of any function), it has been empirically shown to be effective in training deep architectures [341]. The RBM model and training procedure have been discussed in the previous two subsections.

Once an RBM is trained, another RBM is "stacked" on top of it, taking its input from the final already-trained layer. The new visible layer is initialized to a training vector, and values for the units in the already-trained layers are assigned using the current weights and biases. The new RBM is then trained with the procedure above. This whole process is repeated until some desired stopping criterion is met [344].

## Convolutional Neural Networks

Convolutional neural networks (CNNs) have become the method of choice for processing visual and other two-dimensional data [348, 281]. A CNN is composed of one or more convolutional layers with fully connected layers (matching those in typical ANNs) on top. It also uses tied weights and pooling layers. In particular, max-pooling [349] is often used in Fukushima's convolutional architecture [260]. This architecture allows CNNs to take advantage of the 2D structure of input data. In comparison with other deep architectures, CNNs have shown superior results in both image and speech applications. They can also be trained with standard backpropagation. CNNs are easier to train than other regular, deep, feedforward neural networks and have many fewer parameters to estimate, making them a highly attractive architecture to use [350]. Examples of applications in Computer Vision include DeepDream [351]. See the main article on CNNs for numerous additional references.

A recent achievement in deep learning is the use of convolutional DBN (CDBN). CDBNs have structure very similar to a CNN and are trained similar to DBN. Therefore, they exploit the 2D structure of images, like CNNs do, and make use of pretraining like DBN. They provide a generic structure which can be used in many image and signal processing tasks. Recently, many benchmark results on standard image data sets like CIFAR [352] have been obtained using CDBNs [353].

## Recurrent Neural Networks

A recurrent neural network (RNN) is a class of ANN where connections between units form a directed cycle. This creates an internal state of the network which allows it to exhibit dynamic temporal behavior. Unlike feedforward neural networks, RNNs can use their internal memory to process arbitrary sequences of inputs. This makes them applicable to tasks such as unsegmented connected handwriting recognition [354] or speech recognition [273].

Numerous researchers now use a deep learning RNN called the long short-term memory (LSTM) network, published by Hochreiter and Schmidhuber in 1997 [271]. It is a deep learning system that, unlike traditional RNNs, doesn't have the vanishing gradient problem (compare the section on training algorithms below). LSTM is normally augmented by recurrent gates called forget gates [298]. LSTM RNNs prevent backpropagated errors from vanishing or exploding [265]. Instead errors can flow backward through unlimited numbers of virtual layers in LSTM RNNs unfolded in space. That is, LSTM can learn "Very Deep Learning" tasks [262] that require memories of events that happened thousands or even millions of discrete time steps ago. Problem-specific LSTM-like topologies can be evolved [355]. LSTM works even when there are long delays, and it can handle signals that have a mix of low- and high-frequency components.

Today, many applications use stacks of LSTM RNNs [356] and train them by Connectionist Temporal Classification (CTC) [272] to find an RNN weight matrix that maximizes the probability of the label sequences in a training set, given the corresponding input sequences. CTC achieves both alignment and recognition. Around 2007, LSTM started to revolutionize speech recognition, outperforming traditional models in certain speech applications [300]. In 2009, CTC-trained LSTM was the first RNN to win pattern recognition contests, when it won several competitions in connected handwriting recognition [262, 357]. In 2014, the Chinese search giant Baidu used CTC-trained RNNs to break the Switchboard Hub5'00 speech recognition benchmark, without using any traditional speech processing methods [302]. LSTM also improved large-vocabulary speech recognition [273, 274] text-to-speech synthesis [358], also for Google Android [262, 275], and photo-real talking heads [358]. In 2015, Google's speech recognition reportedly experienced a dramatic performance jump of 49% through CTC-trained LSTM, which is now available through Google voice search to all smartphone users [359].

LSTM has also become very popular in the field of natural language processing. Unlike previous models based on HMMs and similar concepts, LSTM can learn to recognize context-sensitive languages [318]. LSTM improved machine translation [291], language modeling [319], and multilingual language processing [320]. LSTM combined with CNNs also improved automatic image captioning [306] and a plethora of other applications.

## Comparisons

When comparing the DBN, convolutional neural network, and recurrent neural network, there are some common properties:

1. They have multiple layers (one can think of a recurrent network as having infinite number of layers).
2. They use backpropagation algorithm for training.
3. They can be combined together to create more powerful networks.

However, there are some special properties of each network:

- Convolutional neural networks:
  1. They are more suitable for data with grid structures such as audio/image/video.
  2. They have a much smaller number of parameters compared to other architectures because it uses small convolutional filters. For this reason people almost always use convolutional layers when it comes to image or video processing.
  3. They are very efficient in training and evaluation using GPUs.
- Recurrent networks:
  1. They have memory of the past and are thus capable of dealing with temporal data. They are very useful for tasks in which current output depends on multiple inputs in the past like speech
  2. They are not able to take big input such as images or videos. Often combined with convolutional neural network in computer vision tasks.

- Deep belief networks:
  1. These are a bit more special, being a generative model having multiple feedforward and feedback passes.
  2. Generative model means that it could generate realistic looking data after initializing at random and performing multiple forward and backward passes. One could use this learning architecture to generate handwritten characters, face images, or even poems.
  3. They are used much less than convolutional networks because the training is difficult and inefficient.

## 7.3  Example 1: Apache Spark for MBD

The proliferation of mobile devices, such as smartphones and Internet of Things (IoT) gadgets, resulted in the recent MBD era. Collecting MBD is unprofitable unless suitable analytics and learning methods are utilized for extracting meaningful information and hidden patterns from data. This example presents an overview and brief tutorial of deep learning in MBD analytics and discusses a scalable learning framework over Apache Spark. Specifically, a distributed deep learning is executed as an iterative MapReduce computing on many Spark workers. Each Spark worker learns a partial deep model on a partition of the overall MBD, and a master deep model is then built by averaging the parameters of all partial models. This Spark-based framework speeds up the learning of deep models consisting of many hidden layers and millions of parameters. We use a context-aware activity recognition application with a real-world data set containing millions of samples to validate our framework and assess its speed-up effectiveness.

### 7.3.1  Introduction

Mobile devices have matured as a reliable and cheap platform for collecting data in pervasive and ubiquitous sensing systems. Specifically, mobile devices are (a) sold in mass-market chains, (b) connected to daily human activities, and (c) supported with embedded communication and sensing modules. According to the latest traffic forecast report by Cisco Systems [360], half a billion mobile devices were globally sold in 2015, and the mobile data traffic grew by 74%, generating 3.7 exabytes (1 exabyte = $10^{18}$ bytes) of mobile data per month. MBD is a concept that describes a massive amount of mobile data which cannot be processed using a single machine. MBD contains useful information for solving many problems such as fraud detection, marketing and targeted advertising, context-aware computing, and healthcare. Therefore, MBD analytics is currently a high-focus topic aiming at extracting meaningful information and patterns from raw mobile data.

Deep learning is a solid tool in MBD analytics. Specifically, deep learning (a) provides high-accuracy results in MBD analytics, (b) avoids the expensive design of handcrafted features, and (c) utilizes the massive unlabeled mobile data for unsupervised feature extraction. Due to the curse of dimensionality and size of MBD, learning deep models in MBD analytics is slow and takes anywhere from a few hours to several days when performed on conventional computing systems. Arguably, most mobile systems are not delay tolerant and decisions should be made as fast as possible to attain high user satisfaction.

To cope with the increased demand on scalable and adaptive mobile systems, this example presents a tutorial on developing a framework that enables time-efficient MBD analytics using deep models with millions of modeling parameters. Our framework is built over Apache Spark [361] which provides an open source cluster computing platform. This enables distributed learning using many computing cores on a cluster where the continuously accessed data is cached to running memory, thus speeding up the learning of deep models by several folds. To prove the viability of the proposed framework, we implement a context-aware activity recognition system [362] on a computing cluster and train deep learning models using millions of data samples collected by mobile crowdsensing. In this test case, a client request includes accelerometer signals and the server is programmed to extract the underlying human activity using deep activity recognition models. We show significant accuracy improvement of deep learning over conventional machine learning methods, improving 9% over random forests and 17.8% over multilayer perceptions from [363]. Moreover, the learning time of deep models is decreased as a result of the paralleled Spark-based implementation compared to a single machine computation. For example, utilizing six Spark workers can accelerate the learning of a five-layer deep model of 20 million parameters by four folds as compared to a single machine computing.

The rest of this example is organized as follows. Section 7.3.2 presents an overview of MBD and discusses the challenges of MBD analytics. Section 7.3.3 discusses the advantages and challenges of deep learning in MBD analytics. Then, Section 7.3.4 proposes a Spark-based framework for learning deep models for time-efficient MBD analytics within large-scale mobile systems. Section 7.3.5 presents experimental analysis using a real-world data set. Interesting research directions are discussed in Section 7.3.6.

## 7.3.2    MBD Concepts and Features

This section first introduces an overview of MBD and then discusses the key characteristics which make MBD analytics challenging.

### The Era of MBD

Figure 7.3 (a) shows a typical architecture of large-scale mobile systems used to connect various types of portable devices such as smartphones, wearable computers, and IoT gadgets. The widespread installation of various types of sensors, such as accelerometer, gyroscope, compass, and GPS sensors, in modern mobile devices allows many new applications. Essentially, each mobile device encapsulates its service request and own sensory data in stateless data-interchange structure (e.g., JavaScript object notation (JSON) format). The stateless format is important as mobile devices operate on different mobile operating systems (e.g., Android, iOS, and Tizen). Based on the collected MBD,

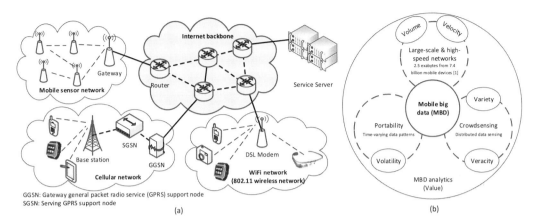

**Figure 7.3** Illustration of the MBD era. (a) Typical architecture of a modern mobile network connecting smartphones, wearable computers, and IoT gadgets. (b) Main technological advances behind the MBD era.

a service server utilizes MBD analytics to discover hidden patterns and information. The importance of MBD analytics stems from its roles in building complex mobile systems that could not be assembled and configured on small data sets. For example, an activity recognition application [362, 364] uses embedded accelerometers of mobile devices to collect proper acceleration data about daily human activities. After receiving a request, the service server maps the accelerometer data to the most probable human activities which are used to support many interactive services (e.g., healthcare, smart building, and pervasive games).

MBD analytics is more versatile than conventional big data problems as data sources are portable and data traffic is crowdsourced. MBD analytics deals with massive amount of data which is collected by millions of mobile devices. Next, we discuss the main characteristics of MBD which complicate data analytics and learning on MBD compared to small data sets.

## Challenges of MBD Analytics

Figure 7.3 (b) shows the main recent technologies that have produced the challenging MBD era: large-scale and high-speed mobile networks, portability, and crowdsourcing. Each technology contributes in forming the MBD characteristics in the following way:

- Large-scale and high-speed mobile networks: The growth of mobile devices and high-speed mobile networks (e.g., WiFi and cellular networks), introduces massive and contentiously increasing mobile data traffic. This has been reflected in the following MBD aspects:
  - *MBD is massive (volume)*. In 2015, 3.7 exabytes of mobile data was generated per month which is expected to increase through the coming years [360].
  - *MBD is generated at increasing rates (velocity)*. MBD flows at a high rate which impacts the latency in serving mobile users. Long queuing time of requests results in less satisfied users and increased cost of late decisions.

- Portability: Each mobile device is free to move independently among many locations. Therefore, *MBD is non-stationary (volatility)*. Due to portability, the time duration for which the collected data is valid for decision-making can be relatively short. MBD analytics should be frequently executed to cope with the newly collected data samples.
- Crowdsourcing: A remarkable trend of mobile applications is crowdsourcing for pervasive sensing which includes a massive data collection from many participating users. Crowdsensing differs from conventional mobile sensing systems as the sensing devices are not owned by one institution but instead by many individuals from different places. This has introduced the following MBD challenges:
  - *MBD quality is not guaranteed (veracity)*. This aspect is critical for assessing the quality uncertainty of MBD as mobile systems do not directly manage the sensing process of mobile devices. Since most mobile data is crowdsourced, MBD can contain low-quality and missing data samples due to noise, malfunctioning or uncalibrated sensors of mobile devices, and even intruders (e.g., badly labeled crowdsourced data). These low-quality data points affect the analytical accuracy of MBD.
  - *MBD is heterogeneous (variety)*. The variety of MBD arises because the data traffic comes from many spatially distributed data sources (i.e., mobile devices). Besides, MBD comes in different data types due to the many sensors that mobile devices support. For example, a triaxial accelerometer generates proper acceleration measurements while a light sensor generates illumination values.

MBD analytics (*value*) is mainly about extracting knowledge and patterns from MBD. In this way, MBD can be utilized for providing better service to mobile users and creating revenues for mobile businesses. The next section discusses deep learning as a solid tool in MBD analytics.

### 7.3.3  Deep Learning in MBD Analytics

Deep learning is a new branch of machine learning which can solve a broad set of complex problem in MBD analytics (e.g., classification and regression). A *deep learning model* consists of simulated neurons and synapses which can be trained to learn hierarchical features from existing MBD samples. The resulting deep model can generalize and process unseen streaming MBD samples.

For simplicity, we present a general discussion of deep learning methods without focusing on the derivations of particular techniques. Nonetheless, we refer interested reader to more technical papers of DBN [293] and stacked denoising autoencoders [365]. A deep model can be scaled to contain many hidden layers and millions of parameters which are difficult to be trained at once. Instead, greedy layer-by-layer learning algorithms [293, 365] were proposed which basically work as follows:

1. *Generative layer-wise pretraining*: This stage requires only unlabeled data which is often abundant and cheap to collect in mobile systems using crowdsourcing. Figure 7.4 shows the layer-wise tuning of a deep model. First, one layer of neurons is trained using the unlabeled data samples. To learn the input data structure, each layer

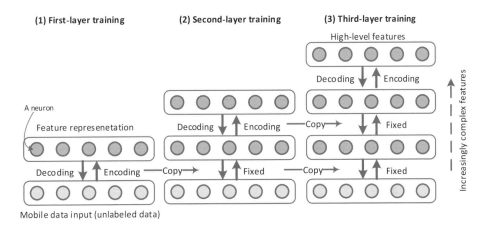

**Figure 7.4** Generative layer-wise training of a deep model. Each layer applies nonlinear transformation to its input vector and produces intrinsic features at its output.

includes encoding and decoding functions: The encoding function uses the input data and the layer parameters to generate a set of new features. Then, the decoding function uses the features and the layer parameters to produce a reconstruction of the input data. As a result, a first set of features is generated at the output of the first layer. Then, a second layer of neurons is added at the top of the first layer, where the output of the first layer is fed as input of the second layer. This process is repeated by adding more layers until a suitable deep model is formed. Accordingly, more complex features are learned at each layer based on the features that were generated at its lower layer.

2. *Discriminative fine-tuning*: The model's parameters which were initialized in the first step are then slightly fine-tuned using the available set of labeled data to solve the problem at hand.

## Deep Learning Advantages in MBD Analytics

Deep learning provides solid learning models for MBD analytics. This argument can be supported with the following advantages of using deep learning in MBD analytics:

- *Deep learning scores high-accuracy results which are a top priority for growing mobile systems.* High-accuracy results of MBD analytics are required for sustainable business and effective decisions. For example, a poor fraud detection results in expensive loss of income for mobile systems. Deep learning models have been reported as state-of-the-art methods in solving many MBD tasks. For example, the authors in [366] propose a method for indoor localization using deep learning and channel state information. In [367], deep learning is successfully applied to inference tasks in mobile sensing (e.g., activity and emotion recognition), and speaker identification.
- *Deep learning generates intrinsic features which are required in MBD analytics.* A *feature* is a measurement attribute extracted from sensory data to capture the underlying phenomenon being observed and enable more effective MBD analytics. Deep

learning can automatically learn high-level features from MBD, eliminating the need for handcrafted features in conventional machine learning methods.

- *Deep Learning can learn from unlabeled mobile data which minimizes the data labeling effort.* In most mobile systems, labeled data is limited, as manual data annotation requires expensive human intervention which is both costly and time consuming. On the other hand, unlabeled data samples are abundant and cheap to collect. Deep learning models utilize unlabeled data samples for generative data exploration during a pretraining stage. This minimizes the need for labeled data during MBD analytics.

- *Multimodal deep learning.* The "variety" aspect of MBD leads to multiple data modalities of multiple sensors (e.g., accelerometer samples, audio, and images). Multimodal deep learning [368] can learn from multiple modalities and heterogeneous input signals.

### Deep Learning Challenges in MBD Analytics

Discussing MBD in terms of volume only and beyond the analytical and profit perspectives is incomplete and restricted. Collecting MBD is unprofitable unless suitable learning methods and analytics are utilized in extracting meaningful information and patterns. Deep learning in MBD analytics is slow and can take a few days of processing time, which does not meet the operation requirements of most modern mobile systems. This is due to the following challenges:

- *Curse of dimensionality*: MBD comes with "volume" and "velocity" related challenges. Historically, data analytics on small amounts of collected data (a.k.a. random sampling) was utilized. Despite the low computational burdens of random sampling, it suffers from poor performance on unseen streaming samples. This performance problem is typically avoided by using the full set of available big data samples which significantly increases the computational burdens.

- *Large-scale deep models*: To fully capture the information on MBD and avoid underfitting, deep learning models should contain millions of free parameters (e.g., a five-layer deep model with 2,000 neurons per layer contains around 20 million parameters). The model free parameters are optimized using gradient-based learning [293, 365] which is computationally expensive for large-scale deep models.

- *Time-varying deep models*: In mobile systems, the continuous adaptation of deep models over time is required due to the "volatility" characteristic of MBD.

To tackle these challenges, we next describe a scalable framework for MBD analytics using deep learning models and Apache Spark.

### 7.3.4     Spark-Based Deep Learning Framework for MBD

Learning deep models in MBD analytics is slow and computationally demanding. Typically, this is due to the large number of parameters of deep models and the large number of MBD samples. Figure 7.5 shows the proposed architecture for learning deep models on MBD with Apache Spark. Apache Spark [361] is an open source platform for scalable MapReduce computing on clusters. The main goal of the proposed framework

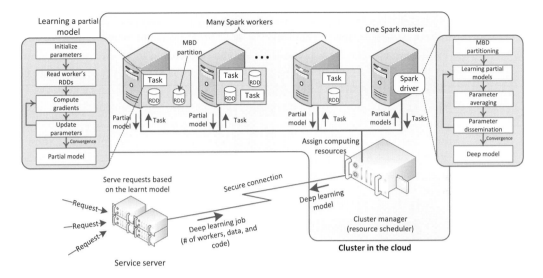

**Figure 7.5** A Spark-based framework for distributed deep learning in MBD analytics.

is speeding up MBD decision-making by parallelizing the learning of deep models to a high-performance computing cluster. In short, the parallelization of a deep model is performed by slicing the MBD into many partitions. Each partition is contained in a resilient distributed data set (RDD) which provides an abstraction for data distribution by the Spark engine. Besides data caching, RDDs of a Spark-program natively support fault-tolerant executions and recover the program operations at worker nodes.

In short, our Spark-based framework consists of two main components: (1) a Spark master and (2) one or more Spark workers. The master machine initializes an instance of the Spark driver that manages the execution of many partial models at a group of Spark workers. At each iteration of the deep learning algorithm (Figure 7.4), each worker node learns a partial deep model on a small partition of the MBD and sends the computed parameters back to the master node. Then, the master node reconstructs a master deep model by averaging the computed partial models of all executor nodes.

## Parallelized Learning Collections

Learning deep models can be performed in two main steps: (1) gradient computation and (2) parameter update (see [293, 365] for the mathematical derivation). In the first step, the learning algorithm iterates through all data batches independently to compute gradient updates (i.e., the rate of change), of the model's parameters. In the second step, the model's parameters are updated by averaging the computed gradient updates on all data batches. These two steps fit the learning of deep models in the MapReduce programming model [369, 370]. In particular, the parallel gradient computation is realized as a Map procedure, while the parameter update step reflects the Reduce procedure. The iterative MapReduce computing of deep learning on Apache Spark is performed as follows:

1. *MBD partitioning*: The overall MBD is first split into many partitions using the *parallelize()* API of Spark. The resulting MBD partitions are stored into RDDs and distributed to the worker nodes. These RDDs are crucial to accelerate the learning of deep models as the memory data access latency is significantly shorter than the disk data operations.

2. *Deep learning parallelism*: The solution of a deep learning problem depends on the solutions of smaller instances of the same learning problem with smaller data sets. In particular, the deep learning job is divided into learning stages. Each learning stage contains a set of independent MapReduce iterations where the solution of one iteration is the input for the next iteration. During each MapReduce iteration, a partial model is trained on a separate partition of the available MBD as follows:

   (a) *Learning partial models*: Each worker node computes the gradient updates of its partitions of the MBD (a.k.a. the Map procedure). During this step, all Spark workers execute the same Map task in parallel but on different partitions of the MBD. In this way, the expensive gradient computation task of the deep model learning is divided into many parallel subtask.

   (b) *Parameter averaging*: Parameters of the partial models are sent to the master machine to build a master deep model by averaging the parameter calculation of all Spark workers (a.k.a. the Reduce procedure).

   (c) *Parameter dissemination*: The resulting master model after the Reduce procedure is disseminated to all worker nodes. A new MapReduce iteration is then started based on the updated parameters. This process is continued until the learning convergence criterion is satisfied.

As a result, a well-tuned deep learning model is generated which can be used to infer information and patterns from streaming requests. In the following, we discuss how the proposed framework helps in tackling the key characteristics of MBD.

## Discussion

The proposed framework is grounded over deep learning and Apache Spark technologies to perform effective MBD analytics. This integration tackles the challenging characteristics of MBD as follows:

- *Deep learning*: Deep learning addresses the "value" and "variety" aspects of MBD. First, deep learning in MBD analytics helps in understanding raw MBD. Therefore, deep learning effectively addresses the "value" aspect of MBD. MBD analytics, as discussed in this example, is integral in providing user-customized mobile services. Second, deep learning enables the learning from multimodal data distributions [368] (e.g., concatenated input from accelerometer and light sensors), which is important for the "variety" issue of MBD.

- *Apache Spark*: The main role of the Spark platform in the proposed framework is tackling the "volume," "velocity," and "volatility" aspects of MBD. Essentially, the Spark engine tackles the "volume" aspect by parallelizing the learning task into many subtasks each performed on a small partition of the overall MBD. Therefore, no single machine is required to process the massive MBD volume as one chunk. Similarly, the Spark engine tackles the "velocity" point through its streaming extensions which enables a fast and high-throughput processing of streaming data. Finally, the "volatility" aspect is addressed by significantly speeding up the training of deep models. This ensures that the learned model reflects the latest dynamics of the mobile system.

The proposed framework does not directly tackle the "veracity" aspect of MBD. This quality aspect requires domain experts to design conditional routines to check the validity of crowdsourced data before being added to a central MBD storage.

## 7.3.5    Prototyping a Context-Aware Activity Recognition Systems

Context-awareness [364, 362] has a high impact on understanding MBD by describing the circumstances during which the data was collected, so as to provide personalized mobile experience to end users (e.g., targeted advertising, healthcare, and social services). A *context* contains attributes of information to describe the sensed environment such as performed human activities, surrounding objects, and locations. A *context learning model* is a program that defines the rules of mapping between raw sensory data and the corresponding context labels (e.g., mapping accelerometer signals to activity labels). This section describes a proof-of-concept case study in which we consider a context-aware activity recognition system (e.g., detect walking, jogging, climbing stairs, sitting, standing, and lying down activities). We use a real-world data set during the training of deep activity recognition models.

### Problem Statement

Accelerometers are sensors which measure proper acceleration of an object due to motion and gravitational force. Modern mobile devices are widely equipped with tiny accelerometer circuits which are produced from electromechanically sensitive elements and generate electrical signals in response to any mechanical motion. The proper acceleration is distinctive from coordinate acceleration in classical mechanics. The latter measures the rate of change of velocity while the former measures acceleration relative to a free fall (i.e., the proper acceleration of an object in a free fall is zero).

Consider a mobile device with an embedded accelerometer sensor that generates proper acceleration samples. Activity recognition is applied to time series data frames which are formulated using a sliding and overlapping window. The number of time-series samples depends on the accelerometer's sampling frequency (in Hertz) and windowing length (in seconds). At time $t$, the activity recognition classifier $f : \mathbf{x}_t \rightarrow \mathcal{S}$ matches the framed acceleration data $\mathbf{x}_t$ with the most probable activity label from the set of supported activity labels $\mathcal{S} = \{1, 2, \ldots, N\}$, where $N$ is the number of supported activities in the activity detection component.

Conventional approaches of recognizing activities require handcrafted features (e.g., statistical features [362]), which are expensive to design, require domain expert knowledge, and generalize poorly to support more activities. To avoid this, a deep activity recognition model learns not only the mapping between raw acceleration data and the corresponding activity label, but also a set of meaningful features which are superior to handcrafted features.

## Experimental Setup

In this section, we use the Actitracker data set [371] which includes accelerometer samples of six conventional activities (walking, jogging, climbing stairs, sitting, standing, and lying down) from 563 crowdsourcing users. Figure 7.6 (a) plots accelerometer signals of the six different activities. Clearly, high-frequency signals are sampled for activities with active body motion (e.g., walking, jogging, and climbing stairs). On the other hand, low-frequency signals are collected during semi-static body motions (e.g., standing, sitting, and lying down). The data is collected using mobile phones with 20Hz of sampling rate, and it contains both labeled and unlabeled data of $2,980,765$ and $38,209,772$ samples, respectively. This is a real-world example of the limited number of labeled data compared with unlabeled data as data labeling requires manual human intervention. The data is framed using a 10-sec windowing function which generates 200 samples of time series samples. We first pretrain deep models on the unlabeled data samples only, and we then fine-tune the models on the labeled data set. To enhance the activity recognition performance, we use the spectrogram of the acceleration signal as input of the deep models. Basically, different activities contain different frequency contents which reflect the body dynamics and movements.

We implemented the proposed framework on a shared cluster system[2] running the load-sharing facility management platform and RedHat Linux. Each node has eight cores (Intel Xeon 5570 CPU with clock speed of 2.93Ghz) and a total of 24GB RAM. In our experiments, we set the cores in multiples of eight to allocate the entire node's resources. One partial model learning task is initialized at each computing core. Each task learns using a data batch consisting of 100 samples for 100 iterations. Clearly, increasing the number of cores results in quicker training of deep models. Finally, it is important to note that distributed deep learning is a strong type of regularization. Thus, regularization techniques, such as the sparsity and dropout constraints, are not recommended to avoid the problem of underfitting.

## Experimental Results

### The impact of deep models

Figure 7.6 (b) shows the activity recognition error under different setups of deep models (number of hidden layers and number of neurons at each layer). Specifically, the capacity of a deep model to capture MBD structures is increased when using deeper models with more layers and neurons. Nonetheless, using deeper models evolves a significant increase in the learning algorithm's computational burdens and time. An accuracy comparison of deep activity recognition models and other conventional methods is shown in Table 7.1. In short, these results clarify that (1) deep models are superior to existing shallow context learning models and (2) the learned hierarchical features of deep models eliminate the need for handcrafted statistical features in conventional methods. In our implementation, we use early stopping to track the model capacity during training, select the best parameters of deep models, and avoid overfitting. The underfitting is typically avoided by using deeper models and more neurons per layer (e.g., 5 layers with

---

[2] www.acrc.a-star.edu.sg.

**Table 7.1** Activity recognition error of deep learning and other conventional methods used in [363]. The conventional methods use handcrafted statistical features.

| Method | Recognition error (%) |
|---|---|
| Multilayer perceptrons | 32.2 |
| Instance-based learning | 31.6 |
| Random forests | 24.1 |
| **Deep learning (5 layers of 2,000 neurons each)** | **14.4** |

2,000 neurons per layer). Next, a speed-up analysis is presented to show the importance of the Spark-based framework for learning deep models on MBD.

*The impact of computing cores*

The main performance metric of cluster-based computing is the task speed-up metric. In particular, we compute the speed-up efficiency as $\frac{T_8}{T_M}$, where $T_8$ is the computing time of one machine with eight cores, and $T_M$ is the computing time under different computing power. Figure 7.6 (c) shows the speed up in learning deep models when the number of computing cores is varied. As the number of cores increases, the learning time decreases. For example, learning a deep model of five layers with 2,000 neurons per layer can be trained in 3.63 hours with six Spark workers. This results in the speed-up efficiency of 4.1 as compared to a single machine computing which takes 14.91 hours.

*MBD veracity*

A normalized confusion matrix of a deep model is shown in Figure 7.7. This confusion matrix shows the high performance of deep models on a per-activity basis (high scores at the diagonal entries). The incorrect detection of the "sitting" activity instead of the "lying down" activity is typically due to the different procedures in performing the activities by crowdsourcing users. This gives a real-world example of the "veracity" characteristic of MBD (i.e., uncertainties in MBD collection).

In the next section, we identify some notable future research directions in MBD collection, labeling, and economics.

## 7.3.6    Future Work

Based on the proposed framework, the following future work can be further pursued.

### Crowd Labeling of MBD

A major challenge facing MBD analysts is the limited amounts of labeled data samples as data labeling is typically a manual process. An important research direction is proposing crowd labeling methods for MBD. The crowd labeling can be designed under two main schemes: (1) paid crowd labeling and (2) embedded crowd labeling. In the paid crowd labeling, the crowdsourcing mobile users annotate mobile data and are accordingly paid based on their labeling performance and speed.

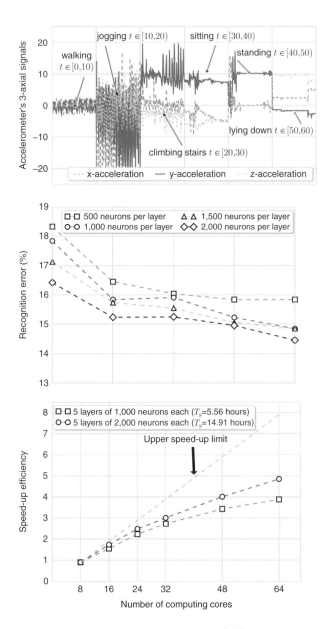

**Figure 7.6** Experimental analysis. (a) Accelerometer signal of different human activities. (b) Recognition accuracy of deep learning models under different deep model setups. (c) Speed-up of learning deep models using the Spark-based framework under different computing cores. The upper speed-up limit is achieved under full CPU utilization and zero communication overhead.

Under this paid scheme, optimal budget allocation methods are required. In the embedded crowd labeling, data labeling can be also achieved by adding labeling tasks within mobile application functional routines (e.g., CAPTCHA-based image labeling [372]). Here, the mobile users can access more functions of a mobile application by indirectly

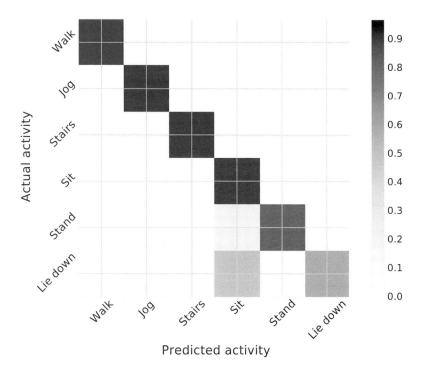

**Figure 7.7** Normalized confusion matrix of a deep model (five layers of 2,000 neurons each).

helping in the data labeling process. More work is required for designing innovative methods for embedded crowd labeling without disturbing the user experience or harming the mobile application's main functionality.

### Economics of MBD

MBD, as discussed in this example, is about extracting meaningful information and patterns from raw mobile data. This information is used during decision-making and to enhance existing mobile services. An important research direction is proposing business models (e.g., pricing and auction design [373]) for selling and buying MBD among mobile organizations and parties.

### Privacy and MBD Collection

As MBD is people-centric, mobile users would be concerned about the risks of sharing their personal mobile data with a service server. Thus, a low percentage of users will opt out of sharing their personal data unless trustworthy privacy mechanisms are applied. Meanwhile, anonymized data collection (i.e., data that could not be used to identify individuals) is adopted by many services. An alternative research direction is proposing fair data exchange models which encourage the sharing of mobile data in return of rewarding points (e.g., premium membership points).

## 7.4     Example 2: User Moving Pattern Extraction

When GPS devices are widely integrated into smartphones, researchers stand a big chance of collecting massive location information that is necessary in studying users' moving behavior and predicting the next location of the users. Once the next location of a user can be determined, it can serve as input for many applications, such as location-based service, scheduling users' access in a mobile network, or even home automation. One important task in predicting the next location is to identify typical users' moving patterns. In this example, we propose a novel method to extract the patterns using deep learning. Experiment results show significant performance improvement of the proposed method compared to the classical principal component analysis method.

### 7.4.1     Introduction

Predicting user movement has been always an interesting topic for researchers in the past decades. However, due to the lack of smart hand-held devices, it has never surged until smartphone with integrated GPS was invented recently. There are many works which focus on location prediction, however, most of them are based on the Markov model [374, 375, 376]. The Markov model is powerful, but it poses some problems such as expressing different patterns over different times or exploiting embedded frequencies in human behavioral patterns [377]. For example, at 6 am during the weekdays, an office worker normally goes to his office, but on the weekend, at the same time, he more likely stays at home. The Markov model treats these two situations in the same way, the probability of going to the office when the user is at home at 6 am during the weekends is as high as that probability during the weekdays. Hence, if typical users' moving patterns can be extracted, it will be useful to combine the Markov model with different moving patterns for different days. The proposed approach can also serve as a stand-alone solution if we want to predict the next possible visiting locations in the rest of the day given location information from midnight to noon [378].

Existing efforts in identifying users' moving patterns can be divided into two major approaches. The first approach is from frequentist statistics [379, 380], in which the most frequent moving sequences of visiting places are extracted. Nevertheless, this approach lacks flexibility. For example, when a user stops by a handy mart on the way going from home to office, the user breaks the stored moving sequence (e.g., [Home → Office]) and consequently, the system fails to predict the next place. In addition to that, locations are only considered to be in the same sequence if and only if they are in the same time slot [380]. This condition does not fit the real scenarios; many factors can affect users' schedules.

The other approach is based on principal component analysis (PCA) to extract eigenbehaviors [378] or eigenplaces [381, 382]. An eigenbehavior is a principal component of our daily moving patterns. In [378], three principal patterns were extracted, including weekday, weekend, and no-signal patterns using PCA. The authors then tried to predict future locations given the observed data and obtained promising experiment results. However, compared to deep learning, which has multiple layers, PCA has a huge disadvantage of shallow structure since it consists of only two layers. With multiple layers, deep learning is more powerful and flexible since it is able to combine many layers to generate observations. In addition to that, only three eigenbehaviors might not be sufficient enough to construct different users' movement patterns. We believe that the results can be further improved by extracting more typical patterns and applying a "deeper" structure.

From the human point of view, given the users' daily trajectories, we can tell approximately how many typical moving patterns there are. Our brain can extract multilayer representations of the trajectories, and at the highest abstract level it concludes a number of typical moving patterns. The question is how to reverse engineer the human brain which has multiple layers of expressions. Theoretical results suggest that we should use a deep structure to model this complicated extraction process [344, 383]. Nevertheless, there was not any closed-form solution for inference in deep learning due to the complicated structure and the lack of an efficient algorithm to set up initial weights until recently. Hinton et al. [384] proposed a method to pretrain the model layer by layer and set the prior based on existing knowledge to cancel out the explain-away effect. The DBN proves to outperform shallow architectures such as PCA in many applications [384]. We distinguish our method from the others in the following aspects.

First, to the best of our knowledge, it is the first attempt to apply deep learning in extracting users' typical moving patterns. Since deep learning outperforms PCA by far, we expect to achieve a better result compared to the result in [378]. Second, in our approach, to extract the patterns, users do not have to manually label their logical places, which was a requirement in [378]. And finally, we apply deep learning to some collected users' trajectories and prove that the deep learning can reconstruct the trajectories with much less error compared to the classical PCA.

The rest of the example is organized as follows. In Section 7.4.2, the process of collecting and representing the trajectories to feed to the input of the deep learning is described. In Section 7.4.3, we discuss the deep learning and its basic components, the RBM. Experiment results are shown in Section 7.4.4.

## 7.4.2 Data Collection

An Android application was written and installed on the HTC Wildfire S smartphones to collect GPS signal every 30 seconds. The data consists of GPS coordinates and recorded timestamps. A group of volunteers living in different cities were chosen to bring the phones on their normal daily routines. The phone GPS is always turned on and locations are collected periodically when the phone is not connected to the wall charger. Data is then automatically uploaded to a Dropbox account at the end of the day. A trajectory of a user collected in 38 days is shown on Figure 7.8.

Based on the timestamps, GPS data in the same day is grouped together to form a data set. The area that includes all the GPS measurements shown on Figure 7.8 is divided into 784 small cells, corresponding to a $(28 \times 28)$ pixels image. One example of a trajectory for a day is illustrated on Figure 7.9 where the raw trace on the left is converted to a binary image on the right. For now, we assume the image pixel values are binary. Within

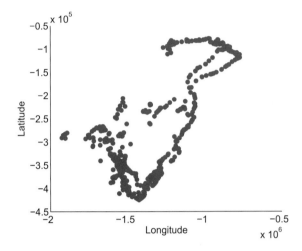

**Figure 7.8** Six-week trajectory.

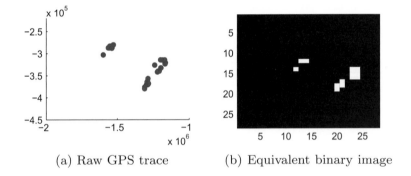

(a) Raw GPS trace    (b) Equivalent binary image

**Figure 7.9** Converting raw trace to binary image.

a day, if the user shows up at a specific cell, then the corresponding cell on the image is assigned a value of 1. Otherwise, the cell value remains at 0. Furthermore, the image is reshaped to create a 1x784 vector. This vector will serve as the input for the deep learning described in the next section.

### 7.4.3    Deep Learning Extraction

Deep learning is a generative model that consists of multiple layers of hidden stochastic latent variables or features. In this example, we use a variant of deep learning called Deep Autoencoder (DAE) [384]. DAE stacks multiple layers of RBM, thus, in the sequel, the generative model of DAE is introduced.

**Restricted Boltzmann Machines**

The Boltzmann machine is a parallel computational model. Units in the Boltzmann machine are stochastic and have energy assigned to them. The RBM is a restricted version of the Boltzmann machine. It does not have any intralayer connection between the hidden nodes, which leads to possible inference algorithms. RBM is an undirected graph that consists of two layers: a hidden layer and a visible layer as well as one bias unit that is always on. Hidden nodes are not connected and are conditionally independent given the observations. This important property makes inference tractable in the RBM. The RBM is illustrated on Figure 7.2, where $v$ is the visible unit or observation and $h$ represents hidden units.

For illustration purpose, we assume that both the observations and the hidden nodes are binary. The activation energy of unit $i$ is

$$E_i = \sum_j w_{ij} x_j, \tag{7.15}$$

where the index of the sum runs over all units connected to hidden unit $i$, $x_j$ represents the $j$th observation and $w_{ij}$ is the weight of the connection between the hidden unit and the observation unit. With activation energy $E_i$, the probability to turn unit $i$ on is a logistic sigmoid function:

$$p_i = 1/[1 + \exp(-E_i)]. \tag{7.16}$$

The larger the value of the activation energy, the more likely unit $i$ will be turned on. The weight of a connection represents a mutual connection between the two units. For example, if both units are very likely to be on or off at the same time, then the weight between them should be positive and large. On the other hand, if they are very unlikely to have the same value then the weight should be small and negative. For example, in our application, we have seven typical patterns or, equivalently, seven hidden units on the top layer. The hidden units are corresponding to seven days in a week and are the output of the algorithm. It is essential to mention that each hidden unit is not a trajectory. It is instead only one hidden node with an activation energy, which is proportional to the probability that the unit is turned on, and a set of related weights. Nevertheless, if the hidden unit is turned on, it can generate a full trajectory by following the weights and activation energies of the lower level. If we apply a trajectory collected on Monday to the visible unit, it is very likely that the Monday hidden unit at the top layer will be turned on and the weight of the connection between them is high.

If we name the vector of visible nodes $\mathbf{v}$ and the vector of hidden nodes $\mathbf{h}$, then the energy of any joint configuration $(\mathbf{v}, \mathbf{h})$ is [293]

$$E(\mathbf{v}, \mathbf{h}) = -\sum_{i \in \mathbf{v}} a_i v_i - \sum_{j \in \mathbf{h}} b_j h_j - \sum_{i,j} v_i h_j w_{ij}, \tag{7.17}$$

where $a_i, b_i$ are the bias associated with visible units and hidden units, respectively, and $w_{ij}$ is the weight between a hidden unit $j$ and a visible unit $i$. The probability of any configuration $(\mathbf{v}, \mathbf{h})$ is

$$p(\mathbf{v}, \mathbf{h}) = \frac{\exp[-E(\mathbf{v}, \mathbf{h})]}{Z}, \tag{7.18}$$

where $Z$ is a normalizing factor to limit the probability in the range of $[0, 1]$. Given the join probability of the configuration, it is easy to find the marginal probability of a set of visible units:

$$p(\mathbf{v}) = \frac{\sum_h \exp[-E(\mathbf{v},\mathbf{h})]}{Z}. \tag{7.19}$$

This probability represents the chance that the model with the determined values of weights will generate the observations set $\mathbf{v}$.

In the training phase, it is expected that the probability is maximized by varying the weights. Thus, the optimum weights can be obtained by taking the derivative of the probability with respect to the weights:

$$\frac{\partial \log p(\mathbf{v})}{\partial w_{ij}} = < vi.hj >_{data} - < vi.hj >_{model}, \tag{7.20}$$

where the brackets $< . >_{data}$ stand for the expectation of the product of the weights and the observed data and $< . >_{model}$ is the same expectation for the model observations that is generated according to the model. Obviously, when the derivative is 0, or in other words when the training data and generated data are similar, we obtain the optimum performance. Hence, the learning algorithm is simply to update the weights with the value in (7.20):

$$\Delta w_{ij} = \epsilon(< vihj >_{data} - < vihj >_{model}), \tag{7.21}$$

where $\epsilon$ is a learning rate. The bigger $\epsilon$ is, the faster the algorithm converges, but the coarser grain of the optimal values the algorithm can achieve. For this reason, $\epsilon$ should be chosen conservatively to achieve the best performance. Experiment results suggest that the value of $\epsilon$ should be set at $0.1$.

### A Deep Autoencoder for Typical Users' Moving Patterns Extraction

A DAE is an artificial neural network used to learn multilayer representations of an input data. Figure 7.10 describes the DAE for our application. It consists of four hidden layers. The lowest hidden layer has 500 hidden units, the next one has 400, the third layer has 300, and the top layer has seven hidden units. Except for the top layer, all the other hidden layers are binary. 0 means the hidden unit is turned off while 1 means the hidden unit is turned on. The top layer has only seven hidden units since human has seven typical moving patterns corresponding to seven days in a week.

Our purpose is to find the weights that minimize the error. However, optimization is over-sensitive to initial chosen weights values. The performance is very poor if a large initial weights are chosen, while with small initial weights, optimization is intractable. Recently, Hinton et al. [384] have shown that by pretraining the network layer by layer, we can obtain a good initial weight values to minimize network errors. After that, a fine-tuning phase is applied to find the local maxima. In the next two subsections, we will describe shortly these two phases.

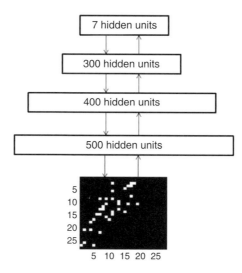

**Figure 7.10** A Deep Autoencoder Network for a user trajectory.

## Pretraining Phase

Figure 7.11 shows the pretraining phase that include two steps. At the first step shown on Figure 7.11(a), for each layer, the hidden units are separated from their upper level hidden units and connected to the lower-level hidden units to form an RBM. Hidden units in every RBM are binary except for the top layer, where the hidden units are drawn from a Gaussian distribution with the mean equal to the activation probability. The binary input of the trajectories is fed as the input of the first RBM. As described in the previous section, the activation probability can be calculated according to (7.16). In turn, the hidden units of the first RBM now become visible units in the second RBM. The same process is repeated until the activation probabilities of the top-layer hidden units are determined.

The step described above is a bottom-up process to achieve activation probabilities for all the hidden units. Now, given the probabilities and the weights, we implement the second step, shown in Figure 7.11(b). In this step, a process from top to bottom is followed to generate the model "visible" units at the bottom layer. The difference between the model "visible" data and the actual data gives us a hint to update the weights according to (7.21). In Figure 7.11(b), the top trajectory is created by the model while the bottom trajectory is the training data.

After this step, we obtain coarse grain optimum values for the weights. To further improve the result, a fine-tuning process is implemented using backpropagation in the next subsection.

## Fine-Tuning Phase

Backpropagation is a supervised method to train deep learning. Since the top hidden units have been learned from the above pretraining phase, we first forward the training data to the input of the network and calculate the difference between the inferred

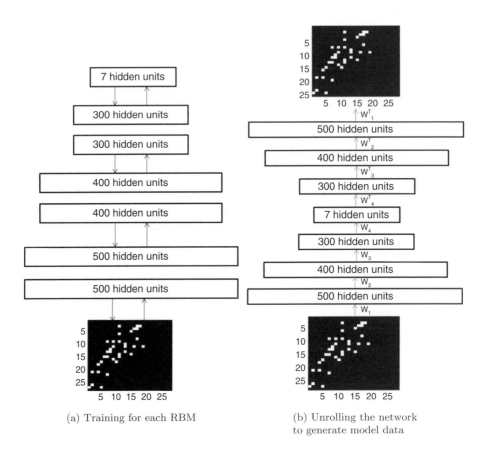

(a) Training for each RBM

(b) Unrolling the network to generate model data

**Figure 7.11**  Pretraining phase.

top-layer hidden unit and the learned top-layer hidden unit. The difference at the top layer is then fed backward to go through the network from the top to the bottom layer using the learned weights, producing differences at every layer. The differences at each layer are later used to update the weights of the corresponding hidden layers. The above process is repeated until the differences are under some threshold or the algorithm reaches its maximum number of iterations.

In this example, we use a backpropagation method called conjugate gradients. Basic backpropagation methods update the weights in the steepest descent direction. While it is sufficient enough to find minima, it is proved that the convergence time is not optimum since the direction of the deepest descent direction may circle around toward a global minima. The conjugate gradients method searches for different directions that are pedicular or conjugate to each other so that optimization in one direction does not affect the optimization in other directions. By searching in different directions, the algorithm can reach the minima much faster. We based on the code provided by [384] to perform the algorithm.

## Typical Pattern Extraction Algorithm

Algorithm 8 summarizes our algorithm to extract typical patterns. Weights are initialized by randomly choosing values from the range of $[-0.1, 0.1]$. Then, the pretraining phase is implemented to find initial weights for the next phase, the fine-tuning phase. After learning all the weights and the typical patterns, a test trace is fed to the network and a corresponding model trace is then regenerated. The difference between the test trace and the regenerated trace is calculated to measure the error, which is numerically described in the next section.

---

**Algorithm 8** Extracting typical patterns algorithm.

---

Initialize: Weights are chosen randomly in the range from [-0.1, 0.1];

**for** $j \leftarrow 1$ to $M$, the number of pretraining iterations **do**

+ Train each RBM from bottom to top

+ Unroll the network to generate data

+ Calculate the difference between the original trajectory and the model trajectory

+ Update the weights

**end**

**for** $j \leftarrow 1$ to $N$, the number of fine-tuning iterations **do**

+ Fine-tune the whole deep network by setting initial weights to the weights obtained from the pretraining phase

**end**

+ Store the top hidden units, which are the typical patterns

---

## 7.4.4    Experiment Results

In this section, we will evaluate the performance of deep learning in reconstructing user trajectory. A trajectory of 38 days from a graduate student was selected, from which 30 days are used for training and the other 8 days are used for testing. After the learning process, weights are learned and then, a test trajectory is applied to the input of the network and used to reconstruct a model trajectory. We will prove that with the learned weights of the hidden units, we can successfully reconstruct the tested trajectories and compare the result with PCA based on the mean square error. The mean square errors of the two algorithms are calculated as the difference between the pixels of the reconstructed trajectory and the input trajectory.

Figure 7.12 shows an input trajectory and its corresponding reconstructed one by deep learning. Notice that the number of dimensions is reduced from $28 \times 28$ to 7. Define the compression ratio as the ratio of the number of the dimensions originally needed to describe the trajectory to the number of dimensions extracted at the top level. We can see that even at a large compression ratio, deep learning performance is still noticeable, especially when compared with PCA. The reconstructed trajectory is not deviated far away from the original one. The mean square error between the two trajectories is 5.3127.

To compare the result with PCA, the same set of training and testing data are used. Seven principal components are extracted to reconstruct the trajectories. As illustrated on Figure 7.13, PCA has much worse performance compared to deep learning. The mean square error of the PCA method is $6.282 * 10^3$, compared to 5.3127 if reconstructed using the deep learning method. Obviously, deep learning beats PCA by far.

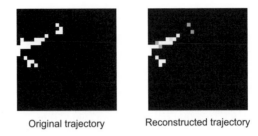

Original trajectory          Reconstructed trajectory

**Figure 7.12**  Reconstructed trajectory using deep learning.

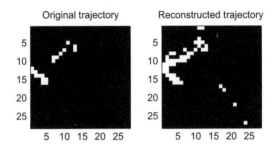

**Figure 7.13**  Reconstructed trajectory using PCA.

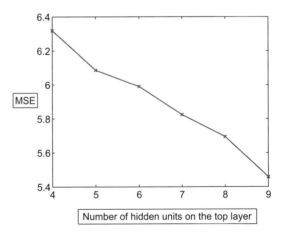

**Figure 7.14**  Varying the number of hidden units on the top layer.

The noticeable improvement in performance that the deep learning can achieve is due to its deep structure while PCA only has a shallow structure of two layers.

We further test the algorithm with various hidden units on the top layer and show the result on Figure 7.14. As we can observe, when the number of hidden units in each layer increases, the performance also increases. Obviously, when the number of hidden units increases, the deep network is more flexible and powerful since it has more branches to generate different sets of trajectories. However, the complexity increases at the same time as a trade-off. Hence, depending on applications, the deep network can be changed to meet both the performance requirements and complexity requirements.

**Possible Future Applications**

In our preliminary work, we only extract typical users' moving patterns. However, many applications can be developed based on our result. For example, at the beginning of the day, we can use the learned extracted moving patterns to generate a set of the most possible locations that a user may visit during the day. In combination with signal strength profile collected at the locations, we can effectively schedule users' access to the network. Delay-tolerant packet transferring can be delayed until the time the user has a strong signal strength instead of transferring the packets immediately.

Another application is to predict next possible locations given the locations observed from beginning of the day until the current time. As soon as new data are collected, they are used to feed to the deep network to learn the weights of the typical patterns. The weights are then matched with the stored weights of different days and the most similar moving pattern will be selected to generate possible future locations for the rest of the day. This application is specifically tested by Eagle et al. in [378]. Since deep learning significantly outperforms PCA, which is used to extract typical patterns in [378], we would believe that our proposed algorithm can significantly improve their results.

## 7.5 Example 3: Combination with Nonparametric Bayesian Learning

In this example, we present an infinite hierarchical nonparametric Bayesian model to extract the hidden factors over observed data, where the number of hidden factors for each layer is unknown and can be potentially infinite. Moreover, the number of layers can also be infinite. We construct the model structure that allows continuous values for the hidden factors and weights, which makes the model suitable for various applications. We use the Metropolis-Hastings (MH) method to infer the model structure. Then the performance of the algorithm is evaluated by the experiments. Simulation results show that the model fits the underlying structure of simulated data.

### 7.5.1 Introduction

Statistical models have been applied to the classification problems in machine learning and data analysis [385]. Some statistical methods make the hypothesis of mathematical models that are controlled by certain parameters to fit the latent structure of observed data [386]. The observed data are assumed to be generated by complex structures which have hierarchical layers and hidden causes [387]. One key challenge is thus the determination of the numbers of layers and hidden variables. However, it is sometimes impractical and challenging to choose any fixed number for the model structure. Therefore, we need flexible nonparametric models that make fewer assumptions and are capable of an unlimited amount of latent structures. The hierarchical nonparametric

Bayesian model [388] assumes unspecified number of latent variables and produces rich kinds of probabilistic structures by constructing cascading layers.

In [389], a nonparametric Bayesian model was proposed with both hidden factors and linking weights being binary. The model accommodates a potentially infinite number of hidden factors and performs well in inferring stroke localizations. In [390], the Indian Buffet Process (IBP) was introduced into factor analysis and enabled their model of handling the infinite case with real-valued weights and factors. In the application realm, nonparametric Bayesian model has been explored to solve various classification and clustering problems. Although the nonparametric Bayesian technique has been advanced by researchers recently, challenges still lie in the problem of constructing the real-valued nonlinear models with the numbers of both hidden layers and hidden factors being infinite.

In this example, we investigate the nonparametric Bayesian model with infinite hierarchical hidden layers and an infinite number of hidden factors in each layer. Our main contributions include: the proposition of the infinity structure both latently and hierarchically; the linking weights are extended from binary values to real values; the proposed model is constructed in a nonlinear fashion; and the employment of the MH algorithm enables the alternative update of the values of hidden factors layer by layer, making the inference procedure recursively. The phantom data are simulated according to our infinite generative model. The inferring algorithm is then applied to the simulated data to extract the data structure. Considering wireless security circumstances, the applications we mainly focus on turn to be the clustering problems. Therefore, the most interest lies in the number of hidden factors which indicates the number of clusters in different hierarchical levels. The simulation results show that this greedy algorithm accomplishes the objective of discovering the number of hidden factors accurately.

## 7.5.2　Generative Model

The objective is to construct a hierarchical Bayesian generative model which allows both infinite layers and infinite components in each layer. To better explain the proposed model, we describe the finite generative model first. Then the infinite model can be obtained by extending the number of hidden factors and the number of layers to infinity.

### Finite Generative Model

Finite generative model is used to model the causal effects among the factors between layers, as those described in [389]. Here we construct the model of one observation layer and two hidden layers. Define the matrix $\mathbf{X} = [\mathbf{x}_1, \mathbf{x}_2, \ldots, \mathbf{x}_T]$ as the data set of $T$ data points with each $\mathbf{x}_t$ being a vector of $N$ dimensions. Accordingly, define the matrix $\mathbf{Y}^1 = [\mathbf{y}_1^1, \mathbf{y}_2^1, \ldots, \mathbf{y}_T^1]$ as the hidden factors of first hidden layer with each $\mathbf{y}_t^1$ being a vector of $K^1$ dimensions. Similarly, we have the definition for the $K^2 \times T$ matrix $\mathbf{Y}^2$ as the hidden factors of the second hidden layer. To express the dependency between two successive layers, we use the $N \times K^1$ weight matrix $\mathbf{W}^1$ and $K^1 \times K^2$ weight matrix $\mathbf{W}^2$, respectively. For instance, if there exists a connection between $\mathbf{Y}_{k,t}^1$ and $\mathbf{X}_{n,t}$, which means the hidden cause $\mathbf{Y}_{k,t}^1$ will influence the generation of data component $\mathbf{X}_{n,t}$, then

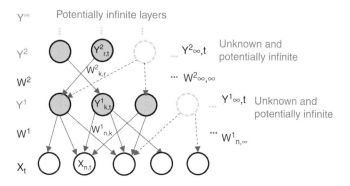

**Figure 7.15** Proposed infinite generative model.

$\mathbf{W}_{n,k}^1 \neq 0$ and $\mathbf{W}_{n,k}^1 \in \mathfrak{R}$. Otherwise, $\mathbf{W}_{n,k}^1 = 0$. The rest hidden vectors $\{\mathbf{y}^i\}$ and weight matrices $\{\mathbf{W}^i\}$ can be derived in the similar way. Figure 7.15 illustrates the proposed infinite generative model structure for a particular instance $t \in \{1, 2, \dots, T\}$.

Within one particular instance $t$, data vector $\mathbf{x}_t$ is generated as follows: First, the hidden vector $\mathbf{y}_t^i$ of the topmost layer is generated according to Gaussian distribution $N(0, \sigma_{y,i}^2)$. Then the weight matrix $\mathbf{W}^i$ is generated according to $\mathbf{W}^i = \mathbf{Z}^i \odot \mathbf{G}^i$, where matrices $\mathbf{Z}^i$ and $\mathbf{G}^i$ are of the same size as $\mathbf{W}^i$ and the symbol $\odot$ indicates the element-wise product operator. We assume each column of matrix $\mathbf{Z}^i$, $\mathbf{Z}_{.,r}^i$, is generated independently as $\mathbf{Z}_{.,r}^i \sim Bernoulli(p_r)$. We further impose a prior distribution for the parameter $p_r \sim Beta(\alpha_i'/K^i, 1)$. It will be demonstrated later that this strategy of constructing matrix $Z^i$ will result in the Indian Buffet Process as the number of variables of $\mathbf{y}_t^i$, $K^i$, approaches the infinity [391]. For the matrix $\mathbf{G}^i$, each column is generated by Gaussian distribution $\mathbf{G}_{.,r}^i \sim N(0, \sigma_r^2)$ with the variance conforming to the inverse gamma prior, $\sigma_r^2 \sim InverseGamma(\alpha_2, \beta_2)$. The matrix $\mathbf{Z}^i$ imposes the selection effect of variables between layers while the matrix $\mathbf{G}^2$ indicates how much influence a variable will receive from its higher-level variables or ancestors. Having obtained the hidden vector $\mathbf{y}_t^i$ and weight matrix $\mathbf{W}^i$, the variables of $\mathbf{y}_t^{i-1}$ are conditionally independently generated given $\mathbf{y}_t^i$ and $\mathbf{W}^i$, and we assume they follow the Gaussian distribution $Y_{k,t}^{i-1} \sim N(0, \sigma_{y,i-1}^2)$, where the parameter $\sigma_{y,i-1}$ is specified by $\sigma_{y,i-1} = \left| \sum_{j=1}^{K^i} W_{k,j}^i Y_{j,t}^i \right|$. It can be verified that the element of weight matrix $\mathbf{W}^i$ follows the distribution:

$$
\begin{aligned}
P(W_{k,r}^i | p_r, \sigma_r^2) = & \left| sgn(W_{k,r}^i) \right| p_r N(W_{k,r}^i; 0, \sigma_r^2) \\
& + (1 - p_r) \delta_0(W_{k,r}^i),
\end{aligned}
\tag{7.22}
$$

where $sgn$ indicates the sign function and symbol $\delta_0$ is a delta function at 0. The downward layers are constructed in the same fashion.

The generation of variables from the first hidden layer $\mathbf{y}_t^1$ to the observed layer $\mathbf{x}_t$ is similar to the procedure above, except for the parameterizations: We assume $\mathbf{Z}_{.,k}^1 \sim Bernoulli(p_k)$ and $p_k \sim Beta(\alpha_1'/K, 1)$ for the matrix $\mathbf{Z}^1$. For the matrix $\mathbf{G}^1$, $\mathbf{G}_{.,k}^1 \sim N(0, \sigma_k^2)$ and $\sigma_k \sim InverseGamma(\alpha_1, \beta_1)$. Hence, the distribution of observed data vector can be expressed as $\mathbf{X}_{n,t} \sim N(0, \sigma_{xn}^2)$ and $\sigma_{xn} = \left| \sum_{j=1}^{K} \mathbf{W}_{n,j}^1 \mathbf{Y}_{j,t}^1 \right|$, where

$$P(\mathbf{W}_{n,k}^1 | p_k, \sigma_k^2) = \left| sgn(\mathbf{W}_{n,k}^1) \right| p_k N(\mathbf{W}_{n,k}^1; 0, \sigma_k^2)$$
$$+ (1 - p_k)\delta_0(\mathbf{W}_{n,k}^1). \tag{7.23}$$

### Infinite Generative Model

The case of an infinite number of layers can be expressed by the recursive equations:

$$\sigma_{y^i j} = \left| \sum_l \mathbf{W}_{j,l}^{i+1} \mathbf{Y}_{l,t}^{i+1} \right|, \tag{7.24}$$

$$\mathbf{Y}_{j,t}^i \sim N(0, \sigma_{y^i j}^2). \tag{7.25}$$

The infinite number of components in each layer can be obtained by taking the limit as $K^i \to \infty$. For example, for our assumptions on $\mathbf{Z}^1$, we have:

$$P(\mathbf{Z}^1 | \mathbf{p}) = \prod_{k=1}^{K} \prod_{n=1}^{N} P(\mathbf{Z}_{n,k}^1 | p_k) = \prod_{k=1}^{K} p_k^{m_k} (1 - p_k)^{N - m_k}, \tag{7.26}$$

where $m_k = \sum_{n=1}^{N} Z_{n,k}^1$ is the number of data points that select hidden factor $\mathbf{Y}_k^1$, and $\mathbf{Z}^1$ is a $N \times K$ matrix. Since we place a prior distribution $Beta(\alpha_1'/K^1, 1)$ on $p_k$, we can integrate out the parameter $\mathbf{p}$ to obtain:

$$P(\mathbf{Z}^1) = \prod_{k=1}^{K} \frac{\frac{\alpha_1'}{K} \Gamma(m_k + \frac{\alpha_1'}{K}) \Gamma(N - m_k + 1)}{\Gamma(N + 1 + \frac{\alpha_1'}{K})}. \tag{7.27}$$

By defining the equivalent class of matrix $\mathbf{Z}^1$ [391], we can find the distribution on $\mathbf{Z}^1$ as $K \to \infty$:

$$P(\mathbf{Z}^1) = \frac{\alpha_1'^{K_h}}{\prod_{n=1}^{N} K_1^n!} e^{-\alpha_1' H_N} \prod_{k=1}^{K_h} \frac{(N - m_k)!(m_k - 1)!}{N!}, \tag{7.28}$$

where $K_1^n$ is the number of first hidden-layer factors being selected by the $n$th variable of the data point $X_{.,t}$, $H_N$ is the harmonic number with $H_N = \sum_{j=1}^{N} \frac{1}{j}$, and $K_h$ is the number of first hidden-layer factors selecting $h$ components of the data point. This distribution corresponds to a stochastic process, the IBP [391].

### 7.5.3    Inference Algorithm

Having constructed the infinite generative model, the goal is to infer $\{\mathbf{W}^1, \mathbf{W}^2, \ldots, \mathbf{Y}^1, \mathbf{Y}^2, \ldots\}$ given observed data $\mathbf{X}$. However, direct estimation of $P(\mathbf{W}^1, \mathbf{W}^2, \ldots, \mathbf{Y}^1, \mathbf{Y}^2, \ldots | \mathbf{X})$ is intractable. Inspired by [293], we perform the inference one layer at a time. Based on this scenario, we use the MH algorithm as an approximate method to infer the first hidden layer $\{\mathbf{Y}^1, \mathbf{W}^1\}$. After inferring the first hidden layer, we use matrix $\mathbf{Y}^1$ as the input data points and perform Bayesian inference at the second hidden layer, and so forth. Since the prior information has been changed during the inference of second hidden layer, we need to re-infer the first hidden layer using the updated upper hidden layers. We iteratively perform inference at each layer until the value of $\{\mathbf{W}^1, \mathbf{W}^2, \ldots, \mathbf{Y}^1, \mathbf{Y}^2, \ldots\}$ converges. It has been proved that this layer-wise inferring strategy is efficient in [392]. Different from [392], the MH algorithm is applied to perform the inference, instead of the CD method.

The MH algorithm defines a Markov chain which allows the change of dimensionality between different states of the model. The new state is generated from the previous state by first generating a candidate state using a specified proposal distribution. Then a decision is made to accept the candidate state or not, based on its probability density relative to that of the previous state, with respect to the desired invariant distribution, $Q$. If the candidate state is adopted, it evolves as the next state of the Markov chain; otherwise, the state of the model stays the same. To better explain the inference algorithm, we specify the problem into one hidden-layer inference. The generalized infinite case can be derived in a similar fashion. In our problem settings, let $\eta$ represent the values of $\mathbf{W}^1, \mathbf{Y}^1, K^1$, where $\mathbf{W}^1$ is the weight matrix connecting the $N \times T$ data matrix $\mathbf{X}$ and the $K^1 \times T$ hidden factors matrix $\mathbf{Y}^1$ and $K^1$ is the dimension of hidden factor $\mathbf{y}_t^1$. Then the change between different states of the model is adopted with probability

$$A(\eta^*, \eta) = min \left[ 1, \frac{P(\mathbf{X}, \eta^*)}{P(\mathbf{X}, \eta)} \frac{Q(\eta|\eta^*)}{Q(\eta^*|\eta)} \right], \tag{7.29}$$

where $\eta^*$ is the proposed new value, $\eta$ is the current value, and $Q(\eta^*|\eta)$ is the probability of proposing $\eta^*$ given $\eta$. The term $P(\mathbf{X}, \eta)$ can be further expressed as

$$P(\mathbf{X}, \eta) = P(\mathbf{X}|\mathbf{W}^1, \mathbf{Y}^1)P(\mathbf{Y}^1|K^1)P(\mathbf{W}^1|K^1)P(K^1). \tag{7.30}$$

The change of dimensionality is completed in this way: Iteratively pick a hidden factor with corresponding column $k$ of $\mathbf{W}^1$ and check the number of linked edges $m_k$. If $m_k = 0$, then remove this hidden factor together with the corresponding column in $\mathbf{W}^1$ and decrease $K^1$. Otherwise, propose a new hidden factor with no linked edges and sample the new values of $\mathbf{Y}^1$ by (7.25).

This new proposed state is accepted with the probability $A(\eta^*, \eta)$. The probability of adding a new hidden factor is approximated by $K_+^1/K^1$ while the probability of generating the new $\mathbf{Y}^1$ is specified by its normal distribution. $Q(\eta^*|\eta)$ is obtained by multiply these two probabilities. To return to the previous configuration, we can delete any hidden factor with the same values as the proposed new row of $\mathbf{Y}^1$. The probability of choosing such a hidden factor is approximated by $1/(K^1 + 1)$. Therefore, we have

$$\frac{Q(\eta|\eta^*)}{Q(\eta^*|\eta)} = \frac{1/(K^1 + 1)}{\frac{K_+^1}{K^1} \prod_t \frac{1}{\sqrt{2\pi}\sigma_{yk}} e^{\frac{y_{k,t}^2}{2\sigma_{yk}^2}}}, \tag{7.31}$$

$$\frac{P(\mathbf{X}, \eta^*)}{P(\mathbf{X}, \eta)} = \frac{\prod_t \frac{1}{\sqrt{2\pi}\sigma_{yk}} e^{\frac{y_{k,t}^2}{2\sigma_{yk}^2}} P(\mathbf{W}^1|K^1 + 1)P(K^1 + 1)}{P(\mathbf{W}^1|K^1)P(K^1)}, \tag{7.32}$$

where $\frac{P(\mathbf{W}^1|K^1+1)}{P(\mathbf{W}^1|K^1)}$ is just the probability of generating a new column of $\mathbf{Z}^1$ with all zero values, specified by (7.23). And $\frac{P(K^1+1)}{P(K^1)}$ can be computed from Poisson distributions as the priors of IBP. As a result, we have

$$A(\eta^*, \eta) = min \left[ 1, \frac{\frac{1}{K^1+1}P(\mathbf{W}^1|K^1+1)P(K^1+1)}{\frac{K^1_+}{K^1}P(\mathbf{W}^1|K^1)P(K^1)} \right]. \tag{7.33}$$

Similarly, the proposal to delete a hidden factor with no linked edges is accepted with the probability

$$A(\eta^*, \eta) = min \left[ 1, \frac{\frac{1}{K^1+1}P(\mathbf{W}^1|K^1-1)P(K^1-1)}{\frac{K^1_+}{K^1}P(\mathbf{W}^1|K^1)P(K^1)} \right]. \tag{7.34}$$

To accomplish the algorithm, we need to sample $\mathbf{W}^1$ and $\mathbf{Y}^1$. Using the Gibbs sampling, we individually infer each variable of the two matrices in turn from the distributions $P(\mathbf{W}^1_{n,k}|\mathbf{X}, \mathbf{W}^1_{-n,k}, \mathbf{Y}^1)$ and $P(\mathbf{Y}^1_{k,t}|\mathbf{X}, \mathbf{Y}^1_{-k,t}, \mathbf{W}^1)$, where $\mathbf{W}^1_{-n,k}$ means all values of $\mathbf{W}^1$ except for $\mathbf{W}^1_{n,k}$ and $\mathbf{Y}^1_{-k,t}$ means all values of $\mathbf{Y}^1$ except for $\mathbf{Y}^1_{k,t}$. From the construction of our generative model and the Bayes' rule, we have

$$P(\mathbf{W}^1_{n,k}|\mathbf{X}, \mathbf{W}^1_{-n,k}, \mathbf{Y}^1) \propto P(\mathbf{X}|\mathbf{W}^1_{n,k}, \mathbf{W}^1_{-n,k}, \mathbf{Y}^1) \cdot P(\mathbf{W}^1_{n,k}|\mathbf{W}^1_{-n,k}), \tag{7.35}$$

where $P(\mathbf{X}|\mathbf{W}^1_{n,k}, \mathbf{W}^1_{-n,k}, \mathbf{Y}^1)$ is specified by the Gaussian likelihood we choose, and the term $P(\mathbf{W}^1_{n,k}|\mathbf{W}^1_{-n,k})$ can be obtained by integrating out the associated priors:

$$P(\mathbf{W}^1_{n,k}|\mathbf{W}^1_{-n,k}) = P(\mathbf{W}^1_{n,k}|\vec{\mathbf{W}}^1_{-n,k}) = \int_{\vec{\theta} \in S} P(\mathbf{W}^1_{n,k}|\vec{\theta})P(\vec{\theta}|\vec{\mathbf{W}}^1_{-n,k})\mathrm{d}\vec{\theta},$$
$$\vec{\theta} = (\sigma_k^2, p_k); S = \{\sigma_k^2 \in \Re^+, p_k \in [0, 1]\}, \tag{7.36}$$

where $P(\mathbf{W}^1_{n,k}|\vec{\theta})$ is specified by (7.23) and $\vec{\mathbf{W}}^1_{-n,k}$ denotes all values of the $k$th column of matrix $\mathbf{W}^1$ except for $\mathbf{W}^1_{n,k}$. Since the columns of $\mathbf{W}^1$ are generated independently, we compute $P(\mathbf{W}^1_{n,k}|\vec{\mathbf{W}}^1_{-n,k})$, instead of $P(\mathbf{W}^1_{n,k}|\mathbf{W}^1_{-n,k})$. Utilizing the Bayes' rule again, $P(\vec{\theta}|\mathbf{W}^1_{-n,k})$ can be computed by

$$P(\vec{\theta}|\vec{\mathbf{W}}^1_{-n,k}) \propto P(\vec{\mathbf{W}}^1_{-n,k}|\sigma_k^2, p_k) \cdot P(\sigma_k^2, p_k). \tag{7.37}$$

The distribution $P(\vec{\mathbf{W}}^1_{-n,k}|\sigma_k^2, p_k)$ can be computed by evaluating each element of the $k$th column of matrix $\mathbf{W}^1$ except for $\mathbf{W}^1_{n,k}$ based on (7.23). The distribution $P(\sigma_k^2, p_k)$ can be computed by multiplication of $P(\sigma_k^2)$ and $P(p_k)$ which are specified by their priors defined in the generative model.

Similarly, we obtain the expression for $P(\mathbf{Y}^1_{k,t}|\mathbf{X}, \mathbf{Y}^1_{-k,t}, \mathbf{W}^1)$:

$$P(\mathbf{Y}^1_{k,t}|\mathbf{X}, \mathbf{Y}^1_{-k,t}, \mathbf{W}^1) \propto P(\mathbf{X}|\mathbf{Y}^1_{k,t}, \mathbf{Y}^1_{-k,t}, \mathbf{W}^1) \cdot P(\mathbf{Y}^1_{k,t}|\mathbf{Y}^1_{-k,t}), \tag{7.38}$$

where $P(X|\mathbf{Y}_{k,t}^1, \mathbf{Y}_{-k,t}^1, \mathbf{W}^1)$ is specified by the Gaussian likelihood we choose. Since each element of $\mathbf{Y}_{.,t}^1$ is generated independently, $P(\mathbf{Y}_{k,t}^1|\mathbf{Y}_{-k,t}^1)$ can be computed by its priors as (7.25).

The inference at the rest hidden layers is similar to the procedure used to infer the first hidden layer. Therefore, we summarize our inference algorithm in Algorithm 7.5.3.

### 7.5.4 Simulations and Discussions for Wireless Applications

We analyzed the performance of the proposed modified MH algorithm for inferring the true number of hidden factors in the first hidden layer. First, we fix the dimension of the observed data points, $N = 16$, and vary the number of hidden factors, $K$, from 3 to 10. For each integer value of $K$, we generate a data set containing $T = 200$ data instances using the proposed generative model. Within one instance, $\mathbf{Y}^1$ is sampled according to its Gaussian prior. Then the weight matrix $\mathbf{W}^1$ is drawn from its distribution specified by (7.23). Finally, the data point $X$ is generated by the Gaussian distribution where the parameters are expressed in terms of $\mathbf{Y}^1$ and $\mathbf{W}^1$. The rest model parameters are fixed at $\alpha_1' = 3$ for the Beta distribution; $\alpha_1 = 2$ and $\beta_1 = 1$ for the InverseGamma distribution. The modified MH algorithm is initialized with three choices of $K$: $K = 2$, $K = 10$ or random positive integer between 3 and 10, and then runs for 200 iterations. Each data set is estimated ten times by the inference procedure described previously. We record the expectation of the estimated number of hidden factors and its variance as the result.

---

**Algorithm 9** MH steps for inferring first-layer hidden factors

---

**for** $r = 1, 2, \ldots,$ number of iterations**do**

**for** $i = 1, 2, \ldots, N$ **do**

+ iteratively select column $k$ of $W$ **if** $m_{-i,k} > 0$

**then**

+ propose adding a new hidden factor with probability specified by (7.33)

+ propose deleting this hidden factor with probability specified by (7.34)

**end**

**for** $k = 1, 2, \ldots, K$ **do**

+ sample $W_{i,k}$ according to (7.35)

**end**

**for** each element of $Y$ **do**

+ sample $Y_{k,t}$ according to (7.38)

**end**

**end**

---

We plotted the results in Figure 7.16. The modified MH algorithm is under the influence of initialization. When initializing $K = 10$, which is much greater than the dimensions of the underlying model, the inferred $K$ values are generally much larger than the true values. However, when initializing $K$ randomly, the results correspondingly

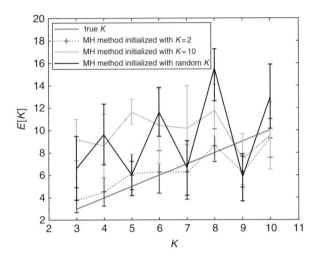

**Figure 7.16** Inferring the number of first-layer hidden factors using Metropolis-Hastings algorithm. Each curve shows the mean and variance of the expected value of the dimensionality $K$.

show some randomness. Another observation is that the MH method tends to overestimate the number of hidden factors. This is because the proposal to add one hidden factor is preferred to be accepted. However, the nominator is usually larger than the denominator because the denominator is composed by the multiplication of probability terms. Hence, the adding proposal is more likely to be accepted.

## 7.6    Summary

Deep learning is the fastest-growing field in machine learning. It uses many-layered DNNs to learn levels of representation and abstraction that make sense of data such as images, sound, and text. In this chapter, we first review the history of deep learning, followed by indicating the milestone papers. The variety of applications are listed. For techniques, we first study the restricted Bolzman machine, and then investigate the algorithms for training. Then the DBN, convoluntional neural networks, and recurrent neural networks are compared. Finally, three practical examples are shown for how to utilize the deep learning in different applications.

In the first example, we have presented and discussed a scalable Spark-based framework for deep learning in MBD analytics. The framework enables the tuning of deep models with many hidden layers and millions of parameters on a computing cluster. Typically, deep learning provides a promising learning tool for adding value by learning intrinsic features from raw MBD. The framework has been validated using a large-scale activity recognition system as a case study. Finally, important research directions on MBD have been outlined.

In the second example, we have implemented a four-layer deep network to learn typical users' moving patterns. At the first step, we pretrain the network by learning it layer by layer using RBM learning algorithm. Then, a backpropagation is deployed to fine-tune the network. Real traces were collected and transferred into binary images. The performance evaluated with the traces is shown to be significantly improved compared to the traditional methods. Our preliminary result also suggests some promising applications and can be extended to many other fields.

In the third example, we developed a deep hierarchical nonparametric Bayesian model to represent the underlying structure of observed data. Correspondingly, we proposed a modified MH algorithm to recover the number of hidden factors. Our simulation results on the hidden layer show that the algorithm discovers the model structure with some estimation errors. However, as shown in the results, our approach is capable of inferring increasing dimensions of hidden structures, which is due to the advantage of the nonparametric Bayesian technique. This indicates that the nonparametric Bayesian approach can be a suitable method for discovering complex structures.

# Part III

## Big Data Applications

# 8 Compressive Sensing-Based Big Data Analysis

Despite the relatively short history of compressive sensing (CS) theory pioneered by the work by Candès, Romberg, and Tao [393, 394, 395] and Donoho [396], the numbers of studies and publications in this area have become amazingly large. On the other hand, the applications of CS just begin to appear. The inborn nature that many signals can be represented by sparse vectors has been recognized in many areas of applications. Examples in wireless communication include the sparse channel impulse response in the time domain, the sparse unitization of the spectrum, and the time and spatial sparsity in the wireless sensor networks. For each of these sparse signals, there are innovative signal acquisition schemes that not only satisfy the requirements by the CS theory, but also are easily realizable on hardware. Efficient signal-recovery algorithms for each system are also available. They guarantee stable signal recovery with high probability.

In this chapter, we provide a concise overview of CS basics and some of its extensions. In subsequent chapters, we focus on CS algorithms and specific areas of CS research related to bid data. This chapter begins with Section 8.1, which gives the motivation of CS, illustrates the typical steps of CS by an example, summarizes the key components CS, and discusses how nearly sparse signals and measurement noise are treated in robust CS. Following these discussions, Section 8.2 compares CS with traditional sensing and examines their advantages and disadvantages. The foundation of CS, sparse representation, is studied in Section 8.3, which briefly covers sparsifying transforms, analytic dictionaries, learned dictionaries, as well as a few extensions of sparse modeling. Next, Section 8.4 discusses a variety of conditions under which one can trust CS for encoding signals during sensing and recovering them faithfully after sensing. Then, two synthetic examples are given to close this chapter. Finally, we outline some applications for CS in Section 8.5.

## 8.1 Background

The Nyquist/Shannon sampling theory has been accepted as the doctrine for signal acquisition and processing ever since it was implied by the work of Nyquist in 1928 [397] and proved by Shannon in 1949 [398]. The theorem says that to exactly reconstruct an *arbitrary* band-limited signal from its samples, the sampling rate needs to be at least twice the bandwidth.

In many applications nowadays, including digital image and video cameras, the Nyquist rate is so high that the excessive number of samples make compression a necessity prior to storage or transmission. In other applications, including medical scanners, radars, and high-speed analog-to-digital converters, increasing the sampling rate is very expensive, either due to the sensor itself being expensive or the measurement process being costly. Often, we are faced with the situation of not being able to collect enough measurements to meet the Nyquist rate.

On the other hand, the chance is that we are often not dealing arbitrary signals. Most signals we are interested in are highly compressible, namely, they can be represented by a set of sparse or nearly sparse coefficients. CS exploits this feature of signals and thus allows a sampling rate significantly lower than the Nyquist rate. Although this book does not cover CS imaging, we use images as an example to illustrate this point. Let us ignore the three channels of the RGB (red, green, blue) colors and imagine a gray-scale image of $n$ pixels represented by a vector $\mathbf{u}^o \in \mathbb{R}^n$ (though when displayed on screens or printed on paper, it has a typical rectangle shape). A computer program can compress $\mathbf{u}^o$ (superscript "$o$" stands for original) into a much smaller vector because only a small amount of information such as objects and their boundaries, certain repeated patterns, textures, and hues in the image are visually important; thence, the rest is "thrown away." In a nutshell, an image compression algorithm compresses $\mathbf{u}^o$ by looking up a dictionary $\Psi$ (e.g., a wavelet basis) for the useful pieces of information in $\mathbf{u}^o$. Mathematically, this is to represent $\mathbf{u}^o$ by a certain *sparse* vector $\mathbf{x}^o$ such that $\Psi\mathbf{x}^o$ contains nearly all the useful information in $\mathbf{u}^o$ and $\|\mathbf{u}^o - \Psi\mathbf{x}^o\|$ is small. The well-known JPEG2000 can compress a megapixel image ($n \approx 1{,}000{,}000$) taken from a camera into just a few thousand wavelet coefficients, reducing its size from a few megabytes to a few dozen kilobytes, leading to economical storage and fast transfer. This fact triggers a question: when only a few thousand coefficients are saved, why are the excessive measurements of a million pixels taken in the first place? A straightforward answer is that because we do not know which are the ones to measure (perhaps also because traditional image sensing cannot directly record wavelet coefficients, or it is very expensive to do so), we must record a sufficient number of pixels so that the subsequent image compression step can compute the wavelet transform at a sufficient resolution and keep the largest wavelet coefficients. In a new sensing paradigm called CS, however, only a small number of measurements of an image (say 1/4 of the number of pixels or fewer) need to be collected; and based upon the compressibility of the image, a sophisticated yet fast algorithm can recover the useful part of it. How is this done?

## 8.1.1    Steps of Compressive Sensing

The entire process of CS consists of three parts, depicted in Figure 8.1:

1. signal sparse representation
2. linear encoding and measurement collection
3. nonlinear decoding (sparse recovery).

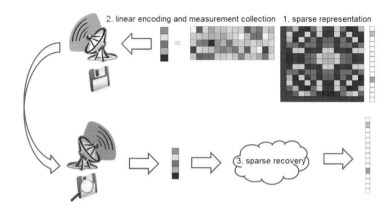

**Figure 8.1** Scheme of compressive sensing.

Part 1 is the foundation of CS, but it can be done after part 2. The importance of a *sparsifying* basis $\Psi$ is clear from the discussions in the last paragraph, and we discuss it in more detail in Section 8.3. So for simplicity, let us ignore $\Psi$ for now and set it to the identity, $\Psi = I$, and thus assume that $\mathbf{u}^o = \mathbf{x}^o$ is itself a $k$-sparse vector, namely, the number of nonzero entries of $\mathbf{x}^o$ is no more than $k$. A common notion for this is

$$\|\mathbf{x}^o\|_0 \leq k.$$

First, $\mathbf{x}^o$ is encoded into a smaller vector $\mathbf{b}$ by a matrix $\mathbf{A} \in \mathbb{R}^{m \times n}$ with $m < n$, which is chosen independently (nonadaptively) of $\mathbf{x}^o$ since $\mathbf{x}^o$ is unknown. This step is written as

$$\mathbf{b} := \mathbf{A}\mathbf{x}^o.$$

In CS applications, this encoding is typically not calculated by a computer but obtained by certain physical, optical, electrical, or electromagnetic means depending on the application. Since $m < n$, $\mathbf{b}$ is a compression of $\mathbf{x}^o$. Next, $\mathbf{b}$ is recorded by a sensor (while $\mathbf{x}^o$ remains unknown) and becomes digitally available. CS theory studies under what condition $\mathbf{x}^o$ can be recovered from $\mathbf{b}$, and when $\mathbf{b}$ is contaminated by noise, how to guarantee the recovery is close to $\mathbf{x}^o$. Clearly, $\mathbf{A}\mathbf{x} = \mathbf{b}$ is an underdetermined equation system and has an infinite number of solutions. This is where the sparsity of $\mathbf{x}^o$ plays its role. Since $\mathbf{x}^o$ is sparse, a natural approach is to find the sparsest solution of $\mathbf{A}\mathbf{x} = \mathbf{b}$, namely, solving

$$\min_{\mathbf{x}} \|\mathbf{x}\|_0 \quad \text{subject to } \mathbf{A}\mathbf{x} = \mathbf{b}. \tag{8.1}$$

However, this combinatorial problem is generally NP-hard [399], and the simple approach of trying all possible supports of cardinality $k$ is computationally intractable. A tractable decoder is needed! Replacing the $\ell_0$ "norm" by the $\ell_1$ norm yields the problem

$$\min_{\mathbf{x}} \|\mathbf{x}\|_1 \quad \text{subject to } \mathbf{A}\mathbf{x} = \mathbf{b}, \tag{8.2}$$

which is a convex program (in particular, it can be transformed to a linear program) and has several fast solvers (see the next chapter). Ideally, we would like (8.2) to recover $\mathbf{x}^o$

when $m$, the number of measurements, is equal to $2k$, which is the "amount of information" that $\mathbf{x}^o$ carries; after all, $\mathbf{x}^o$ is uniquely determined by the $k$ indices and the $k$ values of its nonzero entries. However, we must pay a reasonable price for not knowing the support of $\mathbf{x}^o$ (there are totally $\binom{n}{k}$ possibilities!), yet making sure that a single nonadaptive projection $\mathbf{A}$ good for the reconstruction of all, or at least most, $k$-sparse vectors $\mathbf{x}^o$ by solving a computationally tractable problem (8.2). As we shall see below, there are random matrices $\mathbf{A}$ with $m = O(k \log(n/k))$ rows that, with overwhelming probability, allow any $k$-sparse vector $\mathbf{x}^o$ to be recovered by problem (8.2). Besides solving (8.2) by linear programming algorithms, there are very efficient solvers for (8.2), which run almost as fast as the least-squares solvers; we review several such algorithms in the next chapter. Once the solution $\mathbf{x}^o$ is recovered in a computer, we obtain $\mathbf{u}^o$ and accomplish CS.

## 8.1.2    Robust Compressive Sensing and Signal Recovery

For CS to be useful in real-world sensing applications, it must deal with imperfect sparsity and sensors. Imperfect sparsity occurs when it is difficult to represent the underlying signal by an exactly spare vector (natural images are very good examples), even using a carefully chosen dictionary or transform in Section 8.3. However, it can be much easier to present it using vectors that are *close to being sparse*. To measure how sparse $\mathbf{x}$ is, we have used its number of nonzeros $\|\mathbf{x}\|_0$. We say that $\mathbf{x}$ is (exactly) sparse if $\|\mathbf{x}\|_0 \ll \dim(\mathbf{x})$ – the total number of entries of $\mathbf{x}$. If $\|\mathbf{x}\|_0$ is *not* so small compared to $\dim(\mathbf{x})$ yet $\mathbf{x}$ can be closely approximated by an exactly sparse vector, then $\mathbf{x}$ is called *nearly sparse*. A nearly sparse vector is nearly as useful and powerful as an exactly sparse vector during reconstruction and processing. The quantity below can be used to measure how nearly sparse a signal is.

DEFINITION 11. *Let $\mathbf{x}_{[k]}$ – the best $k$-term approximation of $\mathbf{x}$ – be the vector obtained by setting all but the $k$ largest (in magnitude) entries of $\mathbf{x}$ to zero (ties are broken arbitrarily). The approximate error is*

$$\sigma_{[k]}(\mathbf{x}) := \|\mathbf{x} - \mathbf{x}_{[k]}\|_p, \tag{8.3}$$

*where common choices of $p$ are 1 and 2, and $\|\mathbf{y}\|_p$ is defined as $\left(\sum_i |y_i|^p\right)^{1/p}$.*

For $k$ much smaller than $\dim(\mathbf{x})$, $\sigma_{[k]}$ measures the approximate sparsity of $\mathbf{x}$. A nearly sparse vector $\mathbf{x}$ is one with a small $\sigma_{[k]}(\mathbf{x})$ for some $k \ll \dim(\mathbf{x})$. The so-called *fast-decaying* signals have their $\sigma_{[k]}$ that decay very quickly in $k$, for example, in $O(1/k)$ or even $O(1/c^k)$ with some $c > 1$. As long as the coefficients are fast decaying, a large number of the coefficients can be discarded often without causing too much loss of the information. Figure 8.2 depicts a signal with entries that decay exponentially fast, as well as its best $k$-term approximation error for $k = 1, \dots, 128$.

Imperfect sensors exist in many real-world applications, and they are subject to a variety of errors due to, for example, design/manufacture imperfectness, quantization loss, thermal noise, limited response range, material aging, etc. As a result, the recorded data $\mathbf{b}$ may be contaminated by a certain amount of unknown error. As it is easy to see

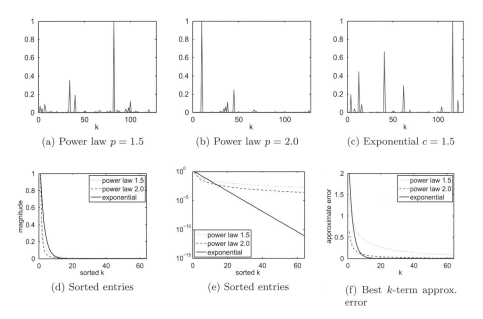

(a) Power law $p = 1.5$          (b) Power law $p = 2.0$          (c) Exponential $c = 1.5$

(d) Sorted entries          (e) Sorted entries          (f) Best $k$-term approx. error

**Figure 8.2** Two power law and one exponentially decaying signals.

noise in an image, we add three different types of noise to image "Cameraman" and demonstrate them in Figure 8.3.

To deal with imperfect sparsity and sensors, we consider the abstract sensing model:

$$\mathbf{b} = \mathbf{A}\mathbf{x}^o + \mathbf{w}, \tag{8.4}$$

where $\mathbf{A}$ is an $m$-by-$n$ matrix and $\mathbf{w}$ is an $n$-dimensional vector of noise or measurement error. The entries of $\mathbf{b}$ are noisy linear measurements of $\mathbf{x}^o$. If $\mathbf{w} \neq \mathbf{0}$, model (8.2) no longer holds. One can consider the $\ell_1$ minimization problem that constraints or penalizes the violation of $\mathbf{b} = \mathbf{A}\mathbf{x}^o$ as follows:

$$\min_{\mathbf{x}} \|\mathbf{x}\|_1 \quad \text{subject to } \|\mathbf{A}\mathbf{x} - \mathbf{b}\|_2 \leq \epsilon, \tag{8.5}$$

$$\min_{\mathbf{x}} \|\mathbf{x}\|_1 + \frac{\mu}{2}\|\mathbf{A}\mathbf{x} - \mathbf{b}\|_2^2. \tag{8.6}$$

The two programs are equivalent in the sense that the solution to one problem is also the solution to the other as long as the parameters $\epsilon$ and $\mu$ are properly set. However, the correspondence between $\epsilon$ and $\mu$ is not generally known *a priori*. Depending on the application and the available problem data, one of $\epsilon$ and $\mu$ might be easier to obtain than the other, so one of (8.5) and (8.6) is preferred to the other. How to choose $\epsilon$ or $\mu$ is an important practical issue. General principles include (i) making statistical assumptions on $\mathbf{w}$ and $\mathbf{x}^o$ and interpreting (8.5) or (8.6) as, for example, MAP estimations; (ii) cross-validation (performing reconstruction on one subset of measurements and validating the recovery on the other subset of measurements); and (iii) finding the best parameters on a test data set and using these parameters on the actual data with proper adjustments to address the differences in scale, dynamic range, sparsity, and noise level.

(a) Original                    (b) Gaussian noise

(c) Random-valued impulsive        (d) Salt-and-pepper

**Figure 8.3** Image "Cameraman" with three different types of noise.

## 8.2    Traditional Sensing vs. Compressive Sensing

In traditional sensing, a signal is first acquired then compressed. Think of taking an image by a digital camera as an example. The CCD or CMOS chip in the camera turns photons into an 2D array of digital signals, which is then compressed and stored in an image file.

Compared to traditional sensing, CS *senses less* and *computes more*. CS "compresses" a signal before it is recorded, whereas traditional sensing compresses a signal after sensing. Since CS senses less, in many of its applications, the save leads to much faster sensing and/or much smaller/cheaper sensors. For instance, in medical applications such as MR imaging, CS is reducing the acquisition time and thus making the imaging experience more comfortable, affordable, and, in CT imaging, CS is reducing the amount of x-ray radiation used to image the subject body and thus making it safer. On the other hand, because traditional sensing senses more and senses before compression, the digital form of the signal is already available during compression; hence, the compression is more flexible, can be adaptive to each signal, and achieves much better compression ratios. Therefore, traditional sensing is more effective at reducing the signal size. The two sensing approaches have their own strengths; neither one will replace the other. They address different bottlenecks, fit different needs, and achieve different performance. They complement each other. However, since CS is a rather new sensing paradigm that senses the information of a signal instead of the signal itself, its theory and computation are less familiar to the communities of researchers and engineers. On the other hand, the principle of CS that simple signals can be recovered from a small number of measurements has stimulated new research results far beyond sensing in areas such as optimization, random matrix theory, statistics, machine learning, medical imaging, etc.

The rest of this chapter provides the reader with brief yet fundamental results of CS in two aspects: sparse representation, and CS encoding and decoding.

## 8.3  Sparse Representation

Sparse representation is the basis of CS. It is concerned with expressing the information of a signal by a small number of real or complex numbers. Mathematically, this is to express a signal $\mathbf{u}^o$ as

$$\mathbf{u}^o = \sum_{i=1}^{p} \psi_i x_i^o,$$

where all but a small number of entries $x_i^o$ are zero (or small enough to safely neglect). $\Psi = [\psi_1\ \psi_2\ \cdots\ \psi_p]$ is called a *dictionary*. A dictionary is sometimes used to sparsely represent just one signal, but more often a set of similar signals. The dictionary atoms encode the most salient information of the signals. On the other hand, it is typical that unwanted artifacts and noise cannot be sparsely represented under the dictionary. Expressing artifacts and noise must involve a large number of $\psi_i$s. So, combined with sparse optimization, denoising is often obtained. Besides using a dictionary, a signal can also become sparse under a certain transform $\Upsilon$, namely, $\Upsilon(\mathbf{u})$ is a sparse vector. Examples include the gradient operator, curvelet transforms, etc. Figure 8.4 presents the DCT and wavelet coefficients, as well as the local variation, of the image "Cameraman." For a given two-dimensional image $\mathbf{u}$, its variation at pixel $(i, j)$ is

$$\|Du_{i,j}\|_2 = \left( (u_{i+1,j} - u_{i,j})^2 + (u_{i,j+1} - u_{i,j})^2 \right)^{1/2}.$$

It is clear from Figure 8.4 that a non-sparse image has nearly sparse representations.

Signal reconstruction and analysis are based on how a signal is represented. The standard representation is discrete sample points in time or space. They are called pixels for images. Some signals such as a stellar map and a wireless channel power vector are sparse under the standard representation, but most signals are not. The standard representation is suitable for operations such as down/upsampling, display, scaling, etc. because these operations are applied to the sample points independently or locally; they do not involve relations between multiple distant sample points. However, for more complicated tasks such as reconstruction, segmentation, and recognition, the standard

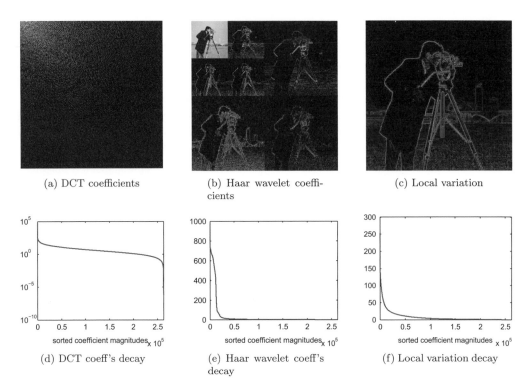

(a) DCT coefficients

(b) Haar wavelet coeffi-
cients

(c) Local variation

(d) DCT coeff's decay

(e) Haar wavelet coeff's
decay

(f) Local variation decay

**Figure 8.4** Sparsity of the image "Cameraman" (the DCT and wavelet coefficients are scaled for better visibility).

representation is often inefficient. We need a meaningful representation that describes the useful characteristics, structures, and features of the signal. For example, in signal denoising, the representation should be sparse only for the wanted part of the signal; for the unwanted parts such as noise and artifacts, the representation should be dense. To recognize the object in a signal, as another example, the representation should be sparse for the most salient features of the object. There are many more other tasks (cf. books [400, 401]) that all look for sparse representations.

Sparse representation involves the choice of dictionary $\Psi$ or sparsifying transform $\Upsilon$. In the simplest cases, $\Psi$ or $\Upsilon$ is orthogonal. Examples include the unitary discrete Fourier transform and various orthogonal wavelet bases. They often equip with fast computation at the complexity of $n \log n$ instead of $n^2$. Well-known non-orthogonal examples include total variation [128], the curvelet transform [402], and various tight frames. These dictionaries are analytic, often designed to handle specific kinds of signals. They have been extensively studied and widely considered in applications as they are easier to analyze and feature faster numerical implementations, compared to the learned (i.e., trained, as opposed to analytic) dictionaries, which we are going to discuss next.

The earliest major work in dictionary learning is due to Olshausen and Field [403], who trained an overcomplete dictionary for sparsely representing small image patches of a set of natural images. Remarkably, the atoms in the trained image were very similar to the simple cell receptive fields in early vision. This result suggests the potential of sparse representation in uncovering fundamental features in complex signals.

Modern training approaches such as MOD [404], kSVD [405], and many others can generate useful dictionaries out of various training sources, from just a single (incomplete) signal, to a few signals of the same kind, and to a database of signals. They can form dictionaries from scratch, with no structural restrictions whatsoever, or from certain functions or distributions with parameters learned from the training set. The training and using of these dictionaries, especially those without any structures, are computationally expensive, but these dictionaries tend to do better on complex signals and complicated tasks for which no analytic dictionaries have been designed.

## 8.3.1 Extensions of Sparse Models

### Joint-Sparse Signals

A set of signals $\mathbf{u}^{(i)}$, $i = 1, \ldots, L$, of the same dimension are jointly sparse if each of them is sparse and their nonzero entries are colocated at (roughly) the same coordinates, or so under dictionaries or transforms. Figure 8.5 depicts joint-sparse signal CS. The model is closely related to group LASSO [406]. Applications of the recovery of such signals [407] include multikernel machine learning [408], source localization [409], neuromagnetic imaging [410], wireless spectrum sensing [411], and many more.

### Low-Rank Matrices

A matrix $\mathbf{M} \in \mathbb{R}^{m \times n}$ of rank $r \ll \min\{m, n\}$ has $mn$ entries but only $r(m + n - r)$ degrees of freedom (consider the singular value decomposition $\mathbf{M} = \mathbf{U}\Sigma\mathbf{V}^\top$; $\mathbf{U}$, $\Sigma$, and $\mathbf{V}$ have $\sum_{i=1}^{r}(m - i)$, $r$, and $\sum_{i=1}^{r}(n - i)$ degrees of freedom, respectively, which sum to $r(m + n - r)$). Figure 8.6 illustrates the approximation of a large low-rank matrix by a pair of skinny and fat matrices. Applications of low-rank matrix recovery [412, 413, 159] include model reduction [414], network Euclidean embedding [415], recovering shape and motion from image streams [416, 417], the Netflix recommendation problem [418], and more. In addition, there are models and applications of decomposing a matrix into the sum of low rank and sparse parts [164].

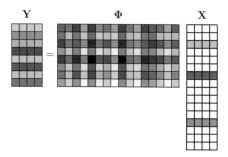

**Figure 8.5** Joint-sparse signal CS.

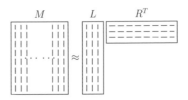

**Figure 8.6** Approximation of a low-rank matrix.

### Unions of Subspaces and Model-Based CS

The support of a $k$-sparse vector is one of the $\binom{n}{k}$ possibilities, but for many signals in practice some or even most of these possibilities are not possible. For example, some transforms' coefficients follow a certain tree structure, some signals' nonzero entries tend to cluster, and some signals must lie in particular linear subspaces. These signals are easier to recover [187], and their structures can be generalized as the union of certain subspaces [419, 420].

## 8.4    CS Encoding and Decoding

In CS, the signal $\mathbf{u}^o = \Psi\mathbf{x}^o$ is encoded to $\mathbf{b} = \mathbf{A}\mathbf{u}^o$ with the hope of recovering $\mathbf{u}^o$ from $\mathbf{A}$ and $\mathbf{b}$. The recovery would be straightforward if $\mathbf{A}$ had full column rank; in this case, $\mathbf{u}^o$ would be the unique solution of

$$\min_{\mathbf{u}} \|\mathbf{b} - \mathbf{A}\mathbf{u}\|_2^2,$$

or $\mathbf{u}^o = (\mathbf{A}^\top\mathbf{A})^{-1}\mathbf{A}^\top\mathbf{b}$. However, CS uses the *downsampled* linear encoding, namely, $\mathbf{A}$ has fewer rows than columns. Such a matrix cannot have full column rank, and $\mathbf{b} = \mathbf{A}\mathbf{u}$ has multiple solutions. Recall that $\mathbf{x}^o$ is sparse or nearly sparse and, prior to CS signal reconstruction, the locations and values of its nonzero entries are unknown. So, the key question is: what kind of matrix $\mathbf{A}$ allows the recovery of $\mathbf{u}^o$ (or a good approximate of $\mathbf{u}^o$) from $\mathbf{b} = \mathbf{A}\mathbf{u}$ given merely that $\mathbf{x}^o$ is sparse? The answer is closely related to a concept called *coherence*.

DEFINITION 12. *In $\mathbb{R}^n$ or $\mathbb{C}^n$, the coherence between the basis $\Phi$, which has elements $\phi_1, \ldots, \phi_n$, and the basis $\Psi$, which has columns $\psi_1, \ldots, \psi_n$, is*

$$\mu(\Phi, \Psi) = \sqrt{n} \max_{1 \le i,j \le n} \frac{|\langle \phi_i, \psi_j \rangle|}{\|\phi_i\|_2 \|\psi_j\|_2}. \tag{8.7}$$

The quantity $\mu(\Phi, \Psi)$ measures how small the closest angle between any two elements of $\Phi$ and $\Psi$ can be. If there exists a pair of elements $\phi_i$ and $\psi_j$ obeying $\phi_i = \alpha\psi_j$ for some scalar $\alpha$ (i.e., they make a zero angle) then $\mu(\Phi, \Psi) = \sqrt{n}$, reaching its maximum, and we say that $\Phi$ and $\Psi$ are coherent. On the other hand, no $\psi_j$ can be orthogonal to all $\phi_i$s since $\Phi = \{\phi_i\}$ can represent any vector including $\psi_j$. It turns out that the minimum of $\mu(\Phi, \Psi)$ is 1. Therefore, $1 \le \mu(\Phi, \Psi) \le \sqrt{n}$. When $\mu(\Phi, \Psi)$ is close to 1, we say that $\Phi$ and $\Psi$ are incoherent. When $\Phi$ and $\Psi$ are coherent, there exists signals that can be sparsely represented by either $\Phi$ or $\Psi$, and in particular, there exist $\psi_i$ that is sparse under $\Phi$ and, conversely, $\phi_i$ that is sparse under $\Psi$. Such signals do not exist when they are incoherent.

How is this related to CS? One can associate the sensing matrix $\mathbf{A}$ with a certain orthogonal basis $\Phi$ in the way that the subspace spanned by the rows of $\mathbf{A}$ equals that by some $m$ elements of $\Phi$. Or, simply speaking, $\mathbf{Au}$ gives just $m$ out of the $n$ coefficients of $\Phi^T \mathbf{u}$. Then the question becomes what relation between two orthogonal bases $\Phi$ and $\Psi$ allows the recovery of $\mathbf{x}^o$ (and thus $\mathbf{u}^o$) from incomplete coefficients of $\Phi^T \Psi \mathbf{x}^o$. The answer is *low coherence* between them, namely, $\mu(\Phi, \Psi)$ is much closer to 1 than $\sqrt{n}$. With low coherence, every element of $\Phi$ is "misaligned" with every element of $\Psi$, so $\langle \phi_i, \psi_j \rangle$s are roughly equal and stay uniformly away from zero. Hence, each coefficient $i$ of $\Phi^T \Psi \mathbf{x}^o$, which equals $\sum_{1 \leq j \leq n} \langle \phi_i, \psi_j \rangle x_j^o$, encodes a guaranteed amount of information of each $x_j^o$. No $x_j^o$ is left out, and no $x_j^o$ dominates. In contrast, high coherence leads to uneven $\langle \phi_i, \psi_j \rangle$, and at least for some $i$, the coefficient $i$ of $\Phi^T \Psi \mathbf{x}^o$ encodes just a few $x_j^o$s while being nearly useless for the others. Since $\mathbf{x}^o$ is sparse, this coefficient $i$ has a good chance to miss the nonzeros of $\mathbf{x}^o$ and thus become useless for recovery $\mathbf{x}^o$. Therefore, low coherence is a guarantee that every measurement does its job of carrying a useful amount of information of all the nonzeros of $\mathbf{x}^o$ no matter where they are. With high coherence, the quality of the measurements depend on the locations of the nonzeros of $\mathbf{x}^o$. The quality of measurements translate the number of measurements required for recovery, which is formally given in the following theorem.

THEOREM 7 ([421]). *For a given* $\mathbf{u}^o = \Psi \mathbf{x}^o$ *where* $\mathbf{x}^o$ *has at most* $k$ *nonzero entries, choose* $m$ *entries of* $\Phi^T \mathbf{u}^o$ *uniformly at random, denoted as vector* $\mathbf{b} = P_\Omega \Phi^T \mathbf{u}^o$, *where* $P_\Omega$ *is the selection operator. As long as*

$$m \geq C \cdot \mu^2(\Phi, \Psi) \cdot (k \log n) \tag{8.8}$$

*for some constant* $C > 0$ *independent of* $k$ *and* $n$, *the solution to* $\min\{\|\mathbf{x}\|_1 : \mathbf{b} = P_\Omega \Phi^T \Psi \mathbf{x}\}$ *is* $\mathbf{x}^o$ *with overwhelming probability. (The result is shown for nearly all possible sign sequences of* $\mathbf{x}^o$.)

The theorem claims that as long as $\Phi$ and $\Psi$ are sufficiently incoherent and a signal is sparse when represented under basis $\Psi$, $\ell_1$-minimization can exactly recover the signal from only $O(\mu^2(\Phi, \Psi) \cdot (k \log n))$ measurements, a number that can be much smaller than the dimension of the signal. Also, the recovery method does not need to know the number of nonzero entries in sparse representation or their locations. On the other hand, there are examples in [422, 423] on which no methods can achieve exact recovery with fewer than $S \log n$ measurements when $\mu(\Phi, \Psi) = 1$. Therefore, the required number of measurements in (8.8) cannot be significantly improved.

Besides low coherence, there are various results telling us when we can trust $\ell_1$-minimization, namely, guarantee a successful CS recovery. To review these conditions, we would simply use the abstract model (8.4). When the measurements are $\mathbf{b} = P_\Omega \Phi^T \mathbf{u}^o$ and the signal is $\mathbf{u}^o = \Psi \mathbf{x}^o$, we can set $\mathbf{A} = P_\Omega \Phi^T \Psi$ in the model (8.4). Among the widely used conditions are those based on the null-space property (NSP), the restricted isometry principle (RIP) [424], the spherical section property (SSP) [425], and a "RIPless" condition [426]. These properties have different strengths, and they together assert that a large number of sensing matrices such as those with entries sampled from sub-Gaussian distributions, the so-called Fourier and Walsh–Hadamard ensembles, and random Toeplitz and circulant ensembles are suitable for CS and have recovery guarantees. Next, we review these conditions and present the corresponding recovery guarantees.

## 8.4.1     The Null-Space Property (NSP)

DEFINITION 13. *Matrix* $\mathbf{A}$ *satisfies the NSP of order k if*

$$\|\mathbf{h}_{\mathcal{S}}\|_1 < \|\mathbf{h}_{\mathcal{S}^c}\|_1, \tag{8.9}$$

*holds for all nonzero* $\mathbf{h}$ *in the null space of* $\mathbf{A}$ *and all coordinate sets* $\mathcal{S} \subset \{1, 2, \cdots, n\}$ *of cardinality* $|\mathcal{S}| \leq k$.

In plain English, the condition examines every nonzero null-space vector of $\mathbf{A}$ and requires it not to concentrate its $\ell_1$-energy on any set of $k$ entries. This is a simple yet natural form of recovery condition since all the feasible solutions of $\mathbf{A}\mathbf{x} = \mathbf{b}$ have the form $\mathbf{x} = \mathbf{x}^o + \mathbf{h}$, where $\mathbf{h}$ is a vector in the null space of $\mathbf{A}$. One of the very first steps toward recovery guarantees is to study the property of the null-space vectors of $\mathbf{A}$. The following theorem relates the null-space property to the exact recovery of $k$-sparse signals by $\ell_1$ minimization.

THEOREM 8 (NSP condition [427, 428, 429]). *Problem* (8.2) *uniquely recovers all* $k$-*sparse vectors* $\mathbf{x}^0$ *from measurements* $\mathbf{b} = \mathbf{A}\mathbf{x}^0$ *if and only if* $\mathbf{A}$ *satisfies the k-NSP.*

The proof is based on the following idea: whenever $\mathbf{x}^o$ has $k$ or fewer nonzero entries, deviating from $\mathbf{x}^o$ to $\mathbf{x}^o + \mathbf{h}$ must cause an increase in $\ell_1$-norm since in (8.9), $\|\mathbf{h}_{\mathcal{S}}\|_1$ is the maximum possible decrease while $\|\mathbf{h}_{\mathcal{S}^c}\|_1$ is the corresponding increase. Figure 8.7 illustrates a simple case where $\mathbf{x}^o$ is exactly recovered by $\ell_1$ minimization since any point $\mathbf{x}^o + \mathbf{h}$ (which is on the gray plane) leads to a larger $\ell_1$-norm.

*Proof. Sufficiency.* Pick any $k$-sparse vector $\mathbf{x}^o$. Let $\mathcal{S} := \mathrm{supp}(\mathbf{x}^o)$ and $\mathcal{Z} = \mathcal{S}^c$. For any *nonzero* $\mathbf{h} \in \mathrm{Null}(\mathbf{A})$, we have $\mathbf{A}(\mathbf{x}^o + \mathbf{h}) = \mathbf{A}\mathbf{x}^o = \mathbf{b}$ and

$$\begin{aligned} \|\mathbf{x}^0 + \mathbf{h}\|_1 &= \|\mathbf{x}_{\mathcal{S}}^0 + \mathbf{h}_{\mathcal{S}}\|_1 + \|\mathbf{h}_{\mathcal{Z}}\|_1 \\ &\geq \|\mathbf{x}_{\mathcal{S}}^0\|_1 - \|\mathbf{h}_{\mathcal{S}}\|_1 + \|\mathbf{h}_{\mathcal{Z}}\|_1 \qquad (8.10) \\ &= \|\mathbf{x}^0\|_1 + (\|\mathbf{h}_{\mathcal{Z}}\|_1 - \|\mathbf{h}_{\mathcal{S}}\|_1), \end{aligned}$$

where we have applied the triangle equality at the second line. Since the $k$-NSP property (8.9) of $\mathbf{A}$ guarantees $\|\mathbf{x}^0 + \mathbf{h}\|_1 > \|\mathbf{x}^0\|_1$, $\mathbf{x}^o$ is the unique solution of (8.2).

*Necessity. The inequality* (8.10) *holds with equality if* $\mathrm{sign}(\mathbf{x}_{\mathcal{S}}^o) = -\mathrm{sign}(\mathbf{h}_{\mathcal{S}})$ *and* $\mathbf{h}_{\mathcal{S}}$ *has a sufficiently small scale. Therefore, for problem* (8.2) *to uniquely recover all* $k$-*sparse vectors* $\mathbf{x}^0$, *property* (8.9) *is also necessary.* □

Note that although the NSP is necessary for exact recovery of all $k$-sparse vectors uniformly, it is no longer necessary if the uniformness is dropped (i.e., if one only considers certain $k$-sparse vector(s)). The NSP is widely used in the proofs of other conditions that

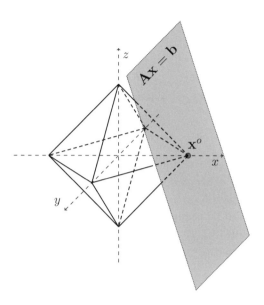

**Figure 8.7** Exact recovery of $\mathbf{x}$ by $\ell_1$ minimization.

guarantee the uniform recovery by (8.2) of sparse vectors. In addition, NSP of order $2k$ is necessary for stable uniform recovery. Consider an arbitrary decoder $\Delta$, computationally tractable or not, which returns a vector from the input $\mathbf{b} = \mathbf{A}\mathbf{x}^o$. If one requires $\Delta$ to be stable in the sense

$$\|\mathbf{x}^o - \Delta(\mathbf{A}\mathbf{x}^o)\|_1 < C \cdot \sigma_{[k]}(\mathbf{x}^o) \tag{8.11}$$

for all vectors $\mathbf{x}^o$, where $C$ is a universal constant and $\sigma_{[k]}$ is defined in (8.3) with $p = 1$ (using $\ell_1$-norm to measure the best $k$-term approximation error), then $\mathbf{A}$ must obey

$$\|\mathbf{h}_{\mathcal{S}}\|_1 < C \cdot \|\mathbf{h}_{\mathcal{S}^c}\|_1, \tag{8.12}$$

for all nonzero $\mathbf{h}$ in the null space of $\mathbf{A}$ and all coordinate sets $\mathcal{S} \subset \{1, 2, \cdots, n\}$ of cardinality $|\mathcal{S}| \leq 2k$. One can show this by applying an argument in [430].

## 8.4.2 The Restricted Isometry Principle (RIP)

Next, we review a property regarding the "orthogonality" of $\mathbf{A}$. This property is more restrictive than the NSP, and several recovery results based on the RIP are proved through establishing the NSP. On the other hand, it gives robustness to measurement noise and has received more attention from researchers. Many recovery algorithms including both optimization and greedy algorithms have been analyzed under the RIP assumptions. Various random matrices satisfy the property with high probability. The property also has consequences beyond signal recovery such as stable embedding, dimension reduction, especially the relation to the Johnson–Lindenstrauss lemma.

DEFINITION 14 ([424]). *Matrix* $\mathbf{A}$ *satisfies the RIP of order k if the smallest constant* $\delta_k$ *obeying*

$$(1 - \delta_k)\|\mathbf{x}\|_2^2 \leq \|\mathbf{Ax}\|_2^2 \leq (1 + \delta_k)\|\mathbf{x}\|_2^2 \tag{8.13}$$

*for all k-sparse vectors* $\mathbf{x} \in \mathbb{R}^n$ *satisfies* $\delta_k < 1$. $\delta_k$ *is called the restricted isometry constant of* $\mathbf{A}$.

Note that it is common practice to say that $\mathbf{A}$ satisfies the RIP as long as $\mathbf{A}$ can satisfy (8.13) after scaling. To apply (8.13) and its recovery results below, one might need to scale $\mathbf{A}$ first. Since CS is concerned with sensing matrices $\mathbf{A}$ with more columns than rows, such $\mathbf{A}$ cannot be orthogonal in the usual sense. On the other hand, the RIP requires $\mathbf{A}$ to be nearly orthogonal only restricted to the set of $k$-sparse vectors. It is sufficient for (8.2) to recover all $k$-sparse vectors uniformly if the property holds for $2k$-sparse vectors with a certain maximum deviation from orthogonality in terms of $\delta_{2k}$. Work [431] shows the sufficiency of $\delta_{2k} < 0.4142$, which is later improved to $\delta_{2k} < 0.4531$ [432], $\delta_{2k} < 0.4652$ [433], $\delta_{2k} < 0.4721$ [434], as well as $\delta_{2k} < 0.4931$ [435]. The last bound is up to date. We present a theorem from [435].

THEOREM 9 (RIP condition for exact and stable recovery. [435]). *Consider model* (8.4). *If* $\mathbf{A}$ *satisfies the RIP with* $\delta_{2k} \leq 0.4931$, *then the solution* $\mathbf{x}^*$ *of* (8.5) *with* $\epsilon$ *set to* $\|\mathbf{w}\|_2$ *satisfies*

$$\|\mathbf{x}^* - \mathbf{x}^o\|_1 \leq C_1 \cdot \sqrt{k}\|\mathbf{w}\|_2 + C_2 \cdot \sigma_{[k]}(\mathbf{x}^o), \tag{8.14}$$

$$\|\mathbf{x}^* - \mathbf{x}^o\|_2 \leq \bar{C}_1 \cdot \|\mathbf{w}\|_2 + \bar{C}_2 \cdot \sigma_{[k]}(\mathbf{x}^o)/\sqrt{k}, \tag{8.15}$$

*where* $C_1$, $C_2$, $\bar{C}_1$, *and* $\bar{C}_2$ *are universal constants*.

The results claim that as long as the sensing matrix $\mathbf{A}$ has the RIP, the $\ell_1$ recovery is stable against both the measurement noise and the signal noise, the latter of which is given in terms of the best $k$-term approximation error. When neither noise exists, namely, $\mathbf{w} = \mathbf{0}$ and $\mathbf{x}^o$ is $k$-sparse (thus $\sigma_{[k]}(\mathbf{x}^o) = 0$), inequalities (8.14) and (8.15) give the exact recovery of $\mathbf{x}^* = \mathbf{x}^o$. When there is measurement noise (i.e., $\mathbf{w} \neq \mathbf{0}$) the order of error $O(\|\mathbf{w}\|_2)$ in the recovered signal cannot be significantly improved since even if the locations of the largest $k$ entries of $\mathbf{x}^o$ are given, the least-squares method will still give an error of $O(\|\mathbf{w}\|_2)$. When the signal is not exactly $k$-sparse (i.e., $\sigma_{[k]}(\mathbf{x}^o) \neq 0$) the error $O(\sigma_{[k]}(\mathbf{x}^o))$ is also tight since it is attained by power law decaying signals. If the decay of $\mathbf{x}^o$ follows the power law: the $i$th largest entry of $\mathbf{x}^o$ in magnitude obeys $|x_{(i)}| = C \cdot i^{-r}$, $r > 1$, then through basic integration one gets $\sigma_{[k]}(\mathbf{x}^o)/\sqrt{k} = O(k^{-r+1/2})$. On the other hand, even if we let $\mathbf{x}^* = \mathbf{x}^o_{[k]}$, we still get $\|\mathbf{x}^* - \mathbf{x}^o\|_2 = O(k^{-r+1/2})$. Therefore, the bounds in (8.14) and (8.15) cannot be essentially improved. However, note that given additional properties of $\mathbf{w}$ and $\mathbf{x}^o$ such as their statistical distributions, one can do better. Bounds (8.14) and (8.15) are optimal up to constant factors when applied to all noise and signals uniformly.

Furthermore, the first inequality in the RIP condition (8.13) is necessary for stable recovery against *arbitrary* unknown noise $\mathbf{w}$. A constructive proof can be found in [436].

Recall that CS requires a sensing matrix that is nonadaptive to the underlying signal, namely, the design of the matrix should be independent of any particular signal or its existing measurements. (The design of adaptive and online sensing matrices is interesting and very useful but out of the scope of our discussions.) Somewhat surprisingly, several kinds of random matrices $\mathbf{A}$ satisfy the RIP with at least high probability up to proper scaling. They include, but are not limited to, sensing matrices $\mathbf{A}$ whose entries are i.i.d. samples from the standard normal distribution, symmetric Bernoulli distribution, as well as other sub-Gaussian distributions, those whose columns are sampled uniformly from the unit sphere $\mathbb{S}^{m-1}$, and those formed by randomly projecting (not randomly picking the rows of) an orthogonal basis. As long as their dimension $m \times n$ obeys $m \geq O(k \log(n/k))$, these matrices obey the RIP of order $k$ with the fail probably decreasing to 0 exponentially fast in $m$; see [437, 438]. Various structured random matrices, which are easier to implement or naturally exist in applications, also satisfy the RIP though the best-known requirements on $m$ are typically $O(k \operatorname{poly}(\log(n/k)))$; see [439]. Common examples of such matrices include those formed by randomly selected rows of a discrete Fourier matrix or matrix $\Phi\Psi$ where $\Phi$ and $\Psi$ are incoherent, as well as random circulant or Toeplitz matrices. There are also works such as [440, 441, 442] that design deterministic matrices with RIPs but require $m$ (sometimes, significantly) more than $O(k \log(n/k))$ yet still much less than $O(n)$.

## 8.4.3    The Spherical Section Property

Next, we present recovery conditions based on the property SSP [425, 443] of $\mathbf{A}$. The advantage of this condition over the RIP is that the SSP is invariant to left-multiplying non-singular matrices to the sensing matrix $\mathbf{A}$ as pointed out in [425]. On the other hand, it does not have robustness against arbitrary unknown measurement noise $\mathbf{w}$.

DEFINITION 15 ($\Delta$-SSP [443]). *Let $m$ and $n$ be two integers such that $m > 0$, $n > 0$, and $m < n$. An $(n - m)$-dimensional subspace $\mathcal{V} \subset \mathbb{R}^n$ has the $\Delta$ spherical section property if*

$$\frac{\|\mathbf{h}\|_1}{\|\mathbf{h}\|_2} \geq \sqrt{\frac{m}{\Delta}} \tag{8.16}$$

*holds for all nonzero $\mathbf{h} \in \mathcal{V}$.*

To see the significance of (8.16), we note that (i) $\frac{\|\mathbf{h}\|_1}{\|\mathbf{h}\|_2} \geq 2\sqrt{k}$ for all $\mathbf{h} \in \text{Null}(\mathbf{A})$ is a sufficient condition for the NSP inequality (8.9) and (ii) due to [444, 445], a uniformly random $(n - m)$-dimensional subspace $\mathcal{V} \subset \mathbb{R}^n$ has the SSP for

$$\Delta = C_0(\log(n/m) + 1)$$

with probability at least $1 - \exp(C_1(n - m))$, where $C_0$ and $C_1$ are universal constants. Hence, $m > 4k\Delta$ guarantees (8.9) to hold, and furthermore, if $\text{Null}(\mathbf{A})$ is uniformly random, $m = O(k \log(n/m))$ is sufficient for (8.9) to hold with overwhelming probability (cf. [425, 443]).

THEOREM 10 (SSP condition for exact recovery [425]). *Suppose* Null($\mathbf{A}$) *satisfies the* $\Delta$-*SSP. If*

$$m \geq 4k\Delta, \tag{8.17}$$

*then the null-space condition (8.9) holds for all* $\mathbf{h} \in$ Null($\mathbf{A}$) *and coordinate sets* $\mathcal{S}$ *of cardinality* $|\mathcal{S}| \leq k$. *By Theorem 8, (8.17) guarantees that problem (8.2) recovers any* $k$-*sparse* $\mathbf{x}^o$ *from measurements* $\mathbf{b} = \mathbf{A}\mathbf{x}^o$.

THEOREM 11 (SSP condition for stable recovery [425]). *Suppose* Null($\mathbf{A}$) *satisfies the* $\Delta$-*SSP. Let* $\mathbf{x}^o \in \mathbb{R}^n$ *be an arbitrary vector. If*

$$m \geq 4k\Delta, \tag{8.18}$$

*then the solution* $\mathbf{x}^*$ *of (8.2) satisfies*

$$\|\mathbf{x}^* - \mathbf{x}^o\|_1 \leq 4\sigma_{[k]}(\mathbf{x}^o). \tag{8.19}$$

## 8.4.4    "RIPless" Analysis

Unlike the NSP, RIP, and SSP, the "RIPless" analysis [426] gives *nonuniform* recovery guarantees in the sense that the recovery guarantee is not given for all (exactly or nearly) $k$-sparse signals uniformly but a single, arbitrary signal. However, it applies to a wide class of CS matrices such as those with i.i.d. sub-Gaussian entries, orthogonal transform ensembles satisfying an incoherence condition, random Teoplitz/circulant ensembles, as well as certain tight and continuous frame ensembles, at $O(k \log(n))$ measurements. This analysis is especially useful in situations where the RIP, as well as the NSP and the SSP, is difficult to check or does not hold.

THEOREM 12 (RIPless for exact recovery [426]). *Let* $\mathbf{x}^o \in \mathbb{R}^n$ *be a fixed* $k$-*sparse vector. With probability at least* $1 - 5/n - e^{-\beta}$, $\mathbf{x}^0$ *is the unique solution to problem (8.2) with* $\mathbf{b} = \mathbf{A}\mathbf{x}^o$ *as long as the number of measurements*

$$m \geq C_0(1 + \beta)\mu(\mathbf{A}) \cdot k \log n,$$

*where* $C_0$ *is a universal constant and* $\mu(\mathbf{A})$ *is an incoherence parameter of* $\mathbf{A}$ *(see [426] for its definition and values for various CS matrices).*

## 8.4.5    Non-$\ell_1$ Decoding Methods

Besides $\ell_1$ and $\ell_1$-like minimization, there exist a large number of other optimization and non-optimization models and algorithms that can efficiently recover sparse or structured signals from their CS measurements. Approaches such as greedy algorithms (e.g., [172, 173, 174, 175, 176]), iterative hard-thresholding algorithms (e.g., [179, 180]), combinatorial algorithms (e.g., [446, 447, 448]), and sublinear-complexity algorithms (e.g., [449, 450]) also enjoy recovery guarantees of different forms under certain assumptions on the sensing matrix and the signal. We review some of these methods in the next chapter but do not present the theorems of their recovery guarantees.

## 8.4.6 Examples

We show two examples of CS encoding and decoding. In the first example, we try to acquire a sparse vector $\mathbf{x}^o \in \mathbb{R}^n$ of length $n = 200$ with ten nonzero entries; it is depicted in Figure 8.8(a). We let $\mathbf{A} \in \mathbb{R}^{m \times n}$ be formed from a random subset of $m = 80$ rows of the $n$-dimensional discrete cosine transform (DCT) $\Phi$. Figure 8.8(b) depicts the full measurements $\Phi\mathbf{x}^o$, and Figure 8.8(c) depicts the CS measurements $\mathbf{b} = \mathbf{A}\mathbf{x}^o$, where missing measurements – those in $\Phi\mathbf{x}^o$ but not in $\mathbf{A}\mathbf{x}^o$ – are replaced by zeros. We solved model (8.2) to recover $\mathbf{x}^o$. The solution $\mathbf{x}^*$ is given in Figure 8.8(d), and it is equal to $\mathbf{x}^o$.

From a different perspective, this is also an example of missing data recovery [451]. Given $\mathbf{b}$, which is a portion of the full data $\mathbf{f} = \Phi\mathbf{x}^o$, one can recover the full data $\mathbf{f}$ by exploiting its sparsity under the DCT dictionary $\Phi$. We plot the recovery $\Phi\mathbf{x}^*$ in Figure 8.8(e), and it is equal to $\Phi\mathbf{x}^o$.

Recall that one would like to take as few measurements as possible, that is, $m = k = 10$, but the use of nonadaptive sensing matrices in CS requires more measurements. To illustrate this point, we performed similar calculations for $m = 10, 11, \ldots, 80$, each with 100 repetitions of randomly chosen $m$ measurements of $\Phi\mathbf{x}^o$. The percentages of successful recovery for all $m$ are plotted in Figure 8.8(f), which shows that it is sufficient to have $m > 6k = 60$ to recover $\mathbf{x}^o$.

In the second example, we simulate ideal magnetic resonance imaging (MRI) of an ideal phantom depicted in Figure 8.9(a). The CS samples are a subset of its 2D Fourier

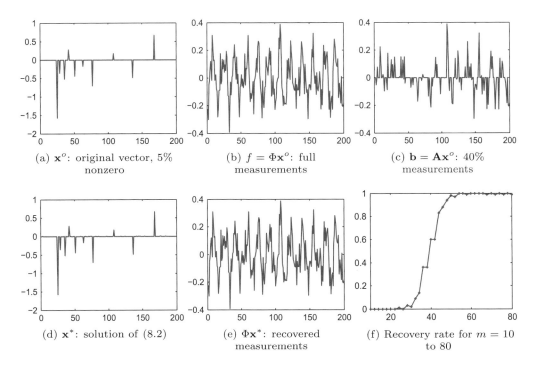

(a) $\mathbf{x}^o$: original vector, 5% nonzero

(b) $f = \Phi\mathbf{x}^o$: full measurements

(c) $\mathbf{b} = \mathbf{A}\mathbf{x}^o$: 40% measurements

(d) $\mathbf{x}^*$: solution of (8.2)

(e) $\Phi\mathbf{x}^*$: recovered measurements

(f) Recovery rate for $m = 10$ to 80

**Figure 8.8** Sparse vector recovery from incomplete measurements. $\Phi$ is a discrete cosine transform.

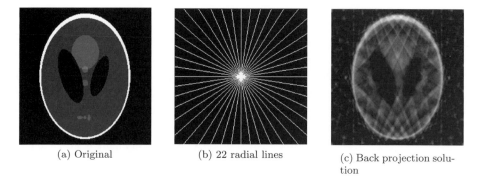

(a) Original                  (b) 22 radial lines                (c) Back projection solution

**Figure 8.9**  Recovering the Shepp–Logan phantom from $k$-space samples on 22 radial lines.

coefficients, which correspond to the points on the 22 radial lines depicted in Figure 8.9(b). (One can further reduce the number of radial lines to as few as 8 by using more accurate algorithms and better models.) Roughly speaking (ignoring the various constraints in MRI sensing), the amount of sensing time required is proportional to the number of samples. CS can potentially reduce the acquisition time by recording only partial Fourier coefficients yet still return faithful reconstructions. Since the phantom image is piece-wise constant and the simulation does not introduce noise, we are able to *exactly* recover the original phantom by total variation minimization given the measurements. On the other hand, if the unsampled Fourier coefficients are replaced by 0, then the inverse Fourier transform gives the image in Figure 8.9(c).

## 8.5    Compressive Sensing-Based Big Data Applications

Compressive sensing is a new signal processing paradigm and aims to encode sparse signals by using far fewer measurements than those in traditional Nyguist setups. It has attracted great attention from researchers in both academia and industry recently. However, there is a significant gap between the theoretical breakthrough of CS and the practical implementation of such a technique, especially in the applications of big data. The previous sections in this chapter introduce the CS techniques and the typical process of using the CS idea as tools in their systems for engineers. In this section, we plan to demonstrate its effective applications and inspire more future possible applications. The Rice University maintains a website for CS related topics [452]. Due to page limitation, we only list a few topics and briefly discuss as follows.

### 8.5.1    CS-Based Analog-to-Digital Converter

An analog-to-digital converter (ADC) is a device that uses sampling to convert a continuous quantity to a discrete time representation in digital form. The reverse operation is performed by a digital-to-analog converter (DAC). ADC and DAC are the gateways between the analog world and digital domain. Most signal processing tasks are implemented in the digital domain. Therefore, ADC and DAC are the key enablers for the digital signal processing and have significant impacts on the system performances.

Next we study two major types of CS-ADC architectures, random demodulator and modulated wideband converter. Then, we investigate the unified framework, Xampling, and briefly explain the other types of implementation.

A data acquisition system, called a random demodulator, is mainly proposed by Rice University [453, 454, 455]. The input signal is assumed to be sparse in some domains such as the frequency domain. In a random demodulator system, the input signal is multiplied by a high-speed sequence from a pseudo-random number generator in the analog domain. As a result, the original signal is modulated and spread out like a CDMA system. Then after a low-pass filter, the modulated signal is sampled with a much slower rate than the Nyquist rate. The intuition behind is that the sparse information can still be recovered from the piece of information obtained through the low-pass filter. In some sense, the random demodulator is implemented from the time domain. An alternative approach is modulated wideband converter [456, 457, 458], which achieves the compression through the frequency domain.

In [459], the comparison of modulated wideband converter and random demodulator is provided. In the following, we try to summarize the several main points:

- Robustness to model mismatch
  The random demodulator is sensitive to inputs with tones slightly displaced from the theoretical grid [459], while the modulated wideband converter is less sensitive to model mismatches.
- Hardware complexity
  The sampling stages of both approaches are similar. So different analog properties of the hardware to ensure accurate mapping to CS cause different design complexities. For random demodulator, time-domain properties of the hardware dictate the necessary accuracy. For modulated wideband converter, the periodicity of the waveforms and low-pass filter need accurate frequency-domain properties. This is especially critical around the cutoff frequency of the low-pass filter.
- Computation load
  For random demodulator, computational loads and memory requirements in the digital domain are the bottleneck. For modulated wideband converter, it is limited for generating the periodic waveforms, which depends on the specific choice of waveform.

Overall, to select the best choice of compressive sensing ADC depends on the analog preprocessing complexity and the input signal properties. In [459], several operative conclusions are drawn for the choice:

1. set system parameters with safeguards to accommodate possible model mismatches
2. incorporate design constraints that suit the technology generating the source signals
3. balance between nonlinear and linear reconstruction complexities.

The Xampling architecture was introduced by Professor Yonina C. Eldar group in Technion, Haifa, Israel [459, 460, 461, 462]. Xampling targets on low-rate sampling and processing of signals lying in a union of subspaces. Xampling consists of the following two blocks:

1. analog compression: narrows down the input bandwidth prior to sampling with commercial devices
2. nonlinear algorithm: detects the input subspace prior to conventional signal processing.

The major difference is that the core DSP algorithm is the subspace based algorithm (in addition to $l_1$ reconstruction for traditional CS algorithms). Based on this framework, there are many analog CS applications such as a general filter-bank scheme for sparse shift-invariant spaces, periodic nonuniform sampling, and modulated wideband conversion for multiband communications with unknown carrier frequencies, acquisition techniques for finite rate of innovation signals with applications to medical and radar imaging, and random demodulation of sparse harmonic tones [463]. Detailed hardware implementation is also discussed for the practical constraints.

## 8.5.2    Communications

To exploit CS in wireless communication, many applications using CS have been proposed in literature [464, 465, 466]. However, CS has more places to be adopted and emphasized. We only list the following topics related to wireless networking, but hope this can motivate the reader to discover more in the future.

- Channel Estimation
  The wireless channels place fundamental limitations on the performance of many wireless communication systems. A transmitted radio signal propagating from a transmitter to a receiver is generally reflected, scattered, and attenuated by the obstacles in the field. At the receiver, two or more slightly delayed versions of the transmitted signal will superpose together. This so-called fading phenomena in wireless communication channels is a double-edged sword to wireless communication systems [467, 468]. On the one hand, this multipath propagation phenomenon will result in severe fluctuations of the amplitudes, phases, and delays of a radio signal over a short period of time or distance. Interpreting the resultant received signals, which vary widely in amplitude and phase, is a critical design challenge to the receiver. On the other hand, multipath will enhance the time, spatial, and frequency diversity of the channel available for communication, which will lead to gains in data transmission and channel reliability. To obtain the diversity gain and alleviate the side effect of the channel fading, the knowledge of channel state information (CSI) is necessary. Thus, channel estimation is of significant importance for the transceiver design.
  The CS theory [396, 395, 437] is a new technology that has emerged in the area of signal processing, statistics, and wireless communication. By utilizing the fact that a

signal is sparse or compressible in some transform domain, CS can powerfully acquire a signal from a small set of randomly projected measurements with the sampling rate much lower than the Nyquist sampling rate. The word "sparse" means that a signal vector only contains very few nonzero entries. Then, efficient methods such as basis pursuit (BP) [469], orthogonal matching pursuit (OMP) [172], and $l_1$ optimization, can be used to reconstruct the original signal with high fidelity. As more and more experimental evidence suggest that broadband or multi-antenna communication channels have an inherent sparse multipath structure, CS has recently become a promising tool to deal with the channel estimation problem [470, 471, 472, 473, 474, 475].

In a nutshell, CSI is of significant importance for high-speed wireless communication over a multipath channel both at transmitters and receivers. Traditional training-based methods usually probe the channel response in time, frequency, and space domains using a sequence of known training symbols, and then the channel response is reconstructed from the received signals by linear reconstruction techniques. This method is known to be optimal for rich multipath channels. However, more and more physical measurements and experimental evidence indicate that many wireless channels encountered in practice tend to exhibit a sparse multipath structure. In this chapter, we present a new approach to estimating sparse (or effectively sparse) multipath channels that is based on recent advances in the theory of CS. The CS-based approaches can potentially achieve a target reconstruction error using far less energy and, in many instances, latency and bandwidth than that dictated by the traditional least squares-based training methods.

- Ultra-Wideband
  Ultra-wideband (UWB) is one of the major breakthroughs in the area of wireless communications, which is highly sparse in the time domain. Hence, it is natural to introduce CS in the design of UWB systems in order to improve the performance of UWB signal acquisition. The key feature of a UWB system is the narrow pulse width in the time domain, which makes the direct sampling in the time domain very difficult. Hence, we can compress the UWB signal such that it is dispersed in the time domain and fewer samples can reconstruct the original UWB signal. The signal compression can be carried out either at the transmitter or at the receiver, as follows:

  – At the transmitter, we can use a filter to compress the UWB signal.
  – At the receiver, we can use either a microstrip circuit or an array of correlators to complete the signal compression.

  Once the UWB signal is compressed, we can reconstruct the original UWB signal from the samples. Traditional reconstruction algorithms in CS, such as BP or OMP, can be applied directly. However, it is better to take the features of UWB signals into account. Hence, we have discussed the following two approaches suitable for UWB systems:

– block CS, in which we fully utilize the feature of blocked locations of nonzero elements in UWB signals
– Bayesian CS, in which the thermal noise is taken into account.

- Precise Positioning
It is easy to imagine that there are many applications of geographical positioning. For example, in a battlefield, it is very important to know the location of a soldier or a tank. In cellular networks, the location of the mobile user can be used for Emergency-911 services. For goods and item tracking, it is of key importance to track their locations. The precision of positions also ranges from subcentimeters (e.g., robotic surgery) to tens of meters (bus information).

Several classifications of positioning technologies are given below [476]:

  – Classified by signaling scheme: In positioning, the target needs to send out signals to base stations or receive signals from base stations in order to determine the target's position. Essentially, the signal needs to be wireless. Radio frequency (RF), infrared, or optical signals can be used.
  – Classified by RF waveforms: Various RF signals can be used for positioning, such as UWB, CDMA, and OFDM.
  – Classified by positioning-related metrics: The metrics include time-of-arrival (TOA), time difference of arrival (TDOA), angle of arrival (AOA), and received signal strength (RSS).
  – Classified by positioning algorithm: When triangulation-based algorithms are used, the positioning is obtained from the intersections of lines, based on metrics such as AOA. In trilateration-based algorithms, the position is obtained from the intersections of circles, based on metrics such as TDOA or TOA. In fingerprinting-based (also called pattern-matching) algorithms, a training period will be spent to establish a mapping between the location and the received signal fingerprinting (or pattern).

Compressive sensing can be used in various types of positioning systems. The application of CS is classified into the following two categories:

- Direct application: In this category, the positioning problem is formulated as a standard CS one. The key challenge is how to find a linear transformation between the position information and observation, as well as how to find the sparsity.
- Indirect application: The positioning is not formulated as a CS problem. Here, the CS is used to improve the precision of signal acquisition, thus increasing the accuracy of positioning. However, the CS for signal acquisition can be jointly carried out with the positioning algorithm.

- Cognitive Radio Network and Sensor Network
Ever since the 1920s, every wireless system has been required to have an exclusive license from the government in order not to interfere with other users of

the radio spectrum. Today, with the emergence of new technologies that enable new wireless services, virtually all usable radio frequencies are already licensed to commercial operators and government entities. According to former US Federal Communications Commission (FCC) chair William Kennard, we are facing a "spectrum drought" [477]. On the other hand, not every channel in every band is in use all the time; even for premium frequencies below 3 GHz in dense, revenue-rich urban areas, most bands are quiet most of the time. The FCC in the United States and the Ofcom in the United Kingdom, as well as regulatory bodies in other countries, have found that most of the precious, licensed radio frequency spectrum resources are inefficiently utilized [478, 479].

In order to increase the efficiency of spectrum utilization, diverse types of technologies have been deployed. Cognitive radio is one of those that leads to the greatest technological gain in wireless capacity. Through the detection and utilization of the spectra that are assigned to the licensed users but standing idle at certain times, cognitive radio acts as a key enabler for spectrum sharing. Spectrum sensing, aiming at detecting spectrum holes (i.e., channels not used by any primary users), is the precondition for the implementation of cognitive radio. The Cognitive Radio (CR) nodes must constantly sense the spectrum in order to detect the presence of the Primary Radio (PR) nodes and use the spectrum holes without causing harmful interference to the PRs. Hence, sensing the spectrum in a reliable manner is of vital importance and constitutes a major challenge in CR networks. However, detection is compromised when a user experiences shadowing or fading effects or fails in an unknown way. To get a better understanding of the problem, consider the following example: a typical digital TV receiver operating in a 6 MHz band must be able to decode a signal level of at least $-83$ dBm without significant errors [480]. The typical thermal noise in such bands is $-106$ dBm. Hence a CR that is 30 dBm more sensitive has to detect a signal level of $-113$ dBm, which is below the noise floor [481]. In such cases, one CR user cannot distinguish between an unused band and a deep fade. In order to combat such effects, recent studies suggest collaboration among multiple CR nodes for improving spectrum sensing performance.

Collaborative spectrum sensing (CSS) techniques have been introduced to improve the performance of spectrum sensing. By allowing different secondary users to collaborate and share their information, PR detection probability can be greatly increased. There are many results that address cooperative spectrum sensing schemes and challenges. The performance of hard-decision combining scheme and soft-decision combining scheme is investigated in [482, 483]. In these schemes, all secondary users send sensing reports to a common decision center. Cooperative sensing can also be done in a distributed way, where the secondary users collect reports from their neighbors and make the decision individually [484, 485, 486]. Optimized cooperative sensing is studied in [487, 488]. When the channel that forwards sensing observations experiences fading, the sensing performance degrades significantly. This issue is investigated in [489, 490]. Furthermore, energy efficiency in CSS is addressed in [491].

- Multiple Access

  In wireless communications, an important task is the multiple access that resolves the collision of the signals sent from multiple users. Traditional studies assume that all users are active and thus the technique of multiuser detection can be applied. However, in many practical systems like wireless sensor networks, only a random and small fraction of users send signals simultaneously. In this chapter, we study the multiple access with sparse data traffic, in which the task is to recover the data packets and the identities of active users. The general problem can be formulated as a CS one due to the sparsity of active users. Particularly, the feature of discrete unknowns will be incorporated into the reconstruction algorithm. The CS-based multiple access scheme will further be integrated with the channel coding. It has been shown to be effective in the application in the advanced metering infrastructure (AMI) in smart grid using real measurement data.

### 8.5.3    Hyperspectral Imaging

Unlike traditional imaging systems, hyperspectral imaging (HSI) [492, 493] acquires a scene with several millions of pixels in up to hundreds of contiguous wavelengths. Such high-resolution spatio-spectral hyperspectral data (i.e., three-dimensional (3D) datacube organized in the spatial and spectral domain) has an extremely large data size and enormous redundancy, which makes CS [393, 494] a promising solution for hyperspectral data acquisition.

Owing to the inherent 3D structure present in the hyperspectral datacube and the two-dimensional nature of optical sensing hardware, CS-based hyperspectral imagers generally capture a group of linear measurements across a spectral strip or a spatial slice of the datacube at a time. In other words, either the 2D spatial extent of the scene for a single spectral band or the spectral extent for a single spatial (pixel) location is compressed, which can be typically categorized as framing acquisition and pixel-based acquisition [495, 496], respectively. The compressed version of the data can then be sent to a fusion station that will recover the original 3D datacube by utilizing a CS reconstruction algorithm.

To date most existing designs for hyperspectral imagers are based on the framing acquisition in the spatial direction and require a 2D array of sensors or digital micromirror device (DMD) matching the spatial resolution [497, 498, 499, 500]. To obtain images with good spatial resolutions, a large dense array of sensors or DMD has to be applied, thereby entailing potentially prohibitive costs. Furthermore, from an applications perspective, one may only be interested in a specific spatial location in a large-scale scene, or objects of interest arise and move in the scene (e.g., target detection and tracking [501, 502, 503]). Also, different spatial areas may have different requirements of imaging precision. These considerations have motivated the need of pixel-based acquisition in order to permit low-cost and flexible HSI acquisition [495, 496].

Furthermore, most existing reconstruction algorithms for both framing acquisition and pixel-based acquisition can only take advantage of the spatial and spectral information of hyperspectral data from the aspect of sparsity. This is because the foundation of these algorithms is built on conventional CS, which reconstructs the signals by solving a convex programming and proceeds without exploiting additional information (aside from sparsity or compressibility) [393, 494]. For hyperspectral data, the spatial correlation and spectral correlations, which not only reflect in the correlation between the sparse structure of the data (i.e., structured sparsity), but also in the correlation between the amplitudes of the data, can be used to provide helpful prior information (or side information) in the reconstruction processed and assist on increasing the compression ratios and/or decreasing distortion.

## 8.5.4 Data Streaming

In connection-oriented communication, a data stream is a sequence of digitally encoded coherent signals (packets of data or data packets) used to transmit or receive information that is in the process of being transmitted. Data stream mining is the process of extracting knowledge structures from continuous, rapid data records. A data stream is an ordered sequence of instances that in many applications of data stream mining can be read only once or a small number of times using limited computing and storage capabilities. Examples of data streams include computer network traffic, phone conversations, ATM transactions, web searches, and sensor data. Data stream mining can be considered a subfield of data mining, machine learning, and knowledge discovery.

In many data stream mining applications, the goal is to predict the class or value of new instances in the data stream given some knowledge about the class membership or values of previous instances in the data stream. Machine learning techniques can be used to learn this prediction task from labeled examples in an automated fashion. Often, concepts from the field of incremental learning, a generalization of incremental heuristic search, are applied to cope with structural changes, online learning, and real-time demands. In many applications, especially operating within non-stationary environments, the distribution underlying the instances or the rules underlying their labeling may change over time (i.e. the goal of the prediction, the class to be predicted or the target value to be predicted, may change over time). This problem is referred to as concept drift.

Some CS related data streaming approaches are listed as heavy hitters, random sampling, histogram maintenance, dimension reduction, and embeddings.

# 9    Distributed Large-Scale Optimization

In this chapter, we discuss distributed algorithms for solving large-scale optimization problems over networks. In particular, the optimization problem is defined over a network of nodes/agents where each of them may have a piece of data, or may be in charge of updating a subset of variables, and they are required to collaboratively perform the desired optimization. The key question to be addressed is, under various problem settings, what protocols should the agents follow so as to efficiently optimize the overall problem. These types of distributed optimizing algorithms have recently become increasingly important in various big data problems, with applications ranging from distributed machine learning, signal processing, networking, multi-agent systems, parallel optimization, etc. There are a few driving forces behind these new trends, as we briefly list below:

1. The advances in sensor technologies makes distributed data acquisition ubiquitous.
2. The increasing size of optimization problems in big data era calls for distributed data storage.
3. Privacy issues in big data problem prevents direct data exchange among different (possibly geographically distributed) data sources.
4. Using multi-core high-performance computing systems (with either shared or distributed memory) to solve big data optimization problems can significantly reduce the solution time.

In this chapter, we present a few recent algorithms for solving such distributed optimization problems. We will cover a wide range of algorithms and discuss their convergence behavior, as well as their performance in terms of scaling with network/problem sizes.

The chapter is organized as follows. In Sections 9.1–9.4, we present a general problem formulation and a few popular first-order methods. In Section 9.5, we discuss a special cases of the general formulation (i.e., the global consensus problem). In Section 9.6, we compare various algorithms presented in this chapter.

## 9.1    Background

### 9.1.1    Problem Formulation

#### The Optimization Problem
Consider a network with $N$ distributed agents, each capable of communicating with its immediate neighbors; see Figure 9.1 for an illustration. We consider the following

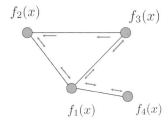

$f_2(x)$    $f_3(x)$

$f_1(x)$    $f_4(x)$

**Figure 9.1** Illustration of a distributed network with $N = 4$ nodes. The arrows in the figure denotes the possible communication directions.

problem of optimizing the sum of possibly nonconvex and nonsmooth objective function having $N$ components

$$\min_{\mathbf{x}} \quad f(\mathbf{x}) := \sum_{i=1}^{N} f_i(\mathbf{x}),$$

(9.1)

$$\text{s.t.} \quad \mathbf{x} \in X.$$

Here $\mathbf{x} \in \mathbb{R}^M$ is the optimization variable; $f_i(\cdot) : \mathbb{R}^M \to \mathbb{R}$ is a component objective function, which represents the local objective function of agent $i$; $X$ is a closed convex set. Let us denote the optimal solution for this problem by $\mathbf{x}^*$, and denote the optimal objective as $f^*$.

### The Network Models

Let us use a graph $\mathcal{G}$ to denote a multi-agent network, which contains a node set $\mathcal{V} = \{1, \cdots, K\}$, and an edge set $\mathcal{E}$. An edge $(ij) \in \mathcal{E}$ if node $i$ can directly communicate with node $j$. Note that depending on whether the underlying communication link is *bidirectional* or not, the graph $\mathcal{G}$ can be either directed or undirected. Let us define the adjacency matrix $\mathbf{A} \in \{0, 1\}^{N \times N}$, where $\mathbf{A}[i,j] = 1$ if $(i,j) \in \mathcal{E}$, and $\mathbf{A}[i,j] = 0$ otherwise. Define the set of in/out neighbors of node $i$ as

$$\mathcal{N}_i^{\text{in}} := \{j \mid (ji) \in \mathcal{E}\}, \quad \mathcal{N}_i^{\text{out}} := \{j \mid (ij) \in \mathcal{E}\}.$$

If the network is undirected, then $\mathcal{N}_i^{\text{in}} = \mathcal{N}_i^{\text{out}} := \mathcal{N}_i$. Similarly, define the in/out degree matrix of node $i$ as

$$\mathbf{D}^{\text{in}} := \text{diag}\{|\mathcal{N}_1^{\text{in}}|, |\mathcal{N}_2^{\text{in}}|, \cdots, |\mathcal{N}_N^{\text{in}}|\} \quad \mathbf{D}^{\text{out}} := \text{diag}\{|\mathcal{N}_1^{\text{out}}|, |\mathcal{N}_2^{\text{out}}|, \cdots, |\mathcal{N}_N^{\text{out}}|\}.$$

If the graph is undirected, then the degree matrix and the Laplacian matrix are given by

$$\mathbf{D} := \text{diag}\{|\mathcal{N}_1|, |\mathcal{N}_2|, \cdots, |\mathcal{N}_N|\}, \quad \mathbf{L} = \mathbf{D} - \mathbf{A}.$$

For a given graph $\mathcal{G}$, let us define an associated weight matrix $\mathbf{W} \in \mathbb{R}^{N \times N}$ as

$$[\mathbf{W}]_{i,j} = \begin{cases} w_{ij} > 0, & \text{if } (ij) \in \mathcal{E}, \text{ or } i = j \\ 0, & \text{otherwise} \end{cases}$$

(9.2)

$$\sum_{j \in \mathcal{V}} [\mathbf{W}]_{i,j} = 1.$$

(9.3)

Clearly the weight matrix is *row-stochastic*. If further it satisfies

$$\sum_{j\in\mathcal{V}}[\mathbf{W}]_{j,i} = 1,$$    (9.4)

then it is a *doubly stochastic matrix*.

Consider a series of graphs $\mathcal{G}(1), \mathcal{G}(2), \cdots$, with respective edge set $\mathcal{E}(t) \subseteq \mathcal{E}$, $t = 1, 2, \cdots$. For each edge set, we will use $\mathbf{W}(t)$ to denote the associated weight matrix whose elements are given by $\{w_{ij}(t)\}$. We call the series of graphs $\{\mathcal{G}(t)\}$ a *B-connected* graph if there exists a constant $B \geq 1$ such that the following graph

$$\left(\mathcal{V}, \mathcal{E}(kB)\bigcup\mathcal{E}(kB + 1),\cdots,\bigcup\mathcal{E}((k + 1)B - 1)\right)$$    (9.5)

is strongly connected[1] for all non-negative integers $k$.

### 9.1.2    Applications

Let us present a few applications of the general formulation (9.1).

**Regression**

Let us consider a formulation that includes classical problems such as LASSO [504], basis pursuit [505], and logistic regression (LR) [506] as special cases. Let $\mathbf{A} = [\mathbf{A}_1^T,\cdots,\mathbf{A}_N^T] \in \mathbb{R}^{KN\times M}$ denote the data matrix, where $KN$ is the number of data samples and $M$ is the dimension of the problem. One can formulate the distributed regression problem by [505, Fig. 1], [507]

$$\min_{\mathbf{x}\in X} \sum_{i=1}^{N} f_i(\mathbf{x}),$$    (9.6)

where $f_i(\mathbf{x})$ is the cost function defined on the local data set $\mathbf{A}_i \in \mathbb{R}^{K\times M}, \mathbf{b}_i \in \mathbb{R}^K$. For example, the LASSO problem has $f_i(\mathbf{x}) = \|\mathbf{b}_i - \mathbf{A}_i\mathbf{x}\|_2^2 + h_i(\mathbf{x})$ where the function $h_i(\mathbf{x})$ is some nonsmooth regularization term. Clearly problem (9.6) is in the form of (9.1).

**Computing Aggregates in P2P Networks**

Consider a data network with a set of $\mathcal{V}$ nodes, where each node $i \in \mathcal{V}$ has a data file of size $\theta_i$, which is the local private information. The nodes are connected via a static undirected graph $\mathcal{G} = \{\mathcal{V}, \mathcal{E}\}$ for a given link set $\mathcal{E}$. The question is how to distributedly compute the averaged size of the file. The problem can be formulated as the following optimization problem [508]:

$$\min_{x\in\mathbb{R}} \sum_{i=1}^{N}(x - \theta_i)^2$$    (9.7)

which is obviously in the form of (9.1).

---

[1]  A strongly connected graph is a graph in which every node $i \in \mathcal{V}$ can be reached by every other node $j \in \mathcal{V}$.

**Distributed Matrix Factorization**

Let us consider the following nonconvex low-rank matrix factorization problem [509]:

$$\min_{L,R} \quad \frac{1}{2}\left\|O - LR^T\right\|_F^2 + \lambda_1\|L\|_F^2 + \lambda_2\|R\|_F^2$$

$$= \min_{L,\{R_i\}} \quad \frac{1}{2}\sum_{i=1}^{N}\left\|O_i - LR_i^T\right\|_F^2 + \lambda_1\|L\|_F^2 + \lambda_2\|R\|_F^2. \tag{9.8}$$

Here $O \in \mathbb{R}^{M \times N}$, $O_i \in \mathbb{R}^{M \times 1}$ is a column of $O$; $L \in \mathbb{R}^{M \times r}$, and $R \in \mathbb{R}^{N \times r}$, and $R_i \in \mathbb{R}^{1 \times r}$ is a row of $R$; $\lambda_1 \geq 0$ and $\lambda_2 \geq 0$ are two constants. Typically the ranks of the matrices $L$ and $R$ are much smaller than $\min\{M, N\}$. Such matrix factorization problem is interesting because we can obtain a low-rank representation of our observation matrix. Consider the case where there are $N$ agents in the system, and each of them has a piece of the data matrix $O$, namely $O_i$. Such scenario arises when $M, N$ are large so that distributed storage is beneficial, or when each piece of $O$ needs to be kept in private. In this case, the local function $f_i$ in (9.1) is given by

$$f_i(L, R_i) = \frac{1}{2}\left\|O_i - LR_i^T\right\|_F^2 + \lambda_1\|L\|_F^2 + \lambda_2\|R_i\|^2.$$

## 9.2 Distributed Gradient/Subgradient Methods

### 9.2.1 Algorithm Description

The distributed gradient/subgradient method is one of the classical methods for distributed optimization. Let us consider the unconstrained version of (9.1), given below:

$$\min_{\mathbf{x} \in \mathbb{R}^M} \quad f(\mathbf{x}) := \sum_{i=1}^{N} f_i(\mathbf{x}), \tag{9.9}$$

where we will assume throughout this section that each $f_i$ is a convex but not necessarily smooth function. We will assume that a subgradient of $f_i(\mathbf{x})$, denoted as $\partial f_i(\mathbf{x})$, exists for all $\mathbf{x} \in \mathbb{R}^M$.

The key idea of distributed gradient/subgradient method is to let each agent $i$ keep a local copy of the global variable $\mathbf{x}$, denoted as $\mathbf{x}_i$, and to perform the following two tasks simultaneously:

1. Ensure consensus over all the agents.
2. Find the global optimal solution.

To this end, let us use $t = 1, 2, 3\cdots$, as the iteration counter, and use $\mathbf{x}_i(t)$ to denote the local copy of $\mathbf{x}$ at $i$th node in $t$th iteration. Let $\mathbf{d}_i(t) \in \partial f_i(\mathbf{x}_i(t))$ denote the subgradient of the local function $f_i$ evaluated at $\mathbf{x}_i(t)$. Let $w_{ij}(t) \geq 0$ denote the weight for the link $(ij)$ at iteration $t$, and let $\gamma > 0$ denote some stepsize parameter. Then the distributed gradient/subgradient method is given by

$$\mathbf{x}_i(t+1) = \sum_{j=1}^{N} w_{ij}(t)\mathbf{x}_j(t) - \gamma \mathbf{d}_i(t), \ \forall \ i \in \mathcal{V}. \tag{9.10}$$

This algorithm is very intuitive; each node performs the classical gradient/subgradient update for its local objective, based on the average of local copies of **x** that belongs to its (in)neighbors. Utilizing the weight matrix defined at the beginning of this chapter, the above iteration can be compactly written as

$$\tilde{\mathbf{x}}(t+1) = (\mathbf{W}(t) \otimes I_M)\tilde{\mathbf{x}}(t) - \gamma \tilde{\mathbf{d}}(t), \ \forall \ i \in \mathcal{V}. \tag{9.11}$$

where $\tilde{\mathbf{x}}(t+1) = \{\mathbf{x}_i(t+1)\}_{i=1}^{N}$, $\tilde{\mathbf{d}}(t) = \{\mathbf{d}_i(t)\}_{i=1}^{N}$, and $I_M$ denote an identity matrix of size $M$.

One can also take an alternative perspective of this algorithm, by expanding the above recursion (9.11). Specifically, let us define a *transition matrix* $\Phi(t, s)$ for any $s$ and $t$ such that $t \geq s$, by

$$\Phi(t, s) = \mathbf{W}(t)\mathbf{W}(t-1), \cdots, \mathbf{W}(s+1)\mathbf{W}(s). \tag{9.12}$$

Then we can expand (9.11), and represent $\mathbf{x}_i(t+1)$ using $x_1(s), \cdots, x_N(s)$ by

$$\mathbf{x}_i(t+1) = \sum_{j=1}^{N} [\Phi(t, s)]_{ij}\mathbf{x}_j(s) - \sum_{r=s}^{t-1} \sum_{j=1}^{N} [\Phi(t, r+1)]_{ij}\mathbf{d}_j(r)\gamma - \gamma\mathbf{d}_i(t), \ \forall \ s \leq t$$

## 9.2.2    Convergence Analysis and Variants

The distributed gradient/subgradient iteration (9.11) was analyzed in [510] by Nedić and Ozdaglar. The main assumptions are listed below.

ASSUMPTION 1. *Assume the following additional assumptions about the weight matrix* **W** *and the graph* $\mathcal{G}$:

1. *The graph is time-varying, and obeys the B-connected rule.*
2. *The weight matrix at a given time instance t follows the definition in (9.2) and (9.3). Further, each* $\mathbf{W}(t)$ *is symmetric.*
3. *The positive entries of* $\mathbf{W}(t)$ *are uniformly bounded away from zero. That is, there exists a scalar* $\eta \in (0, 1)$ *and for all* $i, j, t$

$$\text{if } w_{ij}(t) > 0, \text{ then } w_{ij}(t) > \eta.$$

One of the main results regarding the convergence of the distributed gradient/subgradient iteration given in [510] is stated below.

THEOREM 13. *Assume that Assumption 1 holds true. Further assume that the initial vectors* $\{\mathbf{x}_i(0)\}_{i=1}^{N}$ *satisfy (for some constant $L > 0$)*

$$\max_i \|\mathbf{x}_i(0)\| \leq \gamma L,$$

*and that the subgradients are bounded*

$$\|\mathbf{d}_j(t)\| \leq L, \quad j = 1, \cdots, N, \quad t = 1, 2, \cdots. \tag{9.13}$$

*Let $\hat{\mathbf{x}}(T)$ denote the averaged vector of $\{\mathbf{x}(t)\}_{t=1}^{T}$, that is,*

$$\hat{\mathbf{x}}_i(T) = \frac{1}{T} \sum_{t=0}^{T-1} \mathbf{x}_i(t), \quad i = 1, \cdots, N.$$

*Then we have*

$$f\left(\hat{\mathbf{x}}_i(T)\right) \leq f^* + \frac{N dist^2(\mathbf{y}(0), X^*)}{2\gamma T} + \gamma L \left(\frac{LC}{2} + 2N\hat{L}_1 C_1\right) \tag{9.14}$$

*where $\mathbf{y}(0) = (1/N) \sum_{i=1}^{N} \mathbf{x}_i(0)$, and $C$ is some positive constant; $f^*$ is the optimal objective value; $X^*$ is the optimal solution set; $\hat{L}_1 > 0$ is a constant that bounds the size of the subgradient evaluated at $\{\hat{\mathbf{x}}_i(t)\}$.*

From the above theorem, it is clear that the error term consists of the following two parts:

1. the first term, which is related to the initial solution and is inversely proportional to the stepsize $\gamma$ as well as the number of iterations
2. the second term, which is a constant proportional to the stepsize $\gamma$.

One can see that the first error term decreases with the progress of the algorithm while the second term stays constant. This is not surprising as we have used the constant stepsize $\gamma > 0$ in the algorithm. If a decreasing stepsize is used, then one can show that the algorithm converges to the true global optimal solution.

We briefly mention that the key step in the proof is to characterize the limiting behavior of the transition matrix $\Phi(t, s)$. Indeed, one can show that in the limit, such a transition matrix converges to a uniform steady state distribution for all $s$, that is,

$$\lim_{t \to \infty} \Phi(t, s) = \Phi(s) = \frac{1}{N} \mathbf{1} \mathbf{1}^T, \quad \forall s$$

where $\mathbf{1}$ is the all-one vector. Intuitively, this means that in the limit, the consensus will be easily attainable, as $\mathbf{x} = \Phi(s)\mathbf{x}$. What is perhaps more interesting is that such convergence is in fact *linear*, that is, the entries $[\Phi(t, s)]_{ij}$ converge to $1/N$ with the following rate [511, Proposition 2]):

$$\left|[\Phi(t, s)]_{i,j} - \frac{1}{N}\right| \leq 2\frac{1 + \eta^{-B_0}}{1 - \eta^{B_0}} (1 - \eta^{B_0})^{(t-s)/B_0} \tag{9.15}$$

for all $s$ and $t$ with $t \geq s$, where $\eta$ is the lower bound of the positive components in $\mathbf{W}(t)$, $B_0 = (N - 1)B$.

The above results about the convergence of the distributed subgradient/gradient methods can be generalized significantly in a few directions, when making different (often more relaxed) assumptions on the network as well as the problem instances. For example, Nedić, Ozdaglar, and Parrilo [512] consider using distributed gradient/subgradient iteration to solve certain constrained consensus problem. In particular, consider solving the following feasibility problem over a network of $N$ agents:

$$\text{Find} \quad \{\mathbf{x}_1, \cdots, \mathbf{x}_N\}$$
$$\text{s.t.} \quad \mathbf{x}_i \in X_i, \ \forall \, i \tag{9.16}$$
$$\mathbf{x}_i = \mathbf{x}_j, \ \forall \, i \neq j,$$

where each $X_i \subseteq \mathbb{R}^M$ is a closed convex set known only to agent $i$. Basically the problem can be interpreted as finding the intersection of $N$ convex sets. We assume that there is a vector $\bar{\mathbf{x}} \in \text{int}(\bigcap_i X_i)$; that is, there exists a scalar $\delta > 0$ such that

$$\{\mathbf{z} \mid \|\mathbf{z} - \bar{\mathbf{x}}\| \leq \delta\} \subset \bigcap_i X_i. \tag{9.17}$$

The distributed subgradient/gradient-type method proposed in [512] is given by the following "average and project" iteration:

$$\mathbf{x}_i(t+1) = \text{proj}_{X_i} \left[ \sum_{j=1}^{N} w_{ij}(t) \mathbf{x}_j(t) \right], \quad i = 1, \cdots, N, \tag{9.18}$$

where "$\text{proj}_{X_i}(\cdot)$" denotes the projection onto the set $X_i$. As has been pointed out in [512], there is an interesting connection between the above algorithm and the distributed gradient/subgradient algorithm. Note that problem (9.16) is equivalent to the following *unconstrained* problem:

$$\min_{\mathbf{x}} \quad \frac{1}{2} \sum_{i=1}^{N} \|\mathbf{x} - \text{proj}_{X_i}[\mathbf{x}]\|^2 := \sum_{i=1}^{N} f_i(\mathbf{x}). \tag{9.19}$$

By utilizing the fact that

$$\nabla f_i(\mathbf{x}) = \nabla (1/2) \|\mathbf{x} - \text{proj}_{X_i}[\mathbf{x}]\|^2 = \mathbf{x} - \text{proj}_{X_i}[\mathbf{x}],$$

one can write down the distributed gradient/subgradient iteration below:

$$\mathbf{x}_i(t+1) = \sum_{j=1}^{N} w_{ij}(t) \mathbf{x}_j(t) - \gamma \left( \mathbf{x}_i(t) - \text{proj}_{X_i}[\mathbf{x}_i(t)] \right)$$

while the iterates (9.18) is given by

$$\mathbf{x}_i(t+1) = \sum_{j=1}^{N} w_{ij}(t) \mathbf{x}_j(t) - \left( \sum_{j=1}^{N} w_{ij}(t) \mathbf{x}_j(t) - \text{proj}_{X_i} \left[ \sum_{j=1}^{N} w_{ij}(t) \mathbf{x}_j(t) \right] \right).$$

That is, a constant stepsize $\gamma = 1$ is used, and the gradient $\nabla f_i(\cdot)$ is evaluated on the average $\sum_{j=1}^{N} w_{ij}(t) \mathbf{x}_j(t)$.

Using essentially all the assumptions given in Assumption 1 (except that the symmetricity assumption is replaced by the *doubly stochastic* assumption (9.4)) plus the interior-point assumption (9.17), then one can claim that in the limit, global consensus is achieved, that is, [512, Proposition 2]

$$\lim_{t \to \infty} \|\mathbf{x}_i(t) - \tilde{\mathbf{x}}\| = 0, \quad \forall \, i, \tag{9.20}$$

for some $\tilde{\mathbf{x}} \in \bigcap_i X_i$. Further, one can show that the iterates converge *linearly*, that is,

$$\sum_{i=1}^{N} \|\mathbf{x}_i(t) - \tilde{\mathbf{x}}\|^2 \leq \left(1 - \frac{1}{4R^2}\right)^t \sum_{i=1}^{N} \|\mathbf{x}_i(0) - \tilde{\mathbf{x}}\|^2, \qquad (9.21)$$

where $R = \frac{1}{\delta} \sum_{i=1}^{N} \|\mathbf{x}_i(0) - \bar{\mathbf{x}}\|^2$, and $\delta$ and $\bar{\mathbf{x}}$ is given in (9.17).

Interestingly, the approach here can be further generalized to solve the following constrained optimization problem:

$$
\begin{aligned}
\min \quad & \sum_{i=1}^{N} f_i(\mathbf{x}) \\
\text{s.t.} \quad & \mathbf{x} \in \bigcap_{i=1}^{N} X_i.
\end{aligned}
\qquad (9.22)
$$

The algorithm here is the projected version of (9.10) (named projected subgradient), given below:

$$
\mathbf{v}_i(t) = \sum_{j=1}^{N} w_{ij}(t)\mathbf{x}_i(t)
\qquad (9.23)
$$

$$
\mathbf{x}_i(t+1) = \text{proj}_{X_i} \left[\mathbf{v}_i(t) - \gamma(t)\mathbf{d}_i(t)\right].
$$

First let us analyze the simpler case where $X_i = X$ for all $i$ (the same constraint set assumption). Suppose that the stepsize $\gamma(t)$ is chosen as

$$
\sum_{t=1}^{\infty} \gamma(t) = \infty, \quad \sum_{t=1}^{\infty} \gamma^2(t) < \infty.
\qquad (9.24)
$$

then the projected subgradient converges to the optimal solution of problem (9.22); that is, there exists $\mathbf{x}^* \in X^*$ such that

$$
\lim_{t \to \infty} \|\mathbf{x}_i(t) - \mathbf{x}^*\| = 0, \quad \forall\, i.
\qquad (9.25)
$$

Further, one can show that if the weights $w_{ij}(t)$ are uniform (i.e., $w_{ij}(t) = \frac{1}{N}$ for all $i, j, t$, which implies that the graph is static, and all the agents are connected with the rest of the agents), then under some additional assumptions the projected subgradient algorithm converges to the optimal solution of problem (9.22) (with distinctive $X_i$'s); see [512, Proposition 5].

As another generalization of the classical gradient/subgradient methods, Srivastava and Nedić [513] consider the more practical cases where there is communication noise among the agents, and the communication graph is randomly generated. Such setting closely models practical networks such as the distributed sensor networks, where the nodes in the system are usually activated randomly. In particular, let us use $\mathcal{G} = \{\mathcal{V}, \mathcal{E}\}$ to denote the connected graph that contains all *possible bidirectional* communication links among the agents. The realization of the graph at iteration $t$ is given by a random graph $\mathcal{G}(t) = \{\mathcal{V}, \mathcal{E}(t)\} \subset \mathcal{G}$, which is not required to be bidirectional. The associated weight matrices $\mathbf{W}(t)$'s are i.i.d, with $\bar{\mathbf{W}} = \mathbb{E}[\mathbf{W}(t)] \succ 0$, and $\bar{\mathbf{W}}$ defines a connected graph. The communication noises, denoted as $\xi_{i,j}(t)$ for link $(i, j)$ at iteration $t$, are zero mean, independent among different links, and norm bounded. Consider the following generalization of the gradient/subgradient iteration:

$$\mathbf{v}_i(t+1) = \mathbf{x}_i(t) + \alpha_i(t+1) \sum_{j=1}^{N} [\mathbf{W}(t+1)]_{ij} \left( \mathbf{x}_j(t) + \xi_{ij}(t+1) - \mathbf{x}_i(t) \right)$$

$$\mathbf{x}_i(t+1) = \text{proj}_{X_i} \left[ \mathbf{v}_i(t+1) - \gamma_i(t+1)\mathbf{d}_i(t+1)\chi\{i \in U(t+1)\} \right]$$

where $\alpha_i(t), \gamma_i(t)$ are positive stepsizes; the set $U(t+1)$ is given by

$$U(t+1) := \{i \mid \exists \ell \neq j, \ [\mathbf{W}(t+1)]_{j\ell} > 0\}$$

(i.e., it contains those nodes that have out-neighbors at iteration $t + 1$). Clearly the first step above tries to achieve consensus, while the second step performs the local gradient projection descent. Compared with iteration (9.23), the consensus step is slightly more complicated because of the need to average out the random communication noise.

One version of the convergence claim is the following (cf. [513, Theorem 4]). Suppose that the stepsizes are chosen as follows:

$$\gamma_i(t) = 1/\Gamma_i^{\theta_2}(t), \quad \alpha_i(t) = 1/\Gamma_i^{\theta_1}(t) \tag{9.26}$$

with $\theta_1 \in (1/2, \ 1)$, $\theta_2 > (1 + \theta_1)/2$ and $\Gamma_i(t)$ denotes the total number of times that node $i$ has updated until iteration $t$. Also suppose that the probability $p_i$ of updating any node $i$ at each iteration is strictly positive, and $p_i = p > 0$ for all $i$. Then almost surely $\{\mathbf{x}_i(t)\}$ converges to a common random point in the optimal set $X^*$.

We also briefly discuss the classical problem (which turns out to be a special case of the problems considered in this section) named the "average consensus problem," where $f_i \equiv 0$ for all $i$. Obviously, the only goal here is to reach consensus among all agents. In this case, the iteration (9.11) reduces to the following iteration

$$\tilde{\mathbf{x}}(t+1) = (\mathbf{W}(t) \otimes I_M)\tilde{\mathbf{x}}(t), \ \forall \ i \in \mathcal{V}. \tag{9.27}$$

Intuitively, the agents simply average the most recent iterates of its neighbors. The classical result by Tsitsiklis [514] characterizes the linear convergence of the above iteration.

THEOREM 14 (Convergence for average consensus problem [514]). *Suppose Assumption 1 is satisfied. Then the average consensus iteration (9.27) converges to a consensus with a geometric rate. In particular,*

$$\lim_{t \to \infty} \mathbf{x}_i(t) = \mathbf{x}^*, \quad \forall \ i = 1, \cdots, N,$$

*where $x^*$ is some convex combination of the initial values $\mathbf{x}_1(0), \cdots, \mathbf{x}_N(0)$. Further, we have*

$$\max_i \mathbf{x}_i(t) - \min_j \mathbf{x}_j(t) \leq \left( \max_i \mathbf{x}_i(0) - \min_j \mathbf{x}_j(0) \right) \beta^{\frac{t}{N-1}} \tag{9.28}$$

*with $\beta < 1$ being some constant.*

It is important to note that when only the consensus is sought, the problem becomes easier and one can derive *geometric* rate of convergence. In contrast, such fast rate of convergence is difficult to obtain for general distributed gradient/subgradient iteration (9.11). We will discuss its convergence rate issue in the next subsection.

### 9.2.3 Convergence Rate Analysis

The previous subsection presents a few variants of the distributed gradient/subgradient methods and discusses their convergence results. This section will focus on convergence rate analysis of the related algorithms.

First, it is important to see from Theorem 13 that if a constant stepsize is used, then the algorithm is not going to converge to the global optimal solution. The optimality gap will always be proportional to the size of the stepsize $\gamma$. That is, around the optimal solution set there is a "ball" of size proportional to $\gamma$ to which the algorithm converges. A natural question is then: for a fixed step $\gamma$, how fast should the algorithm should converge to such a "ball"? Recently Yuan, Ling, and Yin [515] have analyzed this question for problem (9.9) with smooth objective functions. They show that the objective error, denoted as $f(1/N \sum_i \mathbf{x}_i(t)) - f^*$, reduces at a speed of $\mathcal{O}(1/t)$ until it reaches $\mathcal{O}(\gamma)$ (i.e., the ball of size proportional to $\gamma$ around the optimal solution set).

More precisely, suppose that the graph $\mathcal{G} = \{\mathcal{V}, \mathcal{E}\}$ is connected, symmetric, and time-invariant, and the associated weight matrix $\mathbf{W}$ is symmetric and doubly stochastic. Suppose that each $f_i$ is a closed and convex function and has Lipschitz gradient with constant $L_i$. Let the stepsize $\gamma$ satisfy

$$\gamma < \mathcal{O}\left(1/\max\{L_1, \cdots, L_N\}\right). \tag{9.29}$$

Let the eigenvalue of $\mathbf{W}$ be sorted as $1 = \lambda_1(\mathbf{W}) \geq \lambda_2(\mathbf{W}) \geq \cdots, \geq \lambda_N(\mathbf{W}) \geq -1$. Define the second largest magnitude of eigenvalues of $\mathbf{W}$ as

$$\beta := \max\{|\lambda_2(\mathbf{W})|, |\lambda_N(\mathbf{W})|\}$$

To bound the convergence speed, we first bound the size of the sequence of gradients. Let $\bar{f}(t) = [\nabla f_1(\mathbf{x}_1(t)); \cdots, \nabla f_N(\mathbf{x}_N(t))] \in \mathbb{R}^{NM \times 1}$, and let $\mathbf{x}(t) = [\mathbf{x}_1(t); \cdots, \mathbf{x}_N(t)] \in \mathbb{R}^{NM \times 1}$. Then [515, Theorem 1] states that if the stepsize $\gamma$ is picked sufficiently small, then the sequence of gradients $\{\bar{f}(t)\}$ is bounded. Specifically, if we have

$$\gamma \leq (1 + \lambda_N(\mathbf{W}))/\max_i L_i$$

then we have

$$\|\bar{f}(t)\| \leq \sqrt{2\max_i L_i \sum_{i=1}^N \left(f_i(0) - f_i^*\right)}.$$

To state the main result, define the error term as $\bar{r}(t) := 1/N \left(f\left(1/N \sum_{i=1}^N \mathbf{x}_i(t)\right) - f^*\right)$. Then we have the following statement [515, Theorem 3]:

THEOREM 15. *Suppose the stepsize $\gamma$ satisfies*

$$\gamma \leq \min \left\{ (1 + \lambda_N(\mathbf{W}))/\max_i L_i, \ N/\sum_{i=1}^{N} L_i \right\}, \tag{9.30}$$

*then while*

$$\bar{r}(t) > C\sqrt{2} \times \frac{\gamma \max_i L_i D}{1 - \beta} = \mathcal{O}\left(\frac{\gamma}{1 - \beta}\right)$$

*where $C, D > 0$ are two constants, the reduction of $\bar{r}(t)$ obeys*

$$\bar{r}(t) \leq \mathcal{O}\left(\frac{1}{\gamma t}\right).$$

Additionally, when the function $f_i$'s are strongly convex, then the convergence rate to the $\mathcal{O}(\gamma)$ ball is *linear*; see [515, Theorem 4].

Let us come back to the convergence rate of the distributed subgradient/gradient iteration (9.10), to the global optimal solution set $X^*$ of problem (9.9). The author of [516] considers the following special case of (9.9) where the nonsmooth terms are explicitly expressed:

$$\min \ \sum_{i=1}^{N} g_i(\mathbf{x}) + h_i(\mathbf{x}). \tag{9.31}$$

Assuming that the proximity operator for the nonsmooth function $h_i$ is easy to compute (cf. (3.11)), the author of [516] proposes the following variants of the distributed subgradient/gradient method, named the distributed proximal gradient method:

$$\mathbf{x}_i(t+1) = \text{prox}_{h_i}^{\gamma(t)} \left[ \mathbf{v}_i(t) - \gamma(t) \nabla g_i(\mathbf{v}_i(t)) \right] \tag{9.32a}$$

$$\mathbf{v}_i(t+1) = \sum_{j=1}^{N} w_{ij}(t) \mathbf{x}_i(t) \tag{9.32b}$$

Clearly, when $h_i \equiv 0$ for all $i$, then the above iteration is simply the distributed gradient iteration (9.10) (with all $\mathbf{d}_i(t) = \nabla f_i(\mathbf{x}_i(t))$). The main assumptions made are given below.

ASSUMPTION 2. *Let us make the following assumptions about problem* (9.31)

1. *Each $g_i$ is convex, continuous differentiable and has Lischitz gradient with constant $L$.*
2. *Each $h_i$ is convex, proper and lower-semicontinuous.*
3. *There exists a scalar $G$ such that for all $\mathbf{x}$, $\|\nabla g_i(\mathbf{x})\| < G$ and $\|\mathbf{z}\| < G$ for each subgradient $\mathbf{z} \in \partial h(\mathbf{x})$.*
4. *$f(\mathbf{x}) = \frac{1}{N}\sum_{i=1}^{N} f_i(\mathbf{x}) = \frac{1}{N}\sum_{i=1}^{N} g_i(\mathbf{x}) + h_i(\mathbf{x})$ has a unique optimizer $\mathbf{x}^*$.*

We have the following convergence rate result [516, Theorem 1]:

THEOREM 16. *Suppose Assumption 1 and 2 hold true, and further assume that* $\mathbf{W}(t)$ *is doubly stochastic for each t. Then if* $\gamma(t) \leq \frac{1}{L}$ *for all t, we have*

$$\min_{1 \leq t \leq T} f_i(t) - f^* \leq \frac{D_1 + D_2 \sum_{t=1}^{T} \gamma^2(t)}{\sum_{t=1}^{T} \gamma(t)} \tag{9.33}$$

*where* $D_1$ *and* $D_2$ *are two scalars.*

According to this result, one can explicitly characterize the convergence rate with different stepsize rules. Let us choose $\gamma(t) = \left(\frac{1}{t}\right)^a$ with some $a > 0$, then one has the following convergence rates [516]:

1. Case 1. $0 < a < 1/2$. In this case, the convergence rate to the global optimal scales with $\mathcal{O}(1/t^a)$.
2. Case 2. $a = 1/2$. In this case, the convergence rate to the global optimal scales with $\mathcal{O}(\ln t/\sqrt{t})$.
3. Case 3. $1/2 < a < 1$. In this case, the convergence rate to the global optimal scales with $\mathcal{O}(1/t^{1-a})$.

Therefore, the fastest convergence rate is the second case, where $\gamma(t) = 1/\sqrt{t}$, and the rate is $\mathcal{O}(\ln t/\sqrt{t})$.

To achieve a faster convergence rate, the author of [516] considers the following iteration (assuming $h_i \equiv 0$ for all $i$):

$$\mathbf{x}_i(t) = \mathbf{v}_i(t-1) - \gamma(t)\nabla g_i(\mathbf{v}_i(t-1)) \tag{9.34a}$$

$$\mathbf{y}_i(t) = \mathbf{x}_i(t) + \frac{t-1}{t+2}(\mathbf{x}_i(t) - \mathbf{x}_i(t-1)) \tag{9.34b}$$

$$\mathbf{v}_i(t) = \sum_{j=1}^{N} w_{ij}(t)\mathbf{y}_j(t) \tag{9.34c}$$

which can be viewed as the distributed version of Nesterov's accelerated gradient method. To see such relation, let us suppose that the last consensus step is done repeatedly (by using a few inner iterations if necessary) so that consensus is achieved (i.e., $\mathbf{v}_i(t) = \mathbf{v}_j(t)$ for all $i \neq j$). Then the above iteration is precisely Nesterov's (centralized) accelerated gradient method.

Of course, it makes no sense to achieve the precise consensus at each iteration of the algorithm, because this will take an arbitrarily long time. Therefore the *distributed accelerated gradient method with multistep consensus* proposes to perform $t$ steps of the consensus iteration (9.34c) at $t$th outer iteration (9.34a)–(9.34c). The intuition is that as the algorithm progresses, more inner consensus steps will be performed, therefore the consensus can be achieved with increased accuracy. It is shown that to reach an $\epsilon$ optimal solution, the algorithm requires $\mathcal{O}(1/\sqrt{\epsilon})$ number of outer iterations, and $\mathcal{O}(1/\epsilon)$ number of communication steps (including the inner consensus iterations). This is significantly faster than the gradient/subgradient iteration with diminishing stepsize discussed above. The same bounds hold for the nonsmooth counterpart of the iteration

(9.34a)–(9.34c), where the proximal gradient steps are taken instead of the gradient steps.

We note briefly that the above acceleration idea has also been considered in the recent work [517] to improve the convergence speed of the distributed subgradient/gradient algorithm. The graph considered therein is a static and undirected graph $\mathcal{G}$, which is simpler than the one discussed above. The first algorithm proposed in [517], which is named the distributed Nesterov gradient algorithm (D-NG), is a single-loop algorithm given below (with $c > 0$ being a positive constant):

$$\mathbf{x}_i(t) = \sum_{j=1}^{N} w_{ij}(t)\mathbf{y}_j(t-1) - \frac{c}{t+1}\nabla g_i(\mathbf{y}_i(t-1)) \tag{9.35a}$$

$$\mathbf{y}_i(t) = \mathbf{x}_i(t) + \frac{t}{t+3}\left(\mathbf{x}_i(t) - \mathbf{x}_i(t-1)\right). \tag{9.35b}$$

The key difference with the iteration (9.34a)–(9.34c) is that this is a *single-loop* algorithm, and the consensus is achieved by using a *diminishing* stepsize $c/(t+1)$. After casting the iteration into the inexact Nesterov gradient method, the authors show that the algorithm converges in the order of $\mathcal{O}(\ln(t)/t)$.

## 9.3    ADMM-Based Methods

In this section, we discuss a very different approach to distributed optimization, which is based on the ADMM method.

### 9.3.1    Problem Formulation

Let us consider solving problem (9.9) over $N$ distributed agents defined over a static and symmetric graph $\mathcal{G} = \{\mathcal{V}, \mathcal{E}\}$. Again assume that if $(i,j) \in \mathcal{E}$, then nodes $i$ and $j$ can exchange information. The graph is symmetric, so we assume that if $(i,j) \in \mathcal{E}$ then $(j,i) \in \mathcal{E}$ as well. Let $|\mathcal{E}| = 2E$.

To introduce the ADMM formulation, let us introduce the link variables $\mathbf{z}_{ij}$ for each $(i,j) \in \mathcal{E}$, and reformulate problem (9.9) by the following *linearly constrained* problem

$$\min_{\{\mathbf{x}_i\},\{\mathbf{z}_{ij}\}} \quad \sum_{i=1}^{N} f_i(\mathbf{x}_i) \tag{9.36}$$

$$\text{s.t.} \quad \mathbf{x}_i - \mathbf{z}_{ij} = \mathbf{0}, \ \mathbf{x}_j - \mathbf{z}_{ij} = \mathbf{0}, \quad \forall (i,j) \in \mathcal{E}.$$

As long as the graph $(\mathcal{N}, \mathcal{E})$ is connected, the constraints of (9.36) enforce $\mathbf{x}_i = \mathbf{z}_{ij} = \mathbf{x}$ for all $i \in \mathcal{N}$ and $(i,j) \in \mathcal{E}$. The ADMM applied to (9.36) is an iterative process consisting of $\mathbf{x}_i$-subproblems, $\mathbf{z}_{ij}$-subproblems, as well as the updates of the dual variables. As will be seen shortly, the $\mathbf{x}_i$-subproblems are independent across the nodes, and the other two are independent across all pairs of neighbors. Therefore, there is no global data exchange and the algorithms belong to the family of *decentralized algorithms* [518]. For applications of the ADMM-based algorithms in distributed signal processing and data analysis, we refer the reader to [519].

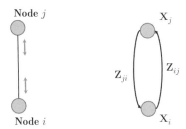

**Figure 9.2** Illustration of the relationship between auxiliary variables $\mathbf{x}$ and $\mathbf{z}$. Left: two nodes $i, j$ connected by a link capable of two-way communication. Right: for a given link and two nodes, introduce the node variables $\mathbf{x}_i$ and $\mathbf{x}_j$, and introduce the link variables $\mathbf{z}_{ij}$ and $\mathbf{z}_{ji}$.

Let us present the consensus ADMM (C-ADMM) algorithm proposed in [518, 520, 504]. First let us introduce two $2EM \times NM$ matrices $\mathbf{A}_s$ and $\mathbf{A}_d$, named block arc source matrix and block destination matrix, and each containing $2E \times N$ blocks $[\mathbf{A}_s]_{e,i}$ and $[\mathbf{A}_d]_{e,i}$, respectively. For the matrix $\mathbf{A}_s$, its block $[\mathbf{A}_s]_{e,i} = \mathbf{I}_M$ if the arc $e$ originates from node $i$, otherwise $[\mathbf{A}_s]_{e,i} = \mathbf{0}_M$. Similarly, for the matrix $\mathbf{A}_d$, its block $[\mathbf{A}_d]_{e,j} = \mathbf{I}_M$ if the arc $e$ ends at node $j$, otherwise $[\mathbf{A}_d]_{e,j} = \mathbf{0}_M$. Using this definition, problem (9.36) can be equivalently written as

$$
\min_{\{\mathbf{x}_i\},\{\mathbf{z}_{ij}\}} \quad \sum_{i=1}^{N} f_i(\mathbf{x}_i) \tag{9.37}
$$
$$
\text{s.t.} \quad \mathbf{A}_s\mathbf{x} = \mathbf{z}, \quad \mathbf{A}_d\mathbf{x} = \mathbf{z}.
$$

Further define $\mathbf{B} = [-\mathbf{I}_{2EM}; -\mathbf{I}_{2EM}] \in \mathbb{R}^{4EM \times 2EM}$ and $\mathbf{A} = [\mathbf{A}_s; \mathbf{A}_d] \in \mathbb{R}^{4EM \times NM}$, the above problem can be compactly expressed as

$$
\min_{\{\mathbf{x}_i\},\{\mathbf{z}_{ij}\}} \quad \sum_{i=1}^{N} f_i(\mathbf{x}_i) \tag{9.38}
$$
$$
\text{s.t.} \quad \mathbf{A}\mathbf{x} + \mathbf{B}\mathbf{z} = \mathbf{0}.
$$

Let us introduce the multipliers $\alpha_{ij}$, $\beta_{ij}$ associated with constraints $\mathbf{x}_i = \mathbf{z}_{ij}$ and $\mathbf{x}_j = \mathbf{z}_{ij}$, respectively. Denote $\boldsymbol{\alpha}$ and $\boldsymbol{\beta}$ as the collection of the multipliers, and they are associated with the constraint $\mathbf{A}_s\mathbf{x} = \mathbf{z}$ and $\mathbf{A}_d\mathbf{x} = \mathbf{z}$, respectively. Define $\boldsymbol{\lambda} = [\boldsymbol{\alpha}; \boldsymbol{\beta}]$, then we can construct the augmented Lagrangian function as

$$
L(\mathbf{x}, \mathbf{z}, \boldsymbol{\lambda}) = f(\mathbf{x}) + \langle \boldsymbol{\lambda}, \mathbf{A}\mathbf{x} + \mathbf{B}\mathbf{z} \rangle + \frac{\rho}{2} \|\mathbf{A}\mathbf{x} + \mathbf{B}\mathbf{z}\|^2, \tag{9.39}
$$

where $\rho > 0$ is a given positive constant. Then the ADMM iteration is given by

$$
\mathbf{x}(t+1) = \arg\min_{\mathbf{x}} L(\mathbf{x}, \mathbf{z}(t), \boldsymbol{\lambda}(t)) \tag{9.40a}
$$
$$
\mathbf{z}(t+1) = \arg\min_{\mathbf{z}} L(\mathbf{x}(t+1), \mathbf{z}, \boldsymbol{\lambda}(t)) \tag{9.40b}
$$
$$
\boldsymbol{\lambda}(t+1) = \boldsymbol{\lambda}(t) + \rho(\mathbf{A}\mathbf{x}(t+1) + \mathbf{B}\mathbf{z}(t+1)) \tag{9.40c}
$$

## 9.3.2      Distributed Implementation

It turns out that the C-ADMM iteration (9.40a)–(9.40c) can be implemented distributedly without requiring any centralized processing, provided that the initial values of $\mathbf{z}$ and $\lambda$, denoted as $\mathbf{z}(0)$ and $\lambda(0)$, are properly chosen (see [518, 520, 504, 519] for the detailed original derivations). To illustrate this point, let us first write down the optimality conditions of (9.40a)–(9.40c):

$$\partial f(\mathbf{x}(t+1)) + \mathbf{A}^T\lambda(t) + \rho\mathbf{A}^T(\mathbf{A}\mathbf{x}(t+1) + \mathbf{B}\mathbf{z}(t)) = 0 \tag{9.41a}$$

$$\mathbf{B}^T\lambda(t) + \rho\mathbf{B}^T(\mathbf{A}\mathbf{x}(t+1) + \mathbf{B}\mathbf{z}(t+1)) = 0 \tag{9.41b}$$

$$\lambda(t+1) - \lambda(t) - \rho\,(\mathbf{A}\mathbf{x}(t+1) + \mathbf{B}\mathbf{z}(t+1)) = 0 \tag{9.41c}$$

which are further equivalent to

$$\partial f(\mathbf{x}(t+1)) + \mathbf{A}^T\lambda(t+1) + \rho\mathbf{A}^T\mathbf{B}(\mathbf{z}(t) - \mathbf{z}(t+1)) = 0 \tag{9.42a}$$

$$\mathbf{B}^T\lambda(t+1) = 0 \tag{9.42b}$$

$$\lambda(t+1) - \lambda(t) - \rho\,(\mathbf{A}\mathbf{x}(t+1) + \mathbf{B}\mathbf{z}(t+1)) = 0 \tag{9.42c}$$

where in the second and third equalities we have utilized the definition of $\lambda(t+1)$ in (9.42c). Recall that we have defined $\lambda = [\alpha; \beta]$, and $\mathbf{B} = [-\mathbf{I}_{2EM}; -\mathbf{I}_{2EM}]$. These facts combined with (9.42b) imply that

$$\alpha(t+1) = -\beta(t+1), \quad \forall\, t = 1, \cdots. \tag{9.43}$$

Utilizing this inequality, the optimality condition for $\mathbf{x}$ (i.e., (9.42a)) is given by

$$0 \in \partial f(\mathbf{x}(t+1)) + \mathbf{M}_-\beta(t+1) - \rho\mathbf{M}_+(\mathbf{z}(t+1) - \mathbf{z}(t)), \tag{9.44}$$

where we have defined

$$\mathbf{M}_+ = \mathbf{A}_s^T + \mathbf{A}_d^T, \quad \mathbf{M}_- = \mathbf{A}_s^T - \mathbf{A}_d^T$$

which are respectively the extended unoriented and oriented incident matrices for graph $\mathcal{G}$. Let us now look at equation (9.42c).

   Let us suppose that $\alpha(0) = -\beta(0)$, therefore combined with (9.43) we must have $\alpha(t) = -\beta(t)$ for all $t = 0, 1, \cdots$. Plugging this into (9.41b) we have

$$\rho\mathbf{B}^T(\mathbf{A}\mathbf{x}(t+1) + \mathbf{B}\mathbf{z}(t+1)) = 0, \ \forall\, t \geq 0, \tag{9.45}$$

or equivalently

$$\mathbf{z}(t+1) = \frac{1}{2}\mathbf{M}_+\mathbf{x}(t+1), \ \forall\, t \geq 0. \tag{9.46}$$

This is a key identity. It says that although we started with an algorithm that has three variables $(\mathbf{x}, \mathbf{z}, \lambda)$, one of them can in fact be replaced by the rest. When we explicitly write down (9.46) we found that

$$\mathbf{z}_{ij}(t+1) = \frac{1}{2}\left(\mathbf{x}_i(t+1) + \mathbf{x}_j(t+1)\right),$$

In words, each link variable is always updated by taking an average between its two adjacent node variables. However, as we will show below, in practice there is no need to keep track of the link variables.

Once the link variables are expressed by the node variables, let us examine the update rule for the remaining two variables. Using the identity (9.46), we have

$$\mathbf{A}\mathbf{x}(t+1) + \mathbf{B}\mathbf{z}(t+1) = \begin{bmatrix} \mathbf{A}_s\mathbf{x}(t+1) \\ \mathbf{A}_d\mathbf{x}(t+1) \end{bmatrix} + \frac{1}{2}\mathbf{B}\mathbf{M}_+^T\mathbf{x}(t+1)$$

$$= \begin{bmatrix} \mathbf{A}_s\mathbf{x}(t+1) \\ \mathbf{A}_d\mathbf{x}(t+1) \end{bmatrix} + \frac{1}{2}\begin{bmatrix} (-\mathbf{A}_s - \mathbf{A}_d)\mathbf{x}(t+1) \\ (-\mathbf{A}_d - \mathbf{A}_s)\mathbf{x}(t+1) \end{bmatrix}$$

$$= \frac{1}{2}\begin{bmatrix} \mathbf{A}_s - \mathbf{A}_d \\ \mathbf{A}_d - \mathbf{A}_s \end{bmatrix}\mathbf{x}(t+1)$$

This implies that

$$\boldsymbol{\alpha}(t+1) = \boldsymbol{\alpha}(t) + \frac{1}{2}(\mathbf{A}_s - \mathbf{A}_d)\mathbf{x}(t+1) = \boldsymbol{\alpha}(t) + \frac{1}{2}\mathbf{M}_-^T\mathbf{x}(t+1). \tag{9.47}$$

Finally utilizing (9.46) and assuming the initialization condition $\mathbf{z}(0) = \frac{1}{2}\mathbf{M}_+^T\mathbf{x}(0)$, we see that the optimality condition for the $\mathbf{x}$ step (9.42a) is given by

$$0 \in \partial f(\mathbf{x}(t+1)) + \mathbf{M}_-\boldsymbol{\alpha}(r+1) + \rho\mathbf{A}^T\mathbf{B}(\mathbf{z}(t) - \mathbf{z}(t+1)). \tag{9.48}$$

Defining a new multiplier $\boldsymbol{\xi} = \mathbf{M}_-\boldsymbol{\alpha}$, we have

$$\nabla f_i(\mathbf{x}(t+1)) + \boldsymbol{\xi}(t+1) + \rho\frac{1}{2}\mathbf{M}_+\mathbf{M}_+^T(\mathbf{x}(t+1) - \mathbf{x}(t)) = 0, \tag{9.49}$$

$$\boldsymbol{\xi}(t+1) = \boldsymbol{\xi}(t) + \rho\frac{1}{2}\mathbf{M}_-\mathbf{M}_-^T\mathbf{x}(t+1). \tag{9.50}$$

Expanding these two conditions component by component, we have

$$0 \in \partial f(\mathbf{x}_i(t+1)) + 2\rho|\mathcal{N}_i|\mathbf{x}_i(t+1) = \rho|\mathcal{N}_i|\mathbf{x}_i(t) + \rho\sum_{j\in\mathcal{N}_i}\mathbf{x}_j(t) - \boldsymbol{\xi}_i(t), \ \forall i \tag{9.51a}$$

$$\boldsymbol{\xi}_i(t+1) = \boldsymbol{\xi}_i(t) + \rho\left(|\mathcal{N}_i|\mathbf{x}_i(t+1) - \sum_{j\in\mathcal{N}_i}\mathbf{x}_j(t+1)\right), \ \forall i. \tag{9.51b}$$

In summary, as long as the following two initial conditions are satisfied

$$\mathbf{z}(0) = \frac{1}{2}\mathbf{M}_+^T\mathbf{x}(0), \quad \boldsymbol{\alpha}(0) = -\boldsymbol{\beta}(0) \tag{9.52}$$

then the ADMM iteration (9.40a)–(9.40c) is equivalent to the iteration (9.51a)–(9.51b).

Finally, let us briefly discuss the distributed implementation of iteration (9.51a)–(9.51b). Let us suppose that each node $i$ is responsible of updating $\mathbf{x}_i$ and $\boldsymbol{\xi}_i$. Let us use $\mathbf{b}_i(t)$ to denote the rhs of (9.51a). It is clear that to obtain this vector the node $i$ only needs $\{\mathbf{x}_j(t)\}_{j\in\mathcal{N}_i}$ from its neighbors. Once $\mathbf{b}_i(t)$ is known, the update $\mathbf{x}_i$ amounts to solving the following (local) optimization problem (which is assumed to be easily solvable):

$$\mathbf{x}_i(t+1) = \arg\min f_i(\mathbf{x}_i) + \frac{1}{2}\|2\rho\,|\mathcal{N}_i|\mathbf{x}_i - \mathbf{b}_i(t)\|^2 \tag{9.53}$$

Similarly, the update of the dual variable is also easy, and only requires the vectors $\{\mathbf{x}_j(t+1)\}_{j\in\mathcal{N}_i}$ from the neighbors $\mathcal{N}_i$.

## 9.3.3     Convergence and Variations

In this section we discuss the convergence behavior of C-ADMM and a few of its variants.

### Convergence Rate of C-ADMM

Because the iteration (9.51a)–(9.51b) is derived directly from the ADMM iteration (9.40a)–(9.40c), its convergence as well as convergence rate follows directly from that of the standard ADMM algorithm (cf. Section 3.6.2). However, if one simply applies the standard results, it is not clear how the rate of C-ADMM depends on the network structure. Intuitively, running the algorithm over a complete graph should be much faster than running it over, say, a ring. Below we discuss the dependence of the rate over the network topology.

To proceed, we make the following assumptions about problem (9.9), as well as the initial multiplier $\alpha(0)$.

ASSUMPTION 3.  *Assume the following:*

1. *Each local objective function $f_i$ is differentiable and strongly convex. That is, for each $i \in \mathcal{N}$ and each $\mathbf{x}, \mathbf{y} \in \mathbb{R}^M$, there exists a constant $m_i > 0$ such that*

$$\langle \nabla f_i(\mathbf{x}) - \nabla f_i(\mathbf{y}), \mathbf{x} - \mathbf{y}\rangle \geq m_i \|\mathbf{x} - \mathbf{y}\|^2.$$

2. *The gradient of $f_i$ is Lipschitz continuous with constant $M_i$, that is,*

$$\|\nabla f_i(\mathbf{x}) - \nabla f_i(\mathbf{y})\| \leq M_i \|\mathbf{x} - \mathbf{y}\|.$$

3. *The initial multiplier $\alpha(0)$ lies in the column space of $\mathbf{M}_-^T$.*

To state the main convergence result, define

$$\mathbf{u} = \begin{pmatrix} \mathbf{z} \\ \alpha \end{pmatrix}, \quad \mathbf{G} = \begin{bmatrix} \rho\mathbf{I}_{2EM} & 0 \\ 0 & \frac{1}{\rho}\mathbf{I}_{2EM} \end{bmatrix} \tag{9.54}$$

The following theorem is from [521, Theorem 1].

THEOREM 17.  *Consider the C-ADMM iteration (9.40a)–(9.40c). The primal variables $\mathbf{x}$ and $\mathbf{z}$ have their unique optimal values $\mathbf{x}^*$ and $\mathbf{z}^*$; the dual variable $\alpha$ has its unique optimal value $\alpha^*$ in the column space of $\mathbf{M}_-^T$. Suppose that Assumption 3 is satisfied, then for any $\mu > 1$ we have*

$$\|\mathbf{u}(t+1) - \mathbf{u}^*\|_G^2 \leq \frac{1}{1+\delta}\|\mathbf{u}(t) - \mathbf{u}^*\|_G^2 \tag{9.55}$$

*where*

$$\delta = \min \left\{ \frac{(\mu - 1)\tilde{\sigma}_{\min}^2(\mathbf{M}_-)}{\mu \sigma_{\max}^2(\mathbf{M}_+)}, \frac{\min_i m_i}{\frac{\rho}{4}\sigma_{\max}^2(\mathbf{M}_+) + \frac{\mu}{\rho} \max_i M_i^2 \tilde{\sigma}_{\min}^{-2}(\mathbf{M}_-)} \right\} > 0 \quad (9.56)$$

*where $\tilde{\sigma}_{\min}(X)$ represents the smallest nonzero singular value of X, and $\sigma_{\max}(X)$ denotes the maximum singular value of X. Further, $\{\mathbf{x}(t)\}$ converges in the following form*

$$\|\mathbf{x}(t+1) - \mathbf{x}^*\|^2 \le \frac{1}{\min_i m_i} \|\mathbf{u}(t) - \mathbf{u}^*\|_G^2. \quad (9.57)$$

The authors of [521] made the following comments about the rate obtained above. Clearly, the theoretical convergence rate is dependent on the constant $\delta$ (large $\delta$ implies fast convergence), which in turn depends on the topology of the graph (i.e., the spectrum of the matrices $\mathbf{M}_-$ and $\mathbf{M}_+$), the property of the objective functions (i.e., $m_i$'s and $M_i$'s), the free parameter $\mu$, as well as the algorithm parameter $\rho$. Let us look at the expression (9.56) closely. First, we observe that it is non-decreasing with the strong convexity constants $m_i$'s, which is as expected. Second, it has a close relationship with the singular values of $\mathbf{M}_+$ and $\mathbf{M}_-$. To understand such dependency, let us define the extended signless Laplacian and extended signed Laplacian matrices as follows:

$$\mathbf{L}_+ = \frac{1}{2}\mathbf{M}_+\mathbf{M}_+^T, \quad \mathbf{L}_- = \frac{1}{2}\mathbf{M}_-\mathbf{M}_-^T. \quad (9.58)$$

We can then define the condition number of the graph $\mathcal{G}$ as

$$\kappa_{\mathcal{G}} = \frac{\sigma_{\max}(\mathbf{M}_+)}{\tilde{\sigma}_{\min}(\mathbf{M}_-)} = \sqrt{\frac{\sigma_{\max}(\mathbf{L}_+)}{\tilde{\sigma}_{\min}(\mathbf{L}_-)}} \quad (9.59)$$

It is known that $\tilde{\sigma}_{\min}(\mathbf{L}_-)$ represents the algebraic connectivity of the underlying graph, and that both $\sigma_{\max}(\mathbf{L}_+)$ and $\tilde{\sigma}_{\min}(\mathbf{L}_-)$ are measures of network connectedness (the larger the value the stronger the connectedness). Also, a larger $\kappa_{\mathcal{G}}$ means the graph has weaker connectedness. Using these facts we see that the better connected of the underlying graph implies larger $\delta$, hence the better convergence rate of the algorithm.

## Convergence of Variants of C-ADMM

In this subsection we discuss a few popular variants of C-ADMM and their convergence guarantees.

One of the main drawbacks of the C-ADMM is that the $\mathbf{x}$-subproblem (9.40a) (or equivalently (9.53)) is required to be solved exactly, at each iteration $t$. This is not always easy for many practical problems. In an effort to reduce the computational complexity of the local updates, recently [85] and [522] independently proposed to perform simple gradient or proximal gradient steps for each $\mathbf{x}$-update, so that the $\mathbf{x}$-subproblem (9.40a) (or equivalently (9.53)) can be solved inexactly. Their algorithms are termed IC-ADMM and DLM, respectively. Below we will discuss the IC-ADMM algorithm proposed in [85] in detail.

To illustrate their scheme, suppose that each local objective function $f_i$ can be expressed as $f_i(\mathbf{x}) = g_i(\mathbf{x}) + h_i(\mathbf{x})$, where $g_i(\mathbf{x})$ is a smooth function with Lipschitz continuous gradient, and $h_i(\mathbf{x})$ is a closed convex lower semi-continuous function, whose proximity operator is easy to compute. Let us denote

$$f(\mathbf{x}) = \sum_{i=1}^{N} g_i(\mathbf{x}) + h_i(\mathbf{x}) := g(\mathbf{x}) + h(\mathbf{x}).$$

The inexact consensus ADMM (IC-ADMM) proposed in [85] replaces the **x**-update step (9.40a) by the following step:

$$\mathbf{x}(t+1) = \arg\min_{\mathbf{x}} \langle \nabla g(\mathbf{x}(t)), \mathbf{x} - \mathbf{x}(t) \rangle + h(\mathbf{x}) + \langle \boldsymbol{\lambda}(t), \mathbf{Ax} + \mathbf{Bz} \rangle$$

$$+ \frac{\rho}{2}\|\mathbf{Ax} + \mathbf{Bz}(t)\|^2 + \frac{1}{2}\|\mathbf{x} - \mathbf{x}(t)\|_{\boldsymbol{\Upsilon}}^2 \qquad (9.60)$$

where $\boldsymbol{\Upsilon} \in \mathbb{R}^{MN \times MN}$ is a block diagonal matrix, with $N$ diagonal blocks $\upsilon_i \mathbf{I}_M$, $i = 1, \cdots, N$, and each $\upsilon_i > 0$ is a positive proximal parameter. We can also reduce this update rule to the following per-node update rule (cf. (9.53) for the definition of $\mathbf{b}_i(t)$):

$$\mathbf{x}_i(t+1) = \arg\min \langle \nabla g_i(\mathbf{x}_i(t)), \mathbf{x}_i - \mathbf{x}_i(t) \rangle + h_i(\mathbf{x}_i)$$

$$+ \frac{1}{2}\|2\rho|\mathcal{N}_i|\mathbf{x}_i - \mathbf{b}_i(t)\|^2 + \frac{\upsilon_i}{2}\|\mathbf{x}_i - \mathbf{x}_i(t)\|^2. \qquad (9.61)$$

Therefore, as long as the proximity operator of $h_i$ is easy to solve, the above subproblem is easy. The following result states the convergence of IC-ADMM [85, Theorem 1].

THEOREM 18. *Suppose that the graph $\mathcal{G}$ is connected, each $g_i(\cdot)$ can be written as $g_i(\mathbf{x}) = \ell_i(\mathbf{A}_i\mathbf{x})$ for some matrix $\mathbf{A}_i$ and a strongly convex function $\ell_i(\cdot)$ with modulus $\sigma_i$, and has Lipschitz continuous gradient $L_i$. Further suppose that $h_i(\cdot)$ is a closed convex lower-semicontinuous function. Then if the proximal parameter $\upsilon_i$ is chosen according to the following rule*

$$\upsilon_i > \frac{L_i^2}{\sigma_i^2}\lambda_{\max}(\mathbf{A}_i^T\mathbf{A}_i) - \rho\lambda_{\min}(\mathbf{L}_+) > 0, \ \forall i \in \mathcal{N}. \qquad (9.62)$$

*Then the following two statements are true:*

1. *The IC-ADMM converges to an optimal primal-dual solution pair.*
2. *If $h_i(\mathbf{x}) \equiv 0$ and $\mathbf{A}_i$ has full column rank for all $i$, then the algorithm converges linearly (in the sense similarly as stated in Theorem 17).*

It is interesting to note that in order to perform an inexact update, there is an additional requirement that the smooth part of the function needs to be a *composite* of a strongly convex function $\ell_i(\cdot)$ and a linear mapping. Further, to ensure convergence, the proximal parameter $\upsilon_i$ should be inversely proportional to the strong convexity modulus $\sigma_i > 0$.

Let us see some numerical results presented in [85], which compare C-ADMM and IC-ADMM on a distributed logistic regression problem. Consider a two-image classification task using images *D*24 and *D*68 from the Brodatz data set (www.ux.uis.no/~tranden/brodatz.html) to generate the regression data matrix **A**. Randomly extract $(NM)/2$ patches with dimension $\sqrt{K} \times \sqrt{K}$ from the two images, respectively, followed by vectorizing the $M$ patches into vectors and stacking them into an $M \times K$ matrix. The rows of the matrix were randomly shuffled and the resultant matrix was used as the data matrix $\mathbf{A}$. The binary labels $\mathbf{b}_i$'s were generated according to the image identity. We use the fast iterative shrinkage thresholding algorithm (FISTA) [156, 523] to solve its $\mathbf{x}$-subproblem of IC-ADMM for each node $i$. The stopping criterion of the FISTA iteration was based on the PG residue (pgr) [156, 523]. For obtaining a high-accuracy solution of the $\mathbf{x}$-subproblem, one may set the stopping criterion as, for instance, $\mathsf{pgr} < 1e^{-5}$.

The following stopping criterion is used, which is based on measuring the solution accuracy $\mathsf{acc} = (\mathsf{obj}(\hat{\mathbf{x}}(t)) - \mathsf{obj}^\star)/\mathsf{obj}^\star$ and variable consensus error $\mathsf{cserr} = \sum_{i=1}^{N} \|\hat{\mathbf{x}}(t) - \mathbf{x}(t)_i\|^2/N$, where $\hat{\mathbf{x}}(t) = \sum_{i=1}^{N} \mathbf{x}_i(t)/N$, $\mathsf{obj}(\hat{\mathbf{x}}(t))$ denotes the objective value of the LR problem evaluated at $\mathbf{x} = \hat{\mathbf{x}}(t)$, and $\mathsf{obj}^\star$ is the optimal value of the LR problem which was obtained by FISTA [156, 523] with a high solution accuracy of $\mathsf{pgr} < 1e^{-6}$. The two algorithms were set to stop whenever $\mathsf{acc}$ and $\mathsf{cserr}$ are both smaller than preset target values.

Consider an example where $N = 10$, $K = 10{,}000$, $M = 10$, $\tau = 0.1$ and $a = 1$. The convergence cures of C-ADMM and IC-ADMM are plotted in Figure 9.3. We have also plotted the consensus subgradient method [512] (cf. (9.23)) where the diminishing stepsize $10/t$ is used. The stopping conditions are $\mathsf{acc} < 10^{-4}$ and $\mathsf{cserr} < 10^{-5}$. For C-ADMM, we considered two cases with $\mathsf{pgr} < 10^{-5}$ (to obtain accurate inner solution) and $\mathsf{pgr} < 10^{-4}$ (to obtain less-accurate inner solution). Clearly the IC-ADMM in general requires more ADMM iterations than the C-ADMM, however, the computational time is usually significantly smaller; see Table 9.1. Also see Figure 9.4 for comparison of computational time between C-ADMM and IC-ADMM.

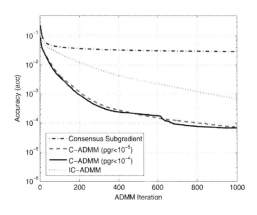

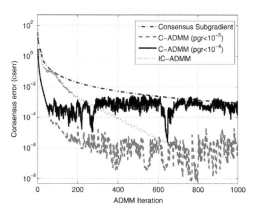

(a) Convergence speed measured by solution accuracy

(b) Convergence speed measured by consensus error

**Figure 9.3** Comparison of the convergence of C-ADMM and IC-ADMM over the distributed logistic regression problem [85].

**Table 9.1** Comparison of C-ADMM and IC-ADMM [85]

| (a) $N = 10, K = 10,000, M = 10, \lambda = 0.1, a = 1$. | | | |
|---|---|---|---|
| | C-ADMM (pgr $< 10^{-5}$) | C-ADMM (pgr $< 10^{-4}$) | IC-ADMM |
| ADMM Ite. | 810 | 675 | 2973 |
| Comput. Time (sec) | **44.56** | 17.86 | **2.14** |
| acc$< 10^{-4}$ | $9.982 \times 10^{-5}$ | $9.91 \times 10^{-5}$ | $9.99 \times 10^{-5}$ |
| cserr$< 10^{-5}$ | $1.53 \times 10^{-6}$ | **$3.425 \times 10^{-4}$** | **$3.859 \times 10^{-9}$** |

| (b) $N = 50, K = 10,000, M = 10, \lambda = 0.15, a = 1$. | | | |
|---|---|---|---|
| | C-ADMM (pgr $< 10^{-5}$) | C-ADMM (pgr $< 10^{-4}$) | IC-ADMM |
| ADMM Ite. | 952 | N/A | 7,251 |
| Comput. Time (sec) | **81.72** | N/A | **9.33** |
| acc$< 10^{-4}$ | $9.99 \times 10^{-5}$ | N/A | $9.999 \times 10^{-5}$ |
| cserr$< 10^{-5}$ | $1.305 \times 10^{-7}$ | N/A | **$1.169 \times 10^{-10}$** |

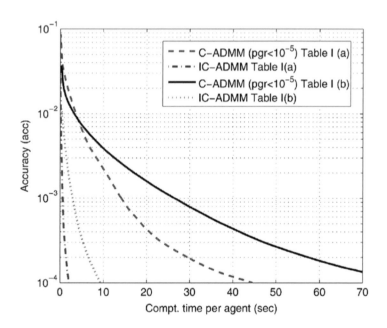

**Figure 9.4** Comparison of the computational time of IC-ADMM and C-ADMM [85].

## 9.4     Other Types of First-Order Method

In this section we discuss a few additional popular distributed optimization algorithms that apparently do not belong to either the ADMM-based or the consensus-based methods.

We first consider the exact first-order algorithm for decentralized consensus optimization (abbreviated as the EXTRA algorithm) proposed by Shi, Ling, Wu, and Yin [524]. Let us consider again solving the problem (9.9) over a network of nodes defined on a symmetric graph $\mathcal{G}$. Assume for simplicity that the global variable $x$ has a single dimension (i.e., $M = 1$), and that each $f_i : \mathbb{R} \to \mathbb{R}$ is a closed convex and differentiable function. Let $x_i \in \mathbb{R}$ denote the local variable belongs to the $i$th node, and let $\mathbf{x} := [x_1, \cdots, x_N]^T$. The classic distributed gradient method, when applied to such problem, can be compactly written as the following iteration (cf. (9.11)):

$$\mathbf{x}(t+1) = \mathbf{W}\mathbf{x}(t) - \gamma(t)\nabla f(\mathbf{x}(t)).$$

The motivation of the EXTRA algorithm is the following [524]: when $t \to \infty$ and when a constant stepsize $\gamma(t) = \gamma > 0$ is used, we have

$$\mathbf{x}(\infty) = \mathbf{W}\mathbf{x}(\infty) - \gamma\nabla f(\mathbf{x}(\infty)).$$

Suppose that in the limit the consensus is achieved, then $\mathbf{W}\mathbf{x}(\infty) = \mathbf{x}(\infty)$, then we must have $\nabla f(\mathbf{x}(\infty)) = \mathbf{0}$, which is not likely to be true for the optimizer of a general smooth function $f$. Therefore, to make sure that consensus is achieved as well as the objective is minimized, one has to have $\gamma \to 0$ (i.e., using diminishing stepsizes).

With the above observation, the EXTRA algorithm adds certain *correction* term to the distributed gradient iteration, in the hope of being able to use constant stepsize $\gamma$. More precisely, the algorithm can be compactly written as

$$\mathbf{x}(1) = \mathbf{W}\mathbf{x}(0) - \gamma\nabla f(\mathbf{x}(0)), \tag{9.63a}$$

$$\mathbf{x}(t+1) = (\mathbf{I} + \mathbf{W})\,\mathbf{x}(t) - \frac{1}{2}(\mathbf{I} + \mathbf{W})\,\mathbf{x}(t-1)$$
$$- \gamma\left[\nabla f(\mathbf{x}(t)) - \nabla f(\mathbf{x}(t-1))\right], \ \forall\, t \geq 1 \tag{9.63b}$$

Note that the matrix $\frac{1}{2}(\mathbf{I} + \mathbf{W})$ can be replaced by more general matrices, but here for simplicity we only use its special form. Clearly, comparing with the distributed gradient iteration, the update in (9.63b) adds an additional term

$$\mathbf{x}(t) - \frac{1}{2}(\mathbf{I} + \mathbf{W})\mathbf{x}(t-1) + \gamma\nabla f(\mathbf{x}(t-1)).$$

We can also write (9.63b) recursively by

$$\mathbf{x}(t+1) = \underbrace{\mathbf{W}\mathbf{x}(t) - \gamma\nabla f(\mathbf{x}(t))}_{\text{distributed gradient iteration}} + \underbrace{\sum_{\ell=0}^{t-1}\left(\mathbf{W} - \frac{1}{2}(\mathbf{I} + \mathbf{W})\right)\mathbf{x}(\ell)}_{\text{correction term}} \tag{9.64}$$

It turns out that by making such modification, one can use constant stepsize $\gamma > 0$ to achieve global convergence. More specifically, let $L_i$ denote the Liptchitz constant of the gradient of $f_i$, and assume that the weight matrix $\mathbf{W}$ is symmetric and doubly stochastic. Then in [524] the authors show that if the stepsize $\gamma > 0$ satisfies

$$\gamma < \frac{\lambda_{\min}(\mathbf{I} + \mathbf{W})}{\max_i L_i} \tag{9.65}$$

then the algorithm will converge to a consensus point $\mathbf{x}^*$, which also optimizes the objective of problem (9.9). By noticing the fact that $\lambda_{\min} (\mathbf{I} + \mathbf{W}) \geq 1$, one can further simplify the above condition to the following sufficient condition:

$$\gamma < \frac{1}{\max_i L_i}. \tag{9.66}$$

Moreover, one can show that the convergence speed is sublinear for general smooth convex objectives, and linear for smooth strongly convex objectives [524, Theorem 3.5, 3.7]. This is a significant improvement compared with the vanilla distributed subgradient methods.

Apparently the EXTRA algorithm is closely related to the distributed gradient iteration, as it can be interpreted as its error-corrected version (cf. (9.64)). However, more careful analysis can reveal that intrinsically it is precisely a distributed inexact ADMM algorithm. This is more or less a technical issue which we will not explore further here.

The second algorithm that we are going to discuss is the dual averaging algorithm proposed by Duchi, Agarwal, and Wainwright [525], which is the distributed version of Nesterov's dual averaging algorithm [526].

Consider solving the following problem:

$$\min \quad \Psi(\mathbf{x}) = \frac{1}{N} \sum_{i=1}^{N} f_i(\mathbf{x}), \quad \text{s.t. } \mathbf{x} \in \mathcal{X} \tag{9.67}$$

over a network of nodes defined on an undirected graph $\mathcal{G}$. Assume that each $f_i$ belongs to a single node $i \in \mathcal{N}$, and it is closed, convex, and lower-semicontinuous (not necessarily differentiable). Further assume that $f_i$ is $L$-Lipschitz, that is,

$$|f_i(\mathbf{x}) - f_i(\mathbf{y})| \leq L\|\mathbf{x} - \mathbf{y}\|, \quad \forall\, i,\, \forall\, \mathbf{x}, \mathbf{y} \in \mathcal{X}.$$

The distributed dual averaging algorithm can be expressed by the following iteration:

$$\mathbf{z}_i(t + 1) = \sum_{j \in \mathcal{N}_i} w_{ji}\mathbf{z}_i(t) + \mathbf{d}_i(t) \tag{9.68a}$$

$$\mathbf{x}_i(t + 1) = \Pi_{\mathcal{X}}^{\Psi} (\mathbf{z}_i(t + 1), \gamma(t)) \tag{9.68b}$$

where $\mathbf{d}_i(t) \in \partial f_i(\mathbf{x}_i(t))$; $w_{ij} \geq 0$ is an element of the doubly stochastic matrix $\mathbf{W}$ satisfying Assumption 1; the function $\Pi_{\mathcal{X}}^{\Psi}(\cdot, \cdot)$ is a projection type operator defined as

$$\Pi_{\mathcal{X}}^{\Psi}(\mathbf{z}, \gamma) := \arg\min_{\mathbf{x} \in \mathcal{X}} \left\{ \langle \mathbf{z}, \mathbf{x} \rangle + \frac{1}{\gamma} \Psi(\mathbf{x}) \right\}$$

where $\Psi(\mathbf{x})$ is a strongly convex function with modulus 1, with $\Psi(\mathbf{x}) \geq 0$ for all $\mathbf{x} \in \mathcal{X}$, and $\Psi(\mathbf{0}) = 0$. At each iteration of the algorithm, each node $i$ first computes the auxiliary vector $\mathbf{z}_i(t + 1)$ which is a combination of its own subgradient as well as the subgradients from its neighbors over all the past iterations. Then it performs a projection of $\mathbf{z}_i(t + 1)$ over the feasible set $\mathcal{X}$. This method is closely related to the projected subgradient method (9.23), with the key difference that the averaging is done over the past subgradients, rather than the past iterates (as is done in the projected subgradient method).

The convergence of the distributed dual averaging method is given below [525, Theorem 1].

THEOREM 19. *Let the sequence $\{\mathbf{x}_i(t)\}$ and $\{\mathbf{z}_i(t)\}$ be generated by the dual averaging method (9.68). Suppose $\Psi(\cdot)$ is strongly convex with respect to the norm $\|\cdot\|$ with dual norm $\|\cdot\|_*$. For any $\mathbf{x}^* \in \mathcal{X}$ and for each node $i \in \mathcal{N}$, we have*

$$f(\hat{\mathbf{x}}_i(T)) - f(\mathbf{x}^*) \leq \mathrm{OPT} + \mathrm{NET},$$

*where*

$$\hat{\mathbf{x}}_i(T) := \frac{1}{T} \sum_{t=1}^{T} \mathbf{x}_i(t), \quad \bar{\mathbf{z}}(t) = \frac{1}{N} \sum_{i=1}^{N} \mathbf{z}_i(t)$$

$$\mathrm{OPT} = \frac{1}{T\gamma(T)} \Psi(\mathbf{x}^*) + \frac{L^2}{2T} \sum_{t=1}^{T} \gamma(t-1)$$

$$\mathrm{NET} = \frac{L}{T} \sum_{t=1}^{T} \gamma(t) \left[ \frac{2}{N} \sum_{j=1}^{N} \|\bar{\mathbf{z}}(t) - \mathbf{z}_j(t)\|_* + \|\bar{\mathbf{z}}(t) - \mathbf{z}_i(t)\|_* \right].$$

Note that the error term OPT consists of two terms related to the errors common to the subgradient-type methods; the error term NET are penalties incurred due to the disagreements among the network nodes. Clearly if $\gamma(t)$ is in the order of $\mathcal{O}(1/\sqrt{t})$, the algorithm should converge to the global optimal solution, provided that $\|\bar{\mathbf{z}}(t) - \mathbf{z}_j(t)\|_*$'s are appropriately bounded.

What is more interesting perhaps is the fact that one can also characterize the convergence behavior of the distributed dual averaging algorithm using the network topology. Let $\sigma_1(\mathbf{W}) = 1$ denote the largest singular value of $\mathbf{W}$, and use

$$\eta(\mathbf{W}) = 1 - \sigma_2(\mathbf{W})$$

denote the spectral gap. Then the convergence rate of the distributed dual averaging can be shown to be inversely proportional to $\eta(\mathbf{W})$; see the following theorem [525, Theorem 2].

THEOREM 20. *Under the conditions and notations of Theorem 19, further assume that $\Psi(\mathbf{x}^*) \leq R^2$. With the stepsize choice of*

$$\gamma(t) = \frac{R\sqrt{1 - \sigma_2(\mathbf{W})}}{4L\sqrt{t}}$$

*we have that*

$$f(\hat{\mathbf{x}}_i(T)) - f(\mathbf{x}^*) \leq 8 \frac{RL}{\sqrt{T}} \frac{\log(T\sqrt{N})}{\sqrt{\eta(\mathbf{W})}}, \quad \forall i \in \mathcal{N}. \tag{9.69}$$

It is known that the connectedness of a given graph is proportional to $\eta(\mathbf{W})$, therefore the above result makes intuitive sense. We refer the interested reader to [525] for the precise bounds derived for a few commonly used graphs, such as the two-dimensional grid, the random geometric graph, and the three-regular expander graph. However, it is important to note that in order to achieve the above network scaling, each node is required to know precisely the spectral gap of the graph, but such information is usually difficult to come by, at least in a distributed manner.

## 9.5     Special Case: Global Consensus Problem

In this section we discuss solving problem (9.9) over a special graph having the so-called "star topology," with a single master node connecting to $N$ distributed computing nodes. The reason that we are especially interested in this type of network structure is that it models a large class of modern high-performance computing architecture, in which the computational tasks and/or pieces of data are distributed to workers in the system, and the overall progress of the computation is centrally managed by some master node.

Let us consider solving problem (9.9), restated below (with $g_i(\cdot)$ being a smooth term and $h(\cdot)$ being a nonsmooth regularizer), over a cluster of $N$ computing nodes:

$$\min_{\mathbf{x}} \quad \sum_{i=1}^{N} g_i(\mathbf{x}) + h(\mathbf{x}), \tag{9.70}$$

$$\text{s.t.} \quad \mathbf{x} \in X \subseteq \mathbb{R}^M.$$

Let each node $i$ keep a local copy of $\mathbf{x}$ as $\mathbf{x}_i \in \mathbb{R}^M$. In this section, we will mainly consider using the ADMM-based method to solve this problem. Later in the next chapter we will also consider different types of randomized algorithm.

Let us introduce a *single* global variable $\mathbf{y}$ and reformulate the above problem as

$$\min_{\{\mathbf{x}_i\}, \mathbf{y} \in X} \left\{ \sum_{i=1}^{N} f_i(\mathbf{x}_i) + h(\mathbf{y}) : \mathbf{x}_i - \mathbf{y} = \mathbf{0}, \quad i = 1, \dots, N \right\}, \tag{9.71}$$

which is referred to as the *global consensus problem* in [86]. Note that compared with the splitting done in (9.36), here due to the special structure of the problem there is no need to introduce the "edge" variables. Each variable $\{\mathbf{x}_i\}$ and $\mathbf{y}$ is associated with a node in the network. The augmented Lagrangian of this problem is given by

$$L(\mathbf{x}, \mathbf{y}, \boldsymbol{\lambda}) = \sum_{i=1}^{N} f_i(\mathbf{x}_i) + h(\mathbf{y}) + \sum_{i=1}^{N} \left( \langle \boldsymbol{\lambda}_i, \mathbf{x}_i - \mathbf{y} \rangle + \frac{\rho_i}{2} \|\mathbf{x}_i - \mathbf{y}\|^2 \right).$$

The subproblems are expressed below

$$\mathbf{x}_i(t+1) = \arg \min_{\mathbf{x}_i} f_i(\mathbf{x}_i) + \langle \boldsymbol{\lambda}_i(t), \mathbf{x}_i - \mathbf{y}(t) \rangle + \frac{\rho_i}{2} \|\mathbf{x}_i - \mathbf{y}(t)\|^2 \tag{9.72a}$$

$$\mathbf{y}_i(t+1) = \arg \min_{\mathbf{y} \in X} h(\mathbf{y}) + \sum_{i=1}^{N} \left( \langle \boldsymbol{\lambda}_i(t+1), \mathbf{x}_i(t+1) - \mathbf{y} \rangle + \frac{\rho_i}{2} \|\mathbf{x}_i(t+1) - \mathbf{y}\|^2 \right) \tag{9.72b}$$

$$\boldsymbol{\lambda}_i(t+1) = \boldsymbol{\lambda}_i(t) + \rho_i(\mathbf{x}_i(t+1) - \mathbf{y}(t+1)). \tag{9.72c}$$

As an example, let us consider the simple LS problem

$$\arg\min_{\mathbf{x}} \frac{1}{2}\|\mathbf{A}\mathbf{x} - \mathbf{b}_i\|_2^2 = \arg\min_{\mathbf{x}} \sum_{i=1}^{N} \frac{1}{2}\|\mathbf{A}_i\mathbf{x} - \mathbf{b}_i\|_2^2$$

where $\mathbf{A}_i$ is the $i$th row of $\mathbf{A}$ and $\mathbf{b}_i$ is the $i$th element of $\mathbf{b}$. Then this problem can be written in the form (9.71) by

$$\min_{\{\mathbf{x}^i\},\mathbf{y}} \left\{ \sum_{i=1}^{N} \frac{1}{2}\|\mathbf{A}_i\mathbf{x}_i - \mathbf{b}_i\|_2^2 : \mathbf{x}_i - \mathbf{y} = 0, i = 1,\dots,N \right\}. \tag{9.73}$$

When implementing the ADMM algorithm on a cluster computer, one can first distribute the problem data $(\mathbf{A}_i, \mathbf{b}_i)$ to different computing nodes, and at each iteration each $\mathbf{x}_i$-subproblems (9.72a) can be solved in parallel, and the $\mathbf{y}$-subproblem and $\lambda$-update are also easy.

It is worth mentioning that a series of recent works [110, 527] have shown that the ADMM is even capable of computing a KKT point for problem (9.70) in the absence of convexity. In particular, let us make the following assumption on problem (9.70).

ASSUMPTION 4. *Assume the following*

1. *There exists a positive constant $L_i > 0$ such that*

$$\|\nabla g_i(\mathbf{x}_i) - \nabla g_i(\mathbf{z}_i)\| \le L_i\|\mathbf{x}_i - \mathbf{z}_i\|, \ \forall \ \mathbf{x}_i, \mathbf{z}_i, \ i = 1,\cdots,N.$$

   *Moreover, $h(\cdot)$ is convex (possible nonsmooth); $X$ is a closed convex set.*
2. *For all $i$, the penalty parameter $\rho_i$ is chosen large enough such that:*
   *(a) For all $i$, the $\mathbf{x}_i$ subproblem (9.72a) is strongly convex with modulus $\gamma_i(\rho_i)$.*
   *(b) For all $i$, $\rho_i\gamma_i(\rho_i) > 2L_i^2$ and $\rho_i \ge L_i$.*
3. *$f(\mathbf{x}) = \sum_{i=1}^{N} g_i(\mathbf{x}) + h(\mathbf{x})$ is bounded from below over $X$.*

Based on the above assumption, we can have the following claim on the convergence of iteration (9.72a)–(9.72c) [110, Theorem 2.4].

THEOREM 21. *Assume that Assumption 3 is satisfied. Then we have the following:*

1. *We have $\lim_{t\to\infty}\|\mathbf{x}_i(t) - \mathbf{y}(t)\| = 0, \ i = 1,\cdots,N$.*
2. *Let $(\{\mathbf{x}_i^*\}, \mathbf{y}^*, \lambda^*)$ denote any limit point of the sequence $\{\{\mathbf{x}_i(t+1)\}, \mathbf{y}(t+1), \lambda(t+1)\}$ generated by the iteration (9.72a)–(9.72c). Then the following statement is true:*

$$0 = \nabla g_i(\mathbf{x}_i^*) + \lambda_i^*, \quad i = 1,\cdots,N.$$

$$\mathbf{y}^* \in \arg\min_{\mathbf{y}\in X} \ h(\mathbf{y}) + \sum_{i=1}^{N} \langle \lambda_i^*, \mathbf{x}_i^* - \mathbf{y} \rangle$$

$$\mathbf{x}_i^* = \mathbf{y}^*, \quad i = 1,\cdots,N.$$

*That is, any limit point of the iteration (9.72a)–(9.72c) is a stationary solution of problem (9.70).*
3. *If $X$ is a compact set, then the sequence of iterates generated by (9.72a)–(9.72c) converges to the set of stationary solutions of problem (9.70). That is,*

$$\lim_{t\to\infty} \text{dist}\left(\left(\{\mathbf{x}_i(t)\}, \mathbf{y}(t), \lambda(t)\right); Z^*\right) = 0, \tag{9.74}$$

where $Z^*$ is the set of primal-dual stationary solutions of problem (9.70); dist$(\mathbf{x}; X)$ denotes the distance between a vector $\mathbf{x}$ and a set $X$, that is,

$$\text{dist}(\mathbf{x}; X) = \min_{\hat{\mathbf{x}} \in X} \|\mathbf{x} - \hat{\mathbf{x}}\|.$$

## 9.6 Comparing Different Algorithms

Now that we have seen a few major algorithms for solving problem (9.9) over a network of nodes, in this section we briefly summarize their features.

### 9.6.1 Problem Types

First, most distributed subgradient/gradient-based algorithms discussed in Section 9.2 have mild requirements on the type of problems to be solved. For example the classical distributed subgradient/gradient iteration (9.10) only requires that $f_i(\cdot)$ be closed and convex and has bounded subgradients. Additional constraint of the type $\mathbf{x} \in \mathcal{X}$ can be easily handled by its constrained version (9.23). On the other hand, the ADMM-based algorithms have stronger requirements on the objective function. The classical C-ADMM require that the problem of minimizing the objective plus certain quadratic proximal term is easily solvable (cf. (9.53)). Recent variants of C-ADMM require that each $f_i(\cdot)$ is either a smooth function [522] and the problem is unconstrained, or it admits the following form $f_i(\cdot) := g_i(\cdot) + h_i(\cdot)$, where $h_i(\cdot)$ is possibly nonsmooth (which can include the indicator function of a convex set), but its proximity operator is easily computable, and $g_i(\cdot)$ has some composite form [85]. Similarly, the EXTRA algorithm also requires that the entire problem is unconstrained, and that $f_i(\cdot)$ is a smooth function. The dual averaging algorithm has the similar requirement as the distributed subgradient algorithm.

### 9.6.2 Graph Types

It is interesting to note that different algorithms are capable of handling very different types of graphs. The classical subgradient/gradient-based algorithm can handle directed, dynamic, B-connected graphs with weight matrices $\mathbf{W}(t)$ being doubly stochastic. If the objective function $f \equiv 0$ (i.e., the nodes are only required to reach consensus), then the graph can be directed, dynamic, B-connected with each $\mathbf{W}(t)$ only being a stochastic matrix [514]. Recently the algorithm has been extended to graphs with randomly activated links, with noisy communications [513], or with only stochastic weight matrices [528]. The ADMM algorithm can only handle undirected and static graphs. To date it is an open question whether the ADMM-based algorithm can handle dynamic time-varying (B-connected and/or randomly activated graphs) as well as directed graphs at the same time. Recent work of [529] shows that certain variants of the ADMM (termed asynchronous distributed ADMM (AD-ADMM)) is capable of sublinear convergence over certain randomly activated graphs, however this result does not extend easily to

**Table 9.2** Comparison of different distributed algorithms

| Algorithms | Problem types | Graph types | Stepsizes | Convergence |
|---|---|---|---|---|
| D-sub/gradient [511] | $f_i$ convex, unconstrained, bounded $\partial f_i$ | W ds, symmetric, directed, B-C | $\gamma$ constant, $\gamma(t)$ diminishing | converge to $\mathcal{O}(\gamma)$ ball, globally convergent |
| D-projected sub/gradient [512] | $f_i$ convex, constrained, bounded $\partial f_i$, $\mathcal{X}_i = \mathcal{X}$ | W ds, directed, B-C | $\gamma(t)$ diminishing | globally convergent |
| D-gradient [515] | $f_i$ smooth, $f_i$ Lip-G | static, undirected | small $\gamma$, (9.30) | converge to $\mathcal{O}(\gamma)$, ball with rate $\mathcal{O}(1/t)$ |
| D-proximal [516] | $f_i = g_i + h_i$, Assumption 2 | W ds, directed, B-C | $\gamma(t) = \sqrt{1/t}$ | converge with rate $\mathcal{O}(\ln(t)/\sqrt{t})$ |
| Accelerated D-proximal [516] | $f_i = g_i + h_i$, Assumption 2 | W ds, directed, B-C | $\gamma \leq 1/L$ | converge with rate $\mathcal{O}(1/t)$ |
| C-ADMM [520, 504] | $f_i$ convex | static, undirected | $\rho$ constant | converge with rate $\mathcal{O}(1/t)$ |
| IC-ADMM [85] | $f_i = g_i + h_i$, $g_i$ composite | static, undirected | $\rho$ constant, $\nu_i$ is (9.62) | converge with rate $\mathcal{O}(1/t)$ |
| DLM [522] | $f_i$ convex, unconstrained, $f_i$ Lip-G | static, undirected | $\rho$ constant, $\nu_i$ large enough | converge with rate $\mathcal{O}(1/t)$ |
| DA-ADMM [529] | $f_i$ convex | randomized, undirected | $\rho$ constant | converge with rate $\mathcal{O}(1/t)$ in expectation |
| EXTRA [524] | $f_i$ convex, unconstrained, $f_i$ Lip-gradient | W ds, static, undirected | $\gamma$ small, (9.65) | converge with rate $\mathcal{O}(1/t)$ |
| D-dual averaging [525] | $f_i$ convex, constrained | W ds, undirected, random | $\gamma(t)$ diminishing, (9.69) | converge with rate $\mathcal{O}(\ln(t)/\sqrt{t})$ |

the $B$-connected and/or directed graphs. The EXTRA algorithm can only handle static undirected graph with doubly stochastic weights. The distributed dual averaging algorithm can handle undirected, possibly randomly activated graphs with doubly stochastic weights.

### 9.6.3     Convergence Rates

The subgradient/gradient-based algorithms typically converge in a rate $\mathcal{O}(\ln(t)/\sqrt{t})$ if appropriate diminishing stepsizes are chosen. If certain sufficiently small constant stepsizes are chosen, then the algorithm (with smooth objective) converges to the ball of size $\mathcal{O}(1/\gamma)$ sublinearly. However in this case since the objective is smooth, it is certainly more appropriate to use either ADMM-based algorithm such as C-ADMM/IC-ADMM/DLM, or the EXTRA algorithm to achieve faster convergence speed of $\mathcal{O}(1/t)$ to a global optimal solution. The distributed dual averaging algorithm achieves a rate similar to the distributed subgradient/gradient algorithms; however, if additional information about the graph topology is known by all the nodes, then the convergence rate can be sharpened. It is worth mentioning that it is possible to accelerate the performance of the distributed subgradient/gradient method by using Nesterov's acceleration scheme; however, to the best of our knowledge it is not known how to accelerate the C-ADMM-based algorithms.

In table 9.2 we summarize most of the algorithms discussed in this section. In this table we use "ds" to denote "doubly stochastic," "B-C" to denote "$B$-connected," and "Lip-G" to denote "Lipschitz gradient."

# 10 Optimization of Finite Sums

In this chapter, we discuss various recent developments for solving the following important problem:

$$\min_{x \in X \in \mathbb{R}^d} f(x) := g(x) + h(x) := \frac{1}{N} \sum_{i=1}^{N} g_i(x) + h(x) \tag{10.1}$$

This problem finds applications in a wide range of machine learning problems, where certain regularized empirical loss function is minimized. Here the $g_i$ function usually represents the smooth cost function evaluated on the $i$th data point; $h_i$ represents convex regularizers; $d$ is the problem dimension and $N$ is the total number of data points. Typically both $N$ and $d$ are large (representing high-dimensional learning over large amount of data). Due to the fact that $n$ is large, the traditional (proximal) gradient-based method that we have introduced in Chapter 3 will be very inefficient and expensive here. The reason is that evaluating the gradient of the smooth function $g(x)$ amounts to evaluating $N$ gradients simultaneously, one for each $g_i$. Can we design algorithms that gradually build upon the gradients of component functions (or data points)? If we can, what is the performance of these methods? These are the key questions to be addressed in this chapter.

In this chapter, we focus on the following families of algorithms:

1. the classical incremental (gradient) algorithms
2. the stochastic gradient-based algorithms
3. the stochastic dual coordinate ascent based algorithms

Our focus will be given to the connections among different algorithms as well as their convergence rate analysis.

The chapter is organized as follows. In Section 10.1 we discuss the properties of the classical incremental gradient methods. In Sections 10.2 and 10.3 we present a few recently developed, stochastic gradient-descent-based algorithms. In Section 10.4, we present the performance lower-bound analysis for optimization with finite sum. In Section 10.5, we present a primal-dual-based algorithmic framework.

## 10.1        The Incremental Gradient Algorithms

First we introduce a family of classical algorithms, the incremental gradient methods (see [530] for a survey). Let us consider a special case of problem (10.1):

$$\min_{x \in X \in \mathbb{R}^d} f(x) := \frac{1}{N} \sum_{i=1}^{N} g_i(x) \tag{10.2}$$

where $g_i$ is a smooth function. The incremental gradient method generates the following iterate:

$$x^{r+1} = \mathrm{proj}_X \left( x^r - \alpha_r \nabla g_{i_r}(x^r) \right), \ r = 1, 2, \cdots . \tag{10.3}$$

Here $r$ is the iteration counter, $\{\alpha_r > 0\}$ is a sequence of stepsizes, and $i_r \in \{1, \cdots, N\}$ is the index picked at iteration $r$. Clearly, for each iteration only a single component is picked out of $n$ components, and the algorithm progresses based on such (rather limited) gradient information. Compared with the classical gradient algorithm, the incremental gradient method can be very fast at the beginning because as soon as *a single* gradient information arrives the algorithm can begin optimize. On the contrary, the gradient method requires the entire gradient information. However, compared with the gradient-based method, the incremental gradient method suffers from slow convergence in the later stages, as to converge to global optimal solution the stepsize sequence $\alpha_r$ is required to be diminishing. Otherwise it is possible that the algorithm will never converge.

**A Example of Divergence** [531]. To understand the behavior of the incremental gradient with constant stepsize, let us consider the following problem:

$$\min f(x) = \frac{1}{2}(x - c_1)^2 + \frac{1}{2}(x - c_2)^2. \tag{10.4}$$

Clearly, $x^* = \frac{c_1 + c_2}{2}$. Let $x^0 = 0$, and suppose a constant stepsize $\alpha$ is used. The incremental gradient method generates the following iterate:

$$x^{r+1}(2) = x^r(1) - \alpha(x^r(1) - c_1) \tag{10.5}$$
$$x^{r+1}(2) = x^r(2) - \alpha(x^r(2) - c_2) \tag{10.6}$$

where $x^r(i)$ denotes the $i$th component at $r$th iteration. Using this iteration, one can check that the following is true [531]:

$$x^{r+1}(1) = (1 - \alpha)^2 x^r(1) + (1 - \alpha)\alpha c_1 + \alpha c_2.$$

For $0 < \alpha < 1$, in the limit we have

$$x^r(1) \to \frac{(1 - \alpha)c_1 + c_2}{2 - \alpha} \tag{10.7}$$

$$x^r(2) \to \frac{(1 - \alpha)c_2 + c_1}{2 - \alpha}. \tag{10.8}$$

Clearly, the sequence of iterates will not converge for a given stepsize $\alpha \in (0, 1)$. Also this result suggests that if $\alpha_r = 1/r$, then the optimal solution can be achieved.

Generally, we have the following convergence results for the incremental gradient methods:

1. If a constant stepsize $\alpha_k = \alpha$ is used, then the iterates will cycle over $N$ points.
2. If $\alpha_r = \mathcal{O}(1/r)$, then the algorithm converges to a global optimal solution of (10.2) (or to stationary solutions if the problem is nonconvex).

However, for big data problems, one does not care very much about whether a given algorithm converges (as convergence should be the minimum criteria), but is more interested in analyzing the convergence rates. It turns out to be difficult to directly analyze the convergence rates of the incremental methods, precisely because of the diminishing stepsize required.

In the seminal work of Blatt, Hero, and Gauchman [532], a variant of the incremental-type algorithm called the "incremental aggregated gradient" (IAG) algorithm is proposed, in which a *constant* stepsize $\alpha$ is used. Consider the unconstrained version of problem (10.2) with $X \equiv R^d$. The algorithm is given as follows (for all $r \geq N$):

$$x^{r+1} = x^r - \alpha \frac{1}{N} d^r \tag{10.9a}$$

$$d^{r+1} = d^r - \nabla f_{i_r}(x^{r+1-N}) + \nabla f_{i_r}(x^{r+1}) \tag{10.9b}$$

where $i_r \in \{1, \cdots, N\}$ is the index picked at iteration $r$, and in this particular form the index is picked *cyclically*. The IAG iteration (10.9a)–(10.9b) is equivalent to

$$x^{r+1} = x^r - \alpha \frac{1}{N} \sum_{n=0}^{N-1} \nabla f_{i_n}(x^{r-n}). \tag{10.10}$$

Clearly, compared with the incremental iteration (10.3), the new algorithm not only uses the current gradient information, but *all* the gradient information that has been evaluated in the latest $N$ iterations. Note that in (10.10), each $\nabla f_{i_n}$ in the sum is evaluated on a different iterate $x^{r-n}$. This is a very reasonable improvement over the incremental gradient algorithm as the past $N$ gradients are readily available. However, it is important to note that to implement the IAG, additional memory is required to store the past gradients. Next let us investigate the convergence and rate of convergence analysis for IAG.

ASSUMPTION 5. *Suppose that the following assumptions hold true:*

1. $f(x) := \sum_{i=1}^{N} g_i(x)$, $h(x) \equiv 0$ and $X \equiv R^d$.
2. *Each component function $g_i$ has a Lipschitz continuous gradient with modulus $M_1$, that is,*

$$\|\nabla g_i(x) - \nabla g_i(z)\| \leq L_i \|x - z\|, \ \forall x, z \in R^d.$$

*Clearly, the function $\sum_{i=1}^{N} g_i$ has a Lipschitz continuous gradient with modulus $L := \sum_{i=1}^{N} L_i$.*

3. $\|\nabla g_i(x)\| \leq M_3$ *for all $i$, and* $\|\nabla f(x)\| \leq M_4$.
4. *Further assume that problem (10.2) has a unique minimizer $x^*$.*

Under Assumption 5, it can be shown that if the stepsize $\alpha$ is small enough, then $\lim_{r \to \infty} x^r = x^*$. For the precise statement of the results, see [532, Proposition 2.7].

Interestingly, the authors of [532] also show that if the objective is a quadratic problem of the form

$$f_i(x) = \frac{1}{2} x^T Q_i x - c_i^T x, \ i = 1, \cdots, N$$

for some matrix $Q_i$ and vector $c_i$, then the IAG converges *linearly* if $\sum_{i=1}^{N} Q_i$ is positive definite.

Recently, Gürbüzbalaban, Ozdaglar, and Parrilo [533] have shown that under similar conditions as discussed above but for general strongly convex objective, the IAG converges globally linearly. In particular, assume that $f(x)$ is strongly convex with modulus $\mu$, and define

$$\kappa = \frac{L}{\mu}.$$

Then we have the following result (see [533]):

THEOREM 22. *Suppose Assumption 5-(1),(2) hold, and $f(x)$ is strongly convex with modulus $\mu$. Suppose the stepsize $\alpha$ is given by*

$$0 < \alpha \le \bar{\alpha} := \frac{a\mu}{NL} \frac{1}{\mu + L}$$

*with $a = 8/25$. Then the IAG iterates $\{x^r\}$ are globally linearly convergent. Further, if $\alpha = \bar{\alpha}/2$, we have*

$$\|x^r - x^*\| \le \left( 1 - \frac{c_N}{(\kappa + 1)^2} \right)^r \|x^0 - x^*\| \tag{10.11a}$$

$$f(x^r) - f(x^*) \le \frac{L}{2} \left( 1 - \frac{c_N}{(\kappa + 1)^2} \right)^{2r} \|x^0 - x^*\|^2. \tag{10.11b}$$

*where $c_N = \frac{2}{25}[N(2N + 1)]^{-1}$.*

We note that the requirement here is that the sum of the functions $f(\cdot)$ is strongly convex; however, the component functions are not required to be individually strongly convex. Later we will see that this assumption may not be sufficient for other related algorithms.

## 10.2   The Stochastic Gradient-Based Algorithms

In this section, we discuss a different family of algorithms for solving problem (10.1), all of which are based on the so-called stochastic gradient algorithms in which the component functions are sampled randomly. The main benefit offered by randomization is the improved convergence rate, as will be shown shortly. Further, certain variations of the stochastic gradient algorithm can achieve linear convergence for strongly convex problem *without* requiring additional memory (as opposed to the IAG-type methods).

## 10.2.1 The SAG-Based Method

In this subsection, we present the SAG (stochastic average gradient) method proposed by Roux, Schmidt, and Bach [534]. Let us still consider problem (10.2) with $X = \mathbb{R}^d$ (i.e., Assumption 5). The SAG algorithm is given in the following table.

---

**The SAG Algorithm**

Randomly pick an index $i_r \in \{1, \cdots, N\}$

$$y_{i_r}^r = x^r, \quad y_j^r = y_j^{r-1}, \ \forall j \neq i_r$$

$$x^{r+1} = x^r - \frac{1}{\eta N} \sum_{i=1}^{N} \nabla g_i(y_i^r)$$

$$= x^r - \frac{1}{\eta N} \sum_{i=1}^{N} \nabla g_i(y_i^{r-1}) + \frac{1}{\eta N} \left( \nabla g_{i_r}(y_{i_r}^{r-1}) - \nabla g_{i_r}(x^r) \right)$$

---

Comparing the SAG with IAG, it is clear that these two algorithms are very closely related. Both algorithms use the averages of the gradients evaluated at the latest $N$ iterations. The only difference perhaps is that in SAG the functions are picked randomly at each iteration, while in IAG the functions are picked deterministically and cyclically.

To analyze the convergence, let us assume that Assumption 5-(1),(2) are true, but replacing the $L_i$ in the second assumption by $L$ for all the component functions. Let us define

$$\sigma^2 := \frac{1}{N} \sum_{i=1}^{N} \| \nabla g_i(x^*) \|^2.$$

Further assume that $g$ is strongly convex with constant $\mu$, that is,

$$g(x) \geq g(y) + \langle \nabla g(y), x - y \rangle + \frac{\mu}{2} \|x - y\|^2, \ \forall x, y \in \mathbb{R}^d.$$

Then the algorithm converges linearly in the sense that, if $N \geq \frac{8L}{\mu}$, $\eta = 2N\mu$, for all $r \geq N$

$$\mathbb{E} \left[ F(x^r) - F^* \right] \leq C \left( 1 - \frac{1}{8N} \right)^r$$

$$C = \frac{16L}{3N} \|x^0 - x^*\|^2 + \frac{4\sigma^2}{3N\mu} \left( 8 \log(1 + \frac{\mu N}{4L} + 1) \right).$$

Or equivalently, to achieve $\epsilon$-optimality, we need the following number of iterations:

$$\mathcal{O}(N \log(1/\epsilon)) = \mathcal{O} \left( (N + \frac{L}{\mu}) \log(1/\epsilon) \right). \tag{10.12}$$

SAG is one of the first stochastic gradient-based algorithms that achieves linear convergence for optimizing a problem with finite sum. However, it has a few drawbacks (listed below) which leave significant room for improvement:

1. It only deals with smooth unconstrained problem.

2. It only deals with strongly convex problem, while many machine learning problems are convex but not strongly convex.
3. The practical convergence is quite slow.
4. It requires the storage of past $N$ gradients.
5. The convergence proof is quite involved.

To alleviate some of the above difficulties, [535], Defazio, Bach, and Lacoste-Julien proposed a SAGA algorithm, listed below, for the general problem formulation (10.1).

---

**The SAGA Algorithm**

Randomly pick an index $i_r \in \{1, \cdots, N\}$

$$y_{i_r}^r = x^r, \quad y_j^r = y_j^{r-1}, \quad \forall j \neq i_r$$

$$w^{r+1} = x^r - \frac{1}{\eta N} \sum_{i=1}^{N} \nabla g_i(y_i^{r-1}) + \frac{1}{\eta} \left( \nabla g_{i_r}(y_{i_r}^{r-1}) - \nabla g_{i_r}(x^r) \right)$$

$$x^{r+1} = \operatorname{prox}_{h+\iota(X)}^{1/\eta}(w^{r+1})$$

---

Here the notation $\iota(X)$ represents the indicator function of the convex set $X$. First note that if the problem is smooth and unconstrained, the last line of the algorithm (the "prox" operation) is not needed. Second, comparing SAGA with SAG, we see that a large "stepsize" is used ($1/\eta$ v.s. $1/(N\eta)$) when updating the variable $w$. This is in fact very reasonable as the term

$$\frac{1}{\eta N} \sum_{i=1}^{N} \nabla g_i(y_i^{r-1}) - \frac{1}{\eta} \left( \nabla g_{i_r}(y_{i_r}^{r-1}) - \nabla g_{i_r}(x^r) \right)$$

represents an unbias estimator of the true gradient at each iteration $r$, while the same term in the SAG represents a *biased* estimate.

The analysis of convergence of SAGA is much simpler than the SAG. However, we have to make the following strong convexity assumption for *each* $g_i$:

$$g_i(x) \geq g_i(y) + \langle \nabla g_i(y), x - y \rangle + \frac{\mu}{2} \|x - y\|^2, \quad \forall x, y \in \mathbb{R}^d, \; \forall i.$$

Then the convergence of SAGA can be proved via the following two cases:

C1: Suppose $h \equiv 0$ and each $g_i$ is strongly convex (modulus $\mu$) and have Lip-gradient (modulus $L$). Choose the following stepsize $\eta = 2(\mu N + L)$. Then we have the following rates

$$\mathbb{E}[\|x^r - x^*\|^2] \leq \left( 1 - \frac{\mu}{2\mu N + L} \right)^r D \tag{10.13}$$

where

$$D := \|x^0 - x^*\|^2 + \frac{N}{\mu N + L} [f(x^0) - \langle f(x^0) - \langle f'(x^*), x^0 - x^* \rangle - f(x^*) \rangle]$$

Therefore, to achieve $\epsilon$-OPT requires $\mathcal{O}\left( (N + \mu L) \log(1/\epsilon) \right)$ iterations.

C2: Suppose $h$ is present and $g_i$ is not strongly convex. Choose $\eta = 3L$. Then we have the following rates

$$\mathbb{E}\left[F(\tilde{x}^R)\right] - F(x^*) \le \frac{4N}{K}D^*$$

where

$$D^* := \|x^0 - x^*\|^2 + \frac{2N}{3L}\left[f(x^0) - \langle f'(x^*), x^0 - x^*\rangle - f(x^*)\right],$$

$$\tilde{x}^R := \frac{1}{R}\sum_{r=1}^{R} x^r.$$

In practice, SAGA is often much faster than SAG. However, its still has a few drawbacks:

1. It requires that each component function to be strongly convex.
2. It still needs the storage of past $N$ gradients.

Next we present an algorithm that can alleviate these two drawbacks.

### 10.2.2  The SVRG Algorithm

In this subsection, we present the SVRG (stochastic variance reduction gradient) method proposed by Johnson and Zhang [536]. The main motivation of this algorithm is to remove the requirement of storing the past gradient. To do so, a "inner loop" is introduced whereby the average of $N$ gradients is reused for a few iterations before updating the new average. The algorithm description is given in the following table. We will refer to the outer iteration (indexed by "$s$") as an *epoch*.

---

**The SVRG Algorithm**

**Iterate**: for $s = 1, 2, \cdots$

   $\tilde{x} = \tilde{x}^{s-1}$

   $\tilde{z} = \dfrac{1}{N}\sum_{i=1}^{N} \nabla g_i(\tilde{x})$

   $x_0 = \tilde{x}$

   **Iterate**: for $t = 1, 2, \cdots, m$

      Randomly pick $i_t$

      $x_t = x_{t-1} - \dfrac{1}{\eta}\left(\nabla g_{i_t}(x_{t-1}) - \nabla g_{i_t}(\tilde{x}) + \tilde{z}\right)$

   **End**

   **Option I**: set $\tilde{x}^s = x_m$

   **Option II**: set $\tilde{x}^s = x_t$ for randomly chosen $t \in \{0, \cdots, m-1\}$

**End**

---

As we have explained above, the inner iteration looks very similar to the SAGA iterations: the "stepsize" $1/\eta$ is used to update the new $x_t$. However, the key difference here is
that the averaged gradient $\tilde{z}$ is never changed within the inner iteration (i.e., it is "reused"
during the entire $m$ inner iterations). Therefore, during the inner iteration no memory of
all the past iterates is needed. Also note that for each epoch, the averaged gradient $\tilde{z}$ is
*recomputed* based on the current iterate $\tilde{x}$, as opposed to the SAG/SAGA/IAG algorithm
where the averaged gradient is formed by a sum of gradients of component functions
*evaluated* at different iterates.

The analysis of the convergence is actually very simple. Suppose that each $g_i$ has Lip-
gradient with modulus $L$, $h \equiv 0$ and $f$ is strongly convex with modulus $\mu$. Also Suppose
the inner iteration number $m$ is sufficiently large so that

$$\alpha := \frac{\eta^2}{\mu(\eta - 2L)m} + \frac{2L}{\eta - 2L} < 1.$$

Then we have the following linear convergence guarantee

$$\mathbb{E}[f(\tilde{x}_s)] - f(x^*) \le \alpha^s[f(\tilde{x}_0) - f(x^*)]. \tag{10.14}$$

To see how different algorithm parameters interact with each other, in [536] the
authors made the following comparison. Suppose the condition number $L/\mu = N$; one
can take

$$\eta = 10L, \quad m = \mathcal{O}(N)$$

to obtain a convergence rate of $\alpha = 1/2$. Then in total we require $N \ln(1/\epsilon)$ number of
iterations (compared with standard batch gradient descent $N^2 \ln(1/\epsilon)$). Note that in this
case, the convergence rate is still

$$\mathcal{O}((N + L/\mu)\log(1/\epsilon)).$$

We briefly mention that the key step of the proof is to bound the "aggregate gradient"
term"

$$v_t := \nabla g_{i_t}(x_{t-1}) - \nabla g_{i_t}(\tilde{x}) + \tilde{z},$$

whose size can be viewed as the variance of the estimation of the gradient. If this term
goes to zero, then the algorithm converges. It is easy to show that the following estimate
holds true:

$$\mathbb{E}[\|v_t\|^2] \le 2\mathbb{E}[\|\nabla g_{i_t}(x_{t-1}) - \nabla g_{i_t}(x^*)\|^2] + 2\mathbb{E}[\|\nabla g_{i_t}(\tilde{x}) - \nabla g_{i_t}(x^*) - \nabla f(\tilde{x})\|^2]$$
$$= 2\mathbb{E}[\|\nabla g_{i_t}(x_{t-1}) - \nabla g_{i_t}(x^*)\|^2] + 2\mathbb{E}[\|\nabla g_{i_t}(\tilde{x}) - \nabla g_{i_t}(x^*) - \mathbb{E}[\nabla g_{i_t}(\tilde{x})$$
$$\quad - \nabla g_{i_t}(x^*)]\|^2]$$
$$\le 2\mathbb{E}[\|\nabla g_{i_t}(x_{t-1}) - \nabla g_{i_t}(x^*)\|^2] + 2\mathbb{E}[\|\nabla g_{i_t}(\tilde{x}) - \nabla g_{i_t}(x^*)]$$
$$\le 4L[f(x_{t_1}) - f(x^*) + f(\tilde{x}) - f(x^*)]$$

where the second inequality uses $\mathbb{E}[\|\xi - \mathbb{E}[\xi]\|^2] \le \mathbb{E}[\|\xi\|^2]$, the last inequality uses
the following key property due to Nestrov [52, Theorem 2.1.5]

If $g_i(x)$ is convex with Lipschitzian gradient (constant $L$), then

$$\frac{1}{L}\|\nabla g_i(x) - \nabla g_i(y)\|^2 \le \langle \nabla g_i(x_i) - \nabla g_i(y_i), x - y \rangle, \ \forall \, x, y \in \mathbb{R}^d.$$

Clearly, if $x_t \to x^*$, then we must have $\|v_t\| \to 0$. This is the reason why the algorithm is designated as a "variance reduction" scheme.

In [536], the authors also mentioned that for convex cases, the algorithm converges with rate $\mathcal{O}(1/\epsilon)$, and that the algorithm also applies to nonconvex cases. The general understanding is that the SVRG trades off time and space, that is, it requires (at least in practice) more gradient evaluations in order to remove the requirement for storing past gradients. This work has been followed by many to extend its applicability to problems with nonsmooth term [537], to strongly convex problem with coordinate descent [538] and so on.

## 10.3  The Stochastic Algorithms in the Dual

Another popular family of algorithms solves (10.1) from the perspective of its dual. In this subsection we present one popular algorithm belonging to this category, the SDCA (stochastic dual coordinate ascent method) algorithm proposed by Shalev-Schwartz and Zhang [539].

Consider the following smooth and unconstrained problem, which is a special case of problem (10.1):

$$\min f(x) = \frac{1}{N} \sum_{i=1}^{N} g_i(b_i^T x) + \frac{\lambda}{2} \|x\|^2 \tag{10.15}$$

where $b_i$ is some vector and the regularizer $h(x) := \frac{\lambda}{2}\|x\|^2$ is a strongly convex and smooth function.

To transform this problem to its dual form, let us introduce $N$ auxiliary variables and write the problem equivalently as

$$\min \quad f(x) := \frac{1}{N} \sum_{i=1}^{N} g_i(c_i) + \frac{\lambda}{2} \|x\|^2,$$

$$\text{s.t.} \quad b_i^T x = c_i, \ \forall \, i.$$

Introduce the dual variables $\{\alpha_i\}_{i=1}^{N}$, one for each linear constraint $b_i^T x = c_i$; the Lagrangian function is given by

$$L(x, c, \alpha) = \frac{1}{N} \sum_{i=1}^{N} g_i(c_i) + \frac{\lambda}{2}\|x\|^2 + \frac{1}{N} \sum_{i=1}^{N} \langle b_i^T x - c_i, \alpha_i \rangle. \tag{10.16}$$

Minimizing with respect to $x$, one obtains

$$\lambda x = \frac{1}{N} b_i^T \alpha. \tag{10.17}$$

Using this expression, the dual problem of (10.15) is given by the following, where $g_i^*(u) := \max_z(zu - g_i(z))$ is the conjugate function for $g_i$:

$$\max_\alpha D(\alpha) := \left[ \frac{1}{N} \sum_{i=1}^{N} -g_i^*(-\alpha_i) - \frac{\lambda}{2} \left\| \frac{1}{\lambda N} \sum_{i=1}^{N} \alpha_i b_i \right\|^2 \right]. \tag{10.18}$$

The SDCA solves the above dual problem using the stochastic coordinate ascent method (cf. Section 3.3, here as we are maximizing the objective, we use "ascent" instead of "descent"). In particular, the algorithm description is given in the following table:

---

**The SDCA Algorithm**

**Iterate**: for $t = 1, 2, \cdots, T$

    Randomly pick $i$

    Find $\nabla \alpha_i$ to maximize

$$- g_i^*(-(\alpha_i^{t-1} + \nabla \alpha_i)) - \frac{\lambda N}{2} \|x^{t-1} + (\lambda N)^{-1} \nabla \alpha_i b_i\|^2$$

    Let $\alpha^t = \alpha^{t-1} + \nabla \alpha_i e_i$

    Let $x^t = x^{t-1} + (\lambda N)^{-1} b_i^T \nabla \alpha_i$

**End**

**Output** :

    Let $\bar{\alpha} = \frac{1}{T-T_0} \sum_{t=T_0+1}^{T} \alpha^{t-1}$

    Let $\bar{w} = \frac{1}{T-T_0} \sum_{t=T_0+1}^{T} w^{t-1}$

---

In the table $e_i$ represents the unit vector in which the $i$th component is 1 and the rest are all zeros. The $\nabla \alpha_i$-maximization step is basically an coordinate ascent over the variable $\alpha_i$'s. The expression for $x^t$ comes from (10.17).

The convergence property of the SDCA can be summarized in the following two cases [539, Theorem 1–Theorem 2]:

C1: If $g_i$ is $Q$ Lipschitz continuous for all $i$, that is,

$$|g_i(x) - g_i(y)| \le Q\|x - y\|, \ \forall \, x, y \in \mathbb{R}^d, \ \forall \, i.$$

Then to obtain $\mathbb{E}[f(\bar{x}) - f(\bar{\alpha})] \le \epsilon$, we need the following number of iterations:

$$T \ge T_0 + N + \frac{4Q^2}{\lambda \epsilon}.$$

C2: If each $g_i$ has Lipschitz continuous gradient with constant $L$, then to obtain $\mathbb{E}[f(\bar{x}) - f(\bar{\alpha})] \le \epsilon$, we need the following number of iterations:

$$T \ge \left(N + \frac{L}{\lambda}\right) \log\left(N + \frac{L}{\lambda} \frac{1}{\epsilon}\right). \tag{10.19}$$

Note that, similar to the SAG and SVRG, here the aggregated objective function $f(x)$ is assumed to be strongly convex, not the component function $g_i$. Therefore the $\lambda$ here corresponds to the constant $\mu$ mentioned in the previous algorithms. However, different from SAG/SAGA/SVRG, here we have to use the particular form where the strongly convex part of the objective is in the form of $\lambda/2\|x\|^2$. The reason is that using this particular form the dual function can be evaluated in closed form (see (10.18)).

## 10.4    Summary

In this chapter we have surveyed a few recent algorithms and their complexity bounds for solving problem (10.1), in which the objective function consists of a sum of finite functions. Overall, most of these methods are quite effective in practice, and they are becoming the state-of-the-art algorithms for large-scale optimization involving big data. Future research directions include extending the algorithms and analysis to convex problems and even for nonconvex problems (e.g., deriving lower bounds, providing lower-bound achieving scheme, etc.). Further, the characterization of generalization error for this family of problems is also worth investigating. We refer the reader to [540] for more discussions along this line.

# 11 Big Data Optimization for Communication Networks

Nowadays, modern communication networks play an important role such as in electric power systems, mobile cloud computing, smart city evolution, and personal healthcare. The employed novel telecommunication technologies make data collection much easier for system operation and control, enable more efficient data transmission for mobile applications, and promise a more intelligent sensing and monitoring for metropolitan city regions. Meanwhile, we are witnessing an unprecedented rise in volume, variety and velocity of information in modern communication networks. A large volume of data are generated by our digital equipments such as mobile devices and computers, smart meters and household appliances, as well as surveillance cameras and sensor-equipped mass rapid transit around the city. The information exposition of big data in modern communication networks makes statistical and computational methods significantly important for data analysis, processing, and optimization. The network operators or service providers who can develop and exploit efficient methods to tackle big data challenges will ensure network security and resiliency, gain market share, increase revenue with distinctive quality of service, as well as achieve intelligent network operation and management.

The unprecedented "big data," reinforced by communication and information technologies, presents us opportunities and challenges. On the one hand, the inferential power of algorithms, which have been shown to be successful on modest-sized data sets, may be amplified by the massive data set. Those data analytic methods for the unprecedented volumes of data promises to improve personalized business model design, intelligent social network analysis, smart city development, efficient healthcare and medical data management, and the smart grid evolution. On the other hand, the sheer volume of data makes it unpractical to collect, store, and process the data set in a centralized fashion. Moreover, the massive data sets are noisy, incomplete, heterogeneous, structured, prone to outliers, and vulnerable to cyber attacks. The error rates, which are part and parcel of any inferential algorithm, may also be amplified by the massive data. Finally, the "big data" problems often come with time constraints, where a medium-quality answer that is obtained quickly can be more useful than a high-quality answer that is obtained slowly. Overall, we are facing a problem in which the classic resources of computation, such as time, space, and energy, are intertwined in complex ways with the massive data resources.

With the era of "big data" comes the need of parallel and distributed algorithms for the large-scale inference and optimization. Numerous problems in statistical and machine learning, compressed sensing, social network analysis, and computational biology formulates optimization problems with millions or billions of variables. Since classical optimization algorithms are not designed to address problems of this size, novel optimization algorithms are emerging to deal with problems in the "big data" setting. An noncomprehensive list of such kind of algorithms includes block coordinate descent method [121, 543, 72],[1] stochastic gradient descent method [544, 545, 546], dual coordinate ascent method [547, 539], alternating direction method of multipliers (ADMM) [518, 548], and Frank-Wolf method (also known as the conditional gradient method) [549, 550]. Each type of the algorithm on the list has its own strength and weakness. The list is sill growing and, due to our limited knowledge and the fast develop nature of this active field of research, many efficient algorithms are not mentioned here.

Though many algorithms can be applied to deal with big data problem in communication systems, we restrict our attention to the class of algorithms based on ADMM. The rest of this chapter is organized as follows. Section 11.1 investigates the mobile data offloading in software-defined networks. Section 11.2 studies the mobile cloud computing algorithms. Section 11.3 illustrates the management schemes in data centers. Section 11.4 tries to conduct the resource allocation for wireless network virtualization. Section 11.5 summarizes this chapter.

## 11.1 Mobile Data Offloading in Software-Defined Networks

We consider a mobile network which consists of $B$ cellular base stations (BSs) and $A$ access points (APs). A BS $b \in \{1,\ldots,B\}$ serves a group of mobile users and has the demand to offload its traffic to APs. An AP $a \in \{1,\ldots,A\}$ is a WiFi or femtocell AP which operates in a different frequency band and supplies its bandwidth for data offloading. The maximum available capacity for data offloading of each AP $a$ is denoted by $C_a$. The software-defined network (SDN) controller manages the BSs and APs through the access network discovery and selection function, and makes the mobile data offloading decisions according to various trigger criteria. Such criteria can be the number of mobile users per BS, available bandwidth/IP address of each BS, or aggregate number of flows on a specific port at a BS.

Let $\mathbf{x}_b = [x_{b1},\ldots,x_{bA}]^\top$ represent the offloaded traffic of BS $b$, where $x_{ba}$ denotes the data of BS $b$ offloaded through AP $a$. Correspondingly, $\mathbf{y}_a = [y_{a1},\ldots,y_{aB}]^\top$ represents the admitted traffic of AP $a$, where $y_{ab}$ represents the admitted data traffic from BS $b$. Generally, a feasible mobile data offloading decision exists when BSs and APs reach an agreement on the amount of offloading data (i.e., $x_{ba} = y_{ab}, \forall a$ and $\forall b$). We assume that the mobile data of BSs can be offloaded to all of the APs without loss of generality. Moreover, we assume that the time is slotted and during each slot duration the offloading demand from BSs is fixed. The SDN controller needs to find a feasible offloading schedule at the beginning of each time slot, while maximizing the utility of BSs at a reasonable cost of APs.

[1] [72] proposes a stochastic block coordinate descent method.

We denote BS $b$'s utility of offloading its traffic to APs by $U_b(\mathbf{x}_b)$, where $U_b(\cdot)$ is designed to be a non-decreasing, non-negative, and concave function in $\mathbf{x}_b, \forall b$. For example, the function can be logarithmic, and the concavity is justified because of diminishing returns of the resources allocated to the offload data. Likewise, we use function $L_a(\mathbf{y}_a)$ to describe the AP $a$'s cost of helping BSs offload data, where $L_a(\cdot)$ is a non-decreasing, non-negative, and convex function in $\mathbf{y}_a, \forall a$. The cost function can be a linear cost function, which means the total cost of APs will increase as the amount of admitted mobile data increases.

For the SDN controller, the total revenue for mobile data offloading is expressed as $\sum_{b=1}^{B} U_b(\mathbf{x}_b) - \sum_{a=1}^{A} L_a(\mathbf{y}_a)$. To maximize the total revenue, the equivalent minimization optimization problem can be formulated as,

$$\min_{\{\mathbf{x}_1,\dots,\mathbf{x}_B\},\{\mathbf{y}_1,\dots,\mathbf{y}_A\}} \sum_{a=1}^{A} L_a(\mathbf{y}_a) - \sum_{b=1}^{B} U_b(\mathbf{x}_b), \tag{11.1}$$

$$\text{s.t} \quad \sum_{b=1}^{B} y_{ab} \leq C_a, \quad \forall a, \tag{11.2}$$

$$x_{ba} = y_{ab}, \quad \forall a, b, \tag{11.3}$$

where (11.2) stands for the capacity constraint at each AP, and (11.3) represents the consensus of BSs and APs on the amount of mobile data.

We propose a fully distributed algorithm to solve the optimization problem (11.1). The computing paradigm of the proposed algorithm is shown in Figure 11.1 and can be described as follows. During each iteration, the BSs and APs update $\mathbf{x}$ and $\mathbf{y}$ concurrently. The updated $\mathbf{x}$ and $\mathbf{y}$ are gathered by the SDN controller, which performs a simple update on $\lambda$ and scatters the dual variables back to the BSs and APs. The iteration goes on until a consensus on the offloading demand and supply is reached.

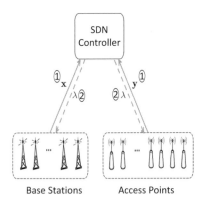

① Gather: BSs and APs concurrently update x and y, which are gathered by controller.
② Scatter: Controller simply updates $\lambda$, which are scattered to BSs and APs

**Figure 11.1** Distributed computing paradigm of proposed algorithm.

Specifically, we first calculate the partial Lagrangian of (11.1), which introduces the Lagrange multipliers only for constraint (11.3),

$$\mathcal{L}_\rho(\mathbf{x}, \mathbf{y}, \boldsymbol{\lambda}) = \sum_{a=1}^{A} L_a(\mathbf{y}_a) - \sum_{b=1}^{B} U_b(\mathbf{x}_b)$$

$$- \sum_{a=1}^{A}\sum_{b=1}^{B} \lambda_{ab}(x_{ba} - y_{ab}) + \frac{\rho}{2}\sum_{a=1}^{A}\sum_{b=1}^{B} \|x_{ba} - y_{ab}\|_2^2, \tag{11.4}$$

where $\boldsymbol{\lambda} \in \mathbb{R}^{AB}$ is the Lagrange multiplier and $\rho$ is the penalty parameter. The updates of BSs and APs can be performed concurrently according to the proximal Jacobian multi-block ADMM. We describe the update procedure of the BSs, APs and SDN controller as follows.

**Base Station Update**: At each BS $b$, the update rule can be expressed as,

$$\mathbf{x}_b^{k+1} = \arg\min_{\mathbf{x}_b}(-U_b(\mathbf{x}_b)+\frac{\rho}{2}\sum_{a=1}^{A}\|x_{ba}-p_{ab}^k\|_2^2+\frac{1}{2}\|\mathbf{x}_b-\mathbf{x}_b^k\|_{\mathbf{P}_i}^2), \tag{11.5}$$

where $\mathbf{P}_i = 0.1\mathbf{I}$ and $\mathbf{I}$ is the identity matrix, and $p_{ab}^k = (y_{ab}^k + \frac{\lambda_{ab}^k}{\rho}), \forall a$ is the "signal" sent from the SDN controller to BS $b$. The update (11.5) is a small-scale unconstrained convex optimization problem. For each round of the update, it sends $\mathbf{x}_b$ of size $A$ to the SDN controller. Note that the update of each BS $b$ is performed independently and can be calculated locally. Once $\mathbf{x}_b$ is updated, it is sent to the SDN controller while the utility function $U_b(\cdot)$ is kept confidential.

---

**Algorithm 10** Distributed Mobile Data Offloading

---

Initialize: $\mathbf{x}^0, \mathbf{y}^0 \boldsymbol{\lambda}^0, \rho > 0, \gamma > 0$;
**for** $k = 0, 1, \ldots$ **do**
  {Update $\mathbf{x}_b$ and $\mathbf{y}_a$ for $b = 1, \ldots, B$ and $a = 1, \ldots, A$, **concurrently**.}
  {**Base station update**, $\forall b$}
  $\mathbf{x}_b^{k+1} = \arg\min_{\mathbf{x}_b} -U_b(\mathbf{x}_b) + \frac{\rho}{2}\sum_{a=1}^{A}\|x_{ba} - y_{ab}^k - \frac{\lambda_{ab}^k}{\rho}\|_2^2 + \frac{1}{2}\|\mathbf{x}_b - \mathbf{x}_b^k\|_{\mathbf{P}_i}^2$;
  {**Access point update**, $\forall a$}
  $\mathbf{y}_a^{k+1} = \arg\min_{\mathbf{y}_b} L_a(\mathbf{y}_a) + \frac{\rho}{2}\sum_{b=1}^{B}\|x_{ba}^k - y_{ab} - \frac{\lambda_{ab}^k}{\rho}\|_2^2 + \frac{1}{2}\|\mathbf{y}_a - \mathbf{y}_a^k\|_{\mathbf{P}_i}^2$;
  {**SDN controller update**}
  $\lambda_{ab}^{k+1} = \lambda_{ab}^k - \gamma\rho\sum_{b=1}^{B}\sum_{a=1}^{A}(x_{ba}^{k+1} - y_{ab}^{k+1})$;
**end for**
Output $\mathbf{x}, \mathbf{y}$;

---

**Access Point Update**: The update rule at each AP $a$ can be expressed as

$$\mathbf{y}_a^{k+1} = \arg\min_{\mathbf{y}_b}(L_a(\mathbf{y}_a) + \frac{\rho}{2}\sum_{b=1}^{B}\|y_{ab} - q_{ba}^k\|_2^2$$

$$+ \frac{1}{2}\|\mathbf{y}_a - \mathbf{y}_a^k\|_{\mathbf{P}_i}^2), \quad \text{s.t.} \quad \sum_{b=1}^{B} y_{ab} \leq C_a, \tag{11.6}$$

where $\mathbf{P}_i = 0.1\mathbf{I}$ and $q_{ba}^k = (x_{ba}^k - \frac{\lambda_{ab}^k}{\rho})$, $\forall b$. $q_{ba}$ is the "signal" from the SDN controller to AP $a$. The update (11.6) is a small-scale convex optimization problem with linear inequality constraints. For each round of the update, it sends $\mathbf{y}_a$ of size $B$ to the SDN controller. The update of $\mathbf{y}$ is also performed independently at each AP. During the update, the information of cost function $L_a(\cdot)$ is kept private. $\mathbf{y}_a$ is sent to the SDN controller once updated.

**SDN Controller Update**: At the SDN controller, the update rule can be expressed as

$$\lambda_{ab}^{k+1} = \lambda_{ab}^k - \gamma\rho \sum_{b=1}^{B} \sum_{a=1}^{A} (x_{ba}^{k+1} - y_{ab}^{k+1}). \tag{11.7}$$

After gathering $\mathbf{x}$ and $\mathbf{y}$ from the BSs and APs, the SDN controller performs a simple update on the dual variable $\lambda$ by a simple algebra operation. After that, the "signal" variables $p_{ba}$ and $q_{ba}$ are scattered back to the corresponding BSs and APs, respectively. For each round of the update, it sends $p_{ba}$, $\forall a$ to each BS b, which is of size $A$, and sends $q_{ba}$, $\forall b$ to each AP a, which is of size $B$.

Note that in the Jacobian type update, the iterations of the BSs and APs are performed concurrently instead of consecutively in the Gauss–Seidel type update. There is no direct communication between the BSs and APs, which kept the intermediated update results of $\mathbf{x}$ and $\mathbf{y}$ confidential to each other. The updates at iteration $k + 1$ only depends on its previous value at iteration $k$, which enables a fully distributed implementation.

At each iteration, the update operations at BSs and APs are quite simple. The update at each BS $b$ and AP $a$ are simple small-scale convex optimization problems, which can be quickly solved by many off-the-shelf tools like CVX [551]. As for the communication overhead, for each iteration the signaling between each BS and SDN controller is of the size $2A$ (size of $\mathbf{x}_b$ and $p_{ba}$, $\forall a$). Likewise, the signaling between each AP and SDN controller is of the size $2B$ (size of $\mathbf{y}_a$ and $q_{ba}$, $\forall b$). The sizes of those signaling messages are quite small compared with the offloading message body and can be communicated in the dedicated control channel. The proposed distributed algorithm is described in Algorithm 10.

We present numerical results by considering a wireless access network which consists of $B = 5$ base stations and $A = \{5, 10\}$ access points coordinated by the SDN controller. The SDN controller will offload mobile data traffic of BSs to APs, and the available capacity of each AP for offloading is $C_a = 10Mbps$. The utility function of BS b is $U_b(\mathbf{x}_b) = \log(\mathbf{x}_b^\top \mathbf{1} + 1)$, where $\mathbf{1}$ is the all-one vector. The cost function of AP a is a linear cost expressed as $L_a(\mathbf{y}_a) = \theta_a * \mathbf{y}_a^\top \mathbf{1}$, where $\theta_a > 0$ is the cost coefficient. The value of $\theta_a$ is application-specific. During numerical tests, we assume $\theta_a$ is a Gaussian random variable which has a distribution $\mathcal{N}(0, 1)$ for simplicity. We perform numerical tests on the offloading decision for one time slot. Figure 11.2 shows that the proposed algorithm converges to the optimal objective in a moderate number of iterations when $B = 5$ and $A = 5$. It takes a longer time for the proposed algorithm to converge when $A = 10$. It indicates that when there are more APs in the access network, it will take a longer time for the SDN controller to coordinate BSs and APs for a consensus on the offloading demand and supply.

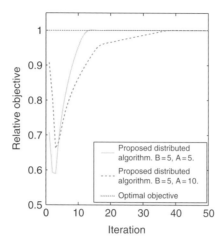

**Figure 11.2** Convergence performance of the proposed algorithm by objective value when $(B = 5, A = 5)$ and $(B = 5, A = 10)$.

## 11.2  Scalable Service Management in Mobile Cloud Computing

In mobile cloud computing (MCC), mobile end users can offload local workload [552] and back up personal data to clouds without explicitly noticed where the service is actually hosted. The service provider needs to dynamically acquire computing resources for service provisioning, and delicately manage online services to optimize the end-to-end performance experienced by its customers. It is known that even a small increase in latency will result in a significant revenue loss for service providers. Thus, mobile service providers usually deploy their services on several cloud-enabled data centers, and perform service management tasks to optimally locate mobile service instances. An illustration of the mobile cloud computing infrastructure is shown in Figure 11.3.

To efficiently manage mobile cloud services, a mobile service provider should appropriately locate client requests to a data center (request allocation), and select an upstream Internet service provider (ISP) link of data center to carry on the traffic back to the client (response routing). Those two tasks are crucial to the success of mobile cloud service, and should be managed adaptively to variations in MCC, such as end user demands, link latency, computation costs, as well as electricity and bandwidth price. Nowadays, the decision of request allocation and response mapping is handled separately, which results in poor service performance and high cost. For example, too many client requests may be allocated to the same data center with limited upstream link bandwidth, or a data center may response to client requests through an expensive ISP link. The management tasks are also computationally intensive due to the large number of mobile devices and the stringent response-time requirement of mobile services. Furthermore, the uncertainty in the wireless link latency of mobile network complicates the problem.

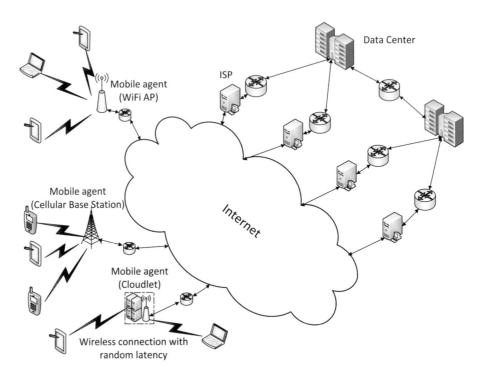

**Figure 11.3** An illustration of mobile cloud computing infrastructure.

## 11.2.1    Literature Review

The service management faced by the mobile service provider can be seen as a network utility maximization (NUM) problem [553], which described a unifying framework for understanding and designing distributed control and resource allocation in communication networks.

The service management in this part is closely related to the mobile service allocation and the traffic engineering in MCC. The framework for offloading mobile computation workload to clouds was proposed in [554], which managed to enhance the energy resource utilization and reduced the computation time of mobile devices. In [555], a mobile service management technology was presented to support novel MCC applications. The network services were reactively relocated to guarantee adequate performance for the client–server communication. A decentralized design for service request allocation was described in [556], which directed client requests to appropriate server replicas to offer better performance. The problem of optimizing the performance of carrying traffic for an online service provider was studied in [557]. The multi-homed traffic engineering for autonomous systems to optimize cost and performance was investigated in [558], and the effect of temperature on cloud service workload management for geo-distributed data centers was analyzed in [559].

The cooperative server selection and traffic engineering between network and content providers who have conflict objectives was proposed in [560], and the concept of the Nash bargaining solution was utilized to enhance the cooperation between ISPs and content providers. [561] extended the optimality result by incorporating practical considerations such as DC-level load balancing and capacity constraints. Recent work [562] considered a coordination of request mapping and response routing for geo-distributed cloud services, and developed a distributed algorithm to solve the large-scale optimization problem. The scalable service management in this part explores and analyzes the effect of random wireless nature on the service management problem, and proposes a distributed stochastic optimization framework with proved convergence property for service management.

In this part, we present a scalable distributed management framework for mobile cloud services, which takes the impact of wireless network characteristics into account. The tasks of clients request allocation and data center response routing are jointly considered, and the management tasks are formulated as a service revenue maximization problem. In particular, the mobile service provider optimally locates client requests to provide qualified service at a reasonable cost under the stochastic wireless link latency. The major contributions are as follows:

1. A stochastic optimization framework for mobile cloud service management is formulated. The clients request allocation and data center response routing are jointly optimized, and the impact of wireless network characteristics on service performance is considered.
2. A distributed approach to solve the large-scale stochastic optimization problem based on ADMM is proposed. The update steps are modified according to the stochastic setting, which can be solved in a parallel fashion on distributed agents and coordinated through dual variables.
3. We prove the convergence of the proposed stochastic distributed optimization algorithm. We evaluate the effectiveness of the proposed algorithm through numerical simulations from both a computation perspective and service management perspective.

## 11.2.2  System Model and Problem Formulation

We consider a set $\mathcal{I}$ of agents in a mobile cloud service. An agent $i \in \{1, \ldots, N\}$ is defined as an access point (AP) of wireless access networks, and the bandwidth capacity of mobile service agent $i$ is $R_i$. A set $\mathcal{J}$ of data centers are indexed by $j \in \{1, \ldots, J\}$. Data centers are interconnected over a backbone network and each data center is multi-homed to $K$ ISP links. The set of ISP links of data center $j$ is denoted by $\mathcal{K}_j$. We assume that all data centers have the same number of ISP links for simplicity. The capacity of ISP links at data center $j$ is denoted by $\mathbf{Q}_j = [Q_{j,1}, \ldots, Q_{j,K}]^\top$. The service provider observes a propagation delay $\mathbf{L}_{j,i} = [L_{j,i,1}, \ldots, L_{j,i,K}]^\top$ over wired connection, where $L_{j,i,k}$ is the average delay between agent $i$ and data center $j$ on the $k^{th}$ ISP link. The wireless link latency between agent $i$ and mobile devices is $\xi_i$.

The requests from mobile devices are first handled by the mobile service agent. After that, one data center at a specific location is assigned to process the request. We use

$a_{i,j}$ as agent $i$'s application requests processed by data center $j$. The request allocation decision variables of mobile service agent $i$ are denoted as $\mathbf{a}_i = [a_{i,1}, \ldots, a_{i,J}]^\top$. In practice, a mobile service agent can be a cloudlet [563] or be implemented on servers that provide mobile network services. Additionally, in this work we assume that an agent has a fine-grained control of the network traffic, which is a reasonable assumption in modern commercial products [556] and techniques like OpenFlow [564]. Once data center $j$ has finished the job, the response traffic will be routed back through ISP links. We use vector $\mathbf{b}_{j,i} = [b_{j,i,1}, \ldots, b_{j,i,K}]^\top$ to denote the traffic routed from the data center $j$ to agent $i$ through $K$ different ISP links, and the matrix $\mathbf{B}_j = [\mathbf{b}_{j,1}, \ldots, \mathbf{b}_{j,N}]^\top$ denote response routing decisions of data center $j$.

## 11.2.3   Mobile Cloud Service Management

In mobile cloud service management, the application requests are allocated to appropriate data centers in order to achieve maximal utility and minimize the cost. The utility and cost functions in the service management can be elaborated as follows:

(1) **Utility of mobile service agents**: The performance objective of agent $i$ is characterized by a utility function $F_i(\cdot)$, which depends on total transmission rate and wireless access network latency. The utility functions can be different among mobile service agents. In this work, $F_i(\cdot)$ is designed to be a non-decreasing, non-negative, and concave function in $\sum_{j \in \mathcal{J}} a_{i,j}$. For example, $F_i(\mathbf{a}_i, \xi_i) = \frac{1}{\xi_i} \log_2(\sum_{j \in \mathcal{J}} a_{i,j} + 1)$, or it can be a more general class of functions that represent the elasticity of service request and/or determine the fairness of resource allocation. Such functions are typically used for the TCP congestion control [565, 566].

(2) **Cost of data centers**: The cost of data center $j$ is characterized by function $G_j$ as

$$G_j(\cdot) = \beta_{j,1}\gamma_{j,1}(\cdot) + \beta_{j,2}\gamma_{j,2}(\cdot) - \beta_{j,3}\gamma_{j,3}(\cdot), \tag{11.8}$$

which has three parts and parameterized by positive coefficients $\beta_{j,1}$, $\beta_{j,2}$, and $\beta_{j,3}$, respectively, to incorporate different degrees of sensitivity to operation cost, link price, and user-perceived latency. The determination of the values of coefficients is service-specific, and the mobile service provider is responsible for choosing the values of parameters based on its service types, data centers it used, and its profit model.

In the first part of $G_j(\cdot)$, $\gamma_{j,1}(\cdot)$ accounts for the operation cost of data center $j$. Here, $\gamma_{j,1}(\cdot)$ is designed to be a non-decreasing, non-negative, and convex function in $\sum_{i \in \mathcal{I}} \mathbf{b}_{j,i}^\top \mathbf{1}$, where $\mathbf{1} \in \mathbb{R}^K$ is an all-one vector. The design of operation cost can incorporate the price of computing resource rental, maintenance cost, and electricity bills [567, 568, 559]. For example, to represent electricity bills [569] at data center $j$, $\gamma_{j,1}(\cdot)$ can be

$$\gamma_{j,1}(\mathbf{B}_j) = Pr_j \times P_e \times [P_{idle} + (P_{peak} - P_{idle}) \sum_{i \in \mathcal{I}} \mathbf{b}_{j,i}^\top \mathbf{1}], \tag{11.9}$$

where $Pr_j$ is the spot electricity price at data center $j$, and $P_e$ is the power usage efficiency. $P_{peak}$ and $P_{idle}$ are server peak power and server idle power, respectively.

In the second part of $G_j(\cdot)$, $\gamma_{j,2}(\cdot)$ stands for the cost of routing traffic from data center $j$ to mobile service agents through ISP links. A linear cost model for ISP links can be adopted:

$$\gamma_{j,2}(\mathbf{B}_j) = \sum_{i \in \mathcal{I}} \mathbf{b}_{j,i}^\top \mathbf{p}_{j,i}, \qquad (11.10)$$

where $\mathbf{p}_{j,i} = (p_{j,i,1}, \ldots, p_{j,i,K})^\top$ is the price vector for ISP links at data center $j$. We assume that the cost of routing traffic on the $k^{th}$ ISP's link from data center $j$ to agent $i$, which is denoted by $p_{j,i,k}$, is known and fixed. Note that nowadays ISPs are adopting sophisticated charging policies (e.g., the 95-percentile charging scheme). It is shown that a linear cost optimization in charging intervals can reduce the 95-percentile cost [557].

In the last part of $G_j(\cdot)$, $\gamma_{j,3}(\cdot)$ captures the user-perceived latency in the response routing from data center $j$ to mobile service agents. $\gamma_{j,3}(\cdot)$ can be a non-decreasing, non-negative, and concave function in $\mathbf{B}_j$:

$$\gamma_{j,3}(\mathbf{B}_j) = \sum_{i \in \mathcal{I}} \mathbf{b}_{j,i}^\top (\mathbf{L}_{max} - \mathbf{L}_{j,i}), \qquad (11.11)$$

where $\mathbf{L}_{max}$ is the maximum tolerable latency. The ISP link delay $\mathbf{L}_{j,i}$ is known and can be obtained through active measurements [570]. We consider the latency between data center $j$ and mobile service agents as a performance metric since user-perceived latency is one of the most important metrics for mobile cloud computing service. Even a small increment can result in significant revenue loss.

(3) **Total revenue**: The goals of maximizing mobile service utility and minimizing data centers' costs usually contradict each other. Allocating users' requests to data centers that offer lower latencies usually incurs higher costs, and over-utilizing the low-cost link for response routing will degrade system performance due to the increased congestion. By jointly considering utilities of mobile service agents and cost of data centers, the total revenue for mobile cloud service management can be formulated as

$$\text{Revenue} = \alpha \sum_{i \in \mathcal{I}} \mathbb{E}_{\xi_i} \{F_i(\mathbf{a}_i, \xi_i)\} - \sum_{j \in \mathcal{J}} G_j(\mathbf{B}_j), \qquad (11.12)$$

where the parameter $\alpha$ is introduced to find a balance between service utility and cost. The mobile service provider needs to customize the cost–performance trade-off to obtain the best revenue.

## 11.2.4 Maximizing Total Revenue

The mobile service provider performs a revenue maximization to improve resource utilization in MCC. The objective of the optimization problem consists of two terms: (i) the service utility from all agents by fulfilling mobile client requests, and (ii) the data center cost for serving service requests. The resulting stochastic optimization problem is presented in its equivalent minimization form as

$$\underset{\{\mathbf{a}_i\}_{i=1}^N, \{\mathbf{B}_j\}_{j=1}^J}{minimize} \quad \sum_{j \in \mathcal{J}} G_j(\mathbf{B}_j) - \alpha \sum_{i \in \mathcal{I}} \mathbb{E}_{\xi_i}\{F_i(\mathbf{a}_i, \xi_i)\} \tag{11.13}$$

$$\text{subject to} \quad \sum_{j \in \mathcal{J}} a_{i,j} \leq R_i, \quad \forall i, \tag{11.14}$$

$$\sum_{i \in \mathcal{I}} a_{i,j} \leq C_j, \quad \forall j, \tag{11.15}$$

$$\sum_{i \in \mathcal{I}} \mathbf{b}_{j,i} \preceq \mathbf{Q}_j, \quad \forall j, \tag{11.16}$$

$$a_{i,j} = \mathbf{b}_{j,i}^\top \mathbf{1}, \quad \forall i,j, \text{ and} \tag{11.17}$$

$$a_{i,j} \geq 0, \quad \mathbf{b}_{j,i} \succeq 0, \quad \forall i,j, \tag{11.18}$$

where (11.14) is the bandwidth capacity constraint for each mobile service agent; (11.15) and (11.16) are data center capacity constraint and link capacity constraint, respectively; (11.17) is the workload conservation constraint between each pair of agent and data center.

The solution to the above stochastic optimization problem ensures the optimal allocation of mobile application requests, while achieving the maximum revenue. Traditionally, this problem is solved in a centralized manner to find the optimal solution. However, the major challenge of the centralized service revenue maximization is the problem size, especially for large systems with an enormous number of agents, communication links, and data centers. Additionally, the randomness of the wireless link latency makes the problem more complicated. To achieve efficient and scalable management of the mobile cloud service, a distributed optimization framework is proposed.

We propose a distributed design to solve the optimization problem (11.13). Specifically, the decision variables $\mathbf{a}_i$ and $\mathbf{B}_j$ are arranged into two groups, which correspond to the mobile service agents request allocation and the data center response routing, respectively. During the optimization, the variables of each group are optimized in a distributed and parallel fashion. In particular, each mobile service agent $i$ solves $\mathbf{a}_i$ and each data center $j$ obtains $\mathbf{B}_j$, and those two groups of decision variables are coordinated through dual variables. The architecture of the proposed mechanism is illustrated in Figure 11.4. We omit detailed derivation due to page limit.

## 11.2.5    Numerical Results

We consider a mobile cloud service which provides applications for $N$ mobile service agents, $N \in \{100, 200, \ldots, 1,000\}$. The service is deployed on ten cloud-enabled geographically distributed data centers, and each data center is multi-homed to three ISP links to deliver services to mobile clients. The capacity of each mobile agent is generated from a uniform distribution $\mathcal{U}(8,000, 10,000)$, with a mean of 9,000 data units. The capacity of each data center is generated in a similar fashion such that the total capacity of data centers is 1.5 times the mobile service agents total capacity. For simplicity, the latency of ISP links is randomly generated from $\mathcal{U}(25,300)$ with a unit of milliseconds, and the stochastic wireless latency $\xi_i$ is generated from an exponential distribution with

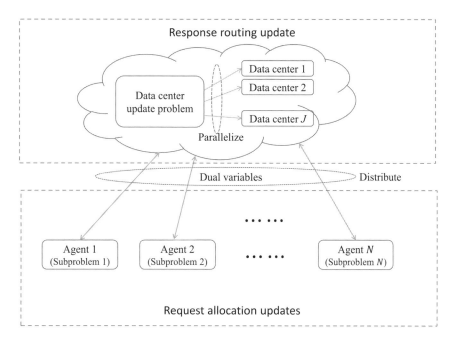

**Figure 11.4** The original service revenue maximization problem is decoupled into two parts, response routing update and request allocation update. The two parts are coordinated through dual variables.

a mean of 5 ms. To describe the service cost at data centers, the 2011 annual average day-ahead on peak prices at ten different local markets are used for data centers [562]. The server peak power and server idle power are set to $200W$ and $100W$, respectively. The power usage efficiently is 1.5. The prices of ISP links are chosen randomly from a finite set of $\{0.005, 0.01, 0.015\}$ monetary units per data unit.

In the following we show the effectiveness of the proposed algorithm on service management. We compare the proposed algorithm with two service management approaches. One is the "cheapest selection" which aims at minimizing the data center cost solely, and the other is the "minimum latency selection" which aims at minimizing the ISP link latency solely. We specify the number of mobile service agents as 100, and the performance comparisons are shown in Figures 11.5 and 11.6.

The cumulative density function (CDF) of the request latency for three mechanisms is shown in Figure 11.5. It is observed that 90% of requests are served with latency less than 100ms for the proposed algorithm, and the latency performance of the proposed algorithm is close to the "minimum latency selection" approach. The "cheapest selection" dose not take the latency performance into consideration explicitly, and thus has the worst performance. For completeness, the comparisons of average latency, revenue, utility, and cost for three service management approaches are shown in Figure 11.6. It is shown in Figure 11.6(a) that the proposed algorithm outperforms other two from perspectives of both average revenue and latency. A detail analysis of the average revenue is shown in Figure 11.6(b). It is shown that the proposed algorithm

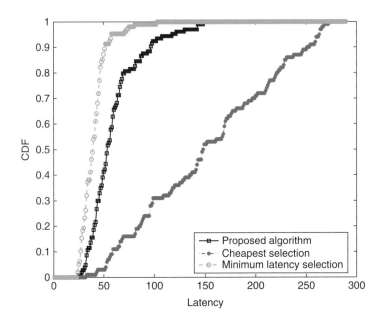

**Figure 11.5** The CDF of the latency for three service management approaches.

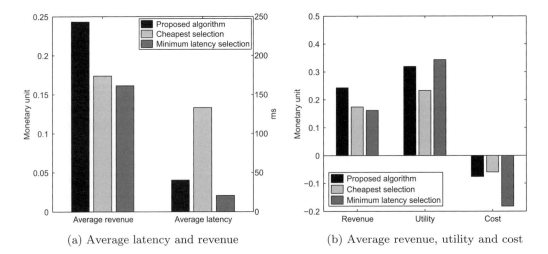

(a) Average latency and revenue                    (b) Average revenue, utility and cost

**Figure 11.6** Comparisons of average latency, revenue, utility and cost for three service management approaches.

chooses the data center neither conservatively to reduce the cost, like the "cheapest selection" approach, nor aggressively to grasp the utility, like the "minimum latency selection." It manages the mobile cloud service strategically to balance utility and cost.

In summary, we have investigated distributed approaches for the service management for mobile cloud computing. We have shown that efficient and scalable management of big data services and data traffic can improve resource utilization and service quality of the could computing.

## 11.3    Scalable Workload Management in Data Centers

Cloud computing has become a part of people's daily lives and has created a large demand for data center computing. Nowadays, people are used to accessing various services such as search engine, e-mail, GPS navigation, streaming video, and social networks deployed on data centers. Meanwhile, small and medium enterprises seize the opportunity to utilize the data center computing ability as a flexible and economically efficient solution for service provisioning. It has great potential for cloud service providers to generate a huge amount of revenue without investing much capital for building and maintaining their own infrastructure. Hence, it is of great incentives for service providers to expand both the number and scale of their data centers worldwide.

The energy consumption of data centers has attracted significant attention from both industry and academia. Currently, significant research efforts have been made to reduce electricity usage and the carbon footprint of data centers. However, the importance of water efficiency, which is a critical concern for future data centers' sustainability, has been undervalued. The AT&T's cooling towers in data center facilities consumed 1 billion gallons of water in 2012, which is approximately 30% of the entire company's water consumption [571]. When operational, the Utah Data Center [572] is expected to consume 1.7 million gallons of water a day in order to cool the people and facilities, which is equivalent to 10,000 households' daily usage. Moreover, the extended droughts have been a potential threat for the daily operation of data centers. On January 17, 2014, California [573] proclaimed a state-wide drought emergency urging a water usage cut of 20% for business and may take mandatory measures if drought conditions continuously worsen. It is imperative to consider water consumption and water efficiency for workload management in data centers.

The workload management problem for data centers can be seen as a NUM problem [553], which describes a unifying framework for understanding and designing distributed control as well as resource allocation in communication networks. [574] presents a novel workload management mechanism which exploits the time variation in electricity price, the availability of renewable energy, and the efficiency cooling. A significant amount of energy cost can be reduced by the proposed method. [559] considers a temperature-aware workload management in geographically distributed data centers for energy efficiency in cooling systems. [562] considers a joint request mapping and response routing for cloud services, which can reduce the energy usage and enhance the quality of cloud services deployed on data centers.

The energy-efficient workload management approaches are insufficient to improve the water efficiency in data centers since the temporal and spatial diversities of data center water effciency are neglected. The significance of water efficiency in data centers has been noticed in recent works [575, 576, 577]. The benefits and possibilities of improvements in power and water usage effectiveness, cost, and operation for data centers are discussed in [575]. [576] presents technologies, design strategies, and operational approaches to improve the energy efficiency and sustainability of data centers. [577] exploits the temporal and spatial diversities of water efficiency, and proposes a novel approach for geographical load balancing.

In this section, we present a distributed workload management framework for data centers, which takes the impact of water consumption into account. In particular, we consider direct water usage in data centers especially for the cooling infrastructure, which is a significant factor for water efficiency in data centers. The management tasks are formulated as a revenue maximization problem, in which the users' service requests are optimally located to provide a qualified service at a reasonable cost. To solve the resultant large-scale optimization problem with manageable complexity, the ADMM is utilized. The optimization problem is broken down into a set of independent subproblems, which can be solved in parallel on distributed nodes and coordinated through dual variables. We evaluate the performance of proposed algorithm by simulations, and the numerical tests validate the effectiveness of the proposed algorithm.

## 11.3.1     Problem Formulation

### Background

Nowadays chillers typically use the most energy in a data center's cooling infrastructure, so large data centers, including AT&T [571] and Google [578], employ evaporative cooling as an energy-efficient substitution. In particular, large cooling towers are used to vent the heat generated by the servers into the environment through water evaporation, and then the evaporated water is replenished with makeup water. Currently, around 40% [579] of large data centers utilize cooling towers for temperature control, and the majority of those data centers rely on fresh water for supplement.

Generally, water consumption in cooling towers consists of two parts: evaporation and blow down [576]. The first one happens when exhaust air leaving a cooling tower also carries away droplets of water. The second one is the dumping of the cooling tower water into the sewage system. Blow down is necessary for a cooling tower due to the water accumulating dissolved minerals as the water evaporates. The data center water efficiency can be described by the water usage effectiveness (WUE), which is defined by the ratio of total water consumption to the information technology (IT) energy usage as follows:

$$\text{WUE} = \frac{\text{Total Water Usage}}{\text{IT Equipment Energy}}. \tag{11.19}$$

The value of WUE depends on the water flow rate, outside humidity, temperature, supply/return water temperature difference, as well as system configurations. Similarly, the data center power usage efficiency (PUE) is defined as:

$$\text{PUE} = \frac{\text{Total Power Usage}}{\text{IT Equipment Energy}}. \tag{11.20}$$

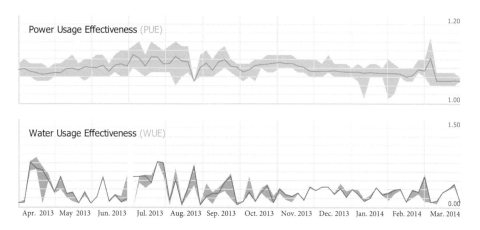

**Figure 11.7** PUE and WUE of Facebook's data center in Prineville, OR, from April 2013 to March 2014 [581].

A record of PUE and WUE of Facebook's data center at Prineville, OR from April 2013 to March 2014 is shown in Fig. 11.7.[2] Note that the WUE of data centers diversifies temporally and spatially, and most of the time the PUE and WUE at the same location demonstrate little correlation to each other.

### System Model

We consider a set $\mathcal{J}$ of geographically distributed data centers, indexed by $j = 1, \ldots, J$. A set $\mathcal{I}$ of users are requesting services from data centers. In this work, a user $i = 1, \ldots, N$ is defined as an aggregated group of customers which have the same IP prefix. The service requests from customers are first handled by the portal server. After that, one data center at a specific location is assigned to process the request. The service request at user $i$ is denoted by $\lambda_i$ and we assume that user's workload is precisely predicted [582].

We use $a_{i,j}$ as user $i$'s application requests processed by data center $j$. The allocation decision variables of user $i$ is denoted as $\mathbf{a}_i = (a_{i,1}, \ldots, a_{i,J})^\top$. In practice, a user can be authoritative DNS servers or HTTP ingress proxies. Additionally, in this work we assume an agent has the fine-grained control of the network traffic, which is a reasonable assumption in nowadays commercial products and techniques [556, 583, 564]. The capacity of data center $j$ is $C_j$, and the total workload from all users is $\sum_{i \in \mathcal{I}} a_{i,j}$. Once data center $j$ finished processing the service request, the response traffic will be routed back to the user. A propagation latency $L_{i,j}$ is observed over wired connection between user $i$ and data center $j$, which can be obtained by active measurements.

The performance objective of user $i$ is characterized by a utility function $F_i(\cdot)$, and the utility functions can be different among different users. In this work, $F_i(\cdot)$ is designed to be a non-decreasing, non-negative, and concave function in $\mathbf{a}_i$. A simple example is $F_i(\mathbf{a}_i) = \sum_{j \in \mathcal{J}} a_{i,j}(L_{max} - L_{i,j})$, where $L_{max}$ is the maximum tolerable latency for

---

[2] Though Facebook utilize the air economizer for cooling, water is still used for temperature and humility control in data centers. For example, the WUE at a cooling tower-based data center is 2.52L/kWh (eBay [580]), and 0.42L/kWh for a "free" air cooling data center (Facebook [581]).

service, or can be a more general class of functions that represent the elasticity of the service request and/or determine the fairness of resource allocation [565].

We characterize the performance objectives of each data center $j$ by function $G_j$, which accounts for the power and water consumptions and parameterized by $\beta_j$ as follows:

$$G_j(\mathbf{a}_i) = \gamma_{j,1}(\mathbf{a}_i) + \beta_j \gamma_{j,2}(\mathbf{a}_i), \tag{11.21}$$

where $\beta_j > 0$ highlights the importance of the water usage in cost function, and can be different across locations. $\gamma_{j,1}(\cdot)$ and $\gamma_{j,2}(\cdot)$ account for the electricity cost and water consumption of data center $j$, respectively, which scale with the total workload assigned to data center $j$. Here, $\gamma_{j,1}(\cdot)$ and $\gamma_{j,2}(\cdot)$ are designed to be non-decreasing, non-negative, and convex function in $\sum_{i\in\mathcal{I}} a_{i,j}$. For example, the electricity bills and water cost at data center $j$ are modeled as follows [567, 569]:

$$\gamma_{j,1}(\mathbf{a}_i) = Pr_j \times \varepsilon_j \times [P_{idle} + (P_{peak} - P_{idle}) \sum_{i\in\mathcal{I}} a_{i,j}], \tag{11.22}$$

$$\gamma_{j,1}(\mathbf{a}_i) = \epsilon_j \times [P_{idle} + (P_{peak} - P_{idle}) \sum_{i\in\mathcal{I}} a_{i,j}], \tag{11.23}$$

where $Pr_j$ is the spot electricity price at the location of data center $j$, and $\varepsilon_j$ is the PUE at data center $j$. $\epsilon_j$ is the WUE at data center $j$. $P_{peak}$ and $P_{idle}$ are the server peak power and server idle power, respectively.

**Workload Management Problem**

The workload management problem can be formulated as an optimization problem to maximize the revenue as follows:

$$\text{minimize}_{\{\mathbf{a}_i\}_{i=1}^N} \quad \sum_{j\in\mathcal{J}} G_j(\mathbf{a}_i) - \alpha \sum_{i\in\mathcal{I}} F_i(\mathbf{a}_i) \tag{11.24}$$

$$\text{subject to} \quad \sum_{j\in\mathcal{J}} a_{i,j} = \lambda_i, \quad \forall i, \tag{11.25}$$

$$\sum_{i\in\mathcal{I}} a_{i,j} \leq C_j, \quad \forall j, \tag{11.26}$$

$$a_{i,j} \geq 0 \quad \forall i,j, \tag{11.27}$$

where (11.25) is the workload conservation constraint for each user, and (11.26) is the data center capacity constraint. The problem (11.24) is not readily solved in a distributed fashion due to coupling of $a_{i,j}$ across data centers in constraint (11.25), and the coupling of $a_{i,j}$ across users in constraint (11.26). In [584], the ADMM is utilized to decouple those constraints and design a distributed approach for (11.24), and we omit the details of the algorithm here.

**Evaluation**

We consider $N$ users in our evaluation, $N \in \{100, 200, \ldots, 1,000\}$. There are ten geographically distributed data centers to deliver services to users. The workload of each user is generated from a uniform distribution $\mathcal{U}(4,000, 8,000)$, with a mean of 6,000 data units. The capacity of each data center is generated in a similar fashion such that the total capacity of data centers is 1.6 times of mobile service agents total capacity. The latency $L_{i,j}$ is randomly generated from $\mathcal{U}(25, 300)$ for simplicity, with a unit of milliseconds.

To describe the service cost at date centers, the 2011 annual average day-ahead on peak prices at ten different local markets as the electricity price for data centers [562] is used. The server peak power and server idle power are set to $200W$ and $100W$, respectively. The power usage efficiently is 1.5. Due to the lack of access to direct WUE for data centers, we generate a synthetic WUE base on the data from Facebook [581].

In the following we show the effectiveness of the proposed algorithm on workload management. We compare the proposed algorithm with the water-oblivious workload management approach. In this case, we specify the number of users as 1,000, and the performance comparison are discussed below.

The comparisons of average latency, revenue, cost, and utility for two workload management approaches are shown in Figure 11.8. It is shown in Figure 11.8(a) that the proposed algorithm outperforms the water-oblivious management approach in average revenue. The proposed algorithm can largely reduce the average operational cost of data centers while only incurring a small lost in average utility. A detailed analysis of the average cost is shown in Figure 11.8(b). It is shown that the proposed algorithm can effectively reduce the water consumptions of data centers, while only leading to a negligible increase in the average power usage.

Next, we compare the average water usage performance of the proposed algorithm with the water-oblivious workload management approach by varying the number of users from 100 to 1,000. The water usage percentage is calculated by dividing the water usage of proposed algorithm by that of the water-oblivious workload management. It is shown in Figure 11.9 that when taking the water usage of data centers into consideration, the proposed workload management algorithm achieves a water usage reduction around 20% at different numbers of users, compared with the water-oblivious workload

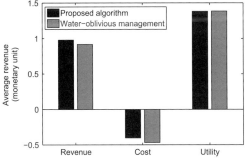

(a) Average revenue, cost and utility

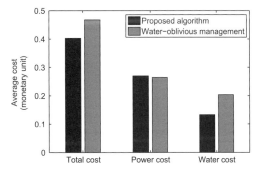

(b) Average total cost, power cost and water cost

**Figure 11.8** Comparisons of average revenue, cost and utility for the proposed algorithm and water-oblivious management approach.

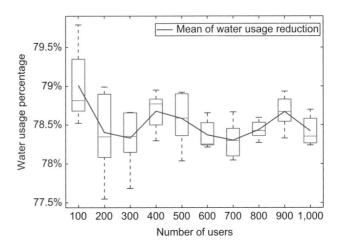

**Figure 11.9** Comparisons of water usage for service management with/without consideration of water efficiency at different number of users.

management approach. The box plot, which depicts groups of numerical data through their quartiles, and the plot of the mean of the water usage reduction in Figure 11.9 shows that such kind of improvement is relatively stable.

In this example, we have presented a distributed workload management mechanism for data centers, and taken the effect of data center water efficiency into account. We have formulated a service revenue maximization problem, in which the users' requests are optimally located to provide qualified service at a reasonable cost. We have used the ADMMs to solve the large-scale stochastic optimization problem with manageable complexity. We have shown that, our algorithm can decompose the optimization problem into a set of independent subproblems, which can be solved in parallel on distributed nodes and coordinated through dual variables. Our numerical tests validate the effectiveness of the proposed algorithm.

## 11.4    Resource Allocation for Wireless Network Virtualization

Wireless network virtualization has been proposed as one of the key enablers to overcome the ossification of the current Internet by allowing diverse services and applications coexist on the same infrastructure [585, 586]. With wireless network virtualization, the traditional Internet service providers are decoupled into infrastructure providers (InPs) who own and manage only infrastructure resources, and service providers (SPs) who concentrate on providing services to subscribers [587]. The physical resources that belong to different InPs are virtualized into a single physical substrate network. Consequently, multiple wireless virtual networks are deployed and operated on top of the single substrate network [588]. By sharing the same infrastructure resources, expenses of wireless network expansion and operation can be significantly reduced [589].

Despite wireless network virtualization having been advocated as a viable technology to enhance resource utilization [590], several research challenges remain to be studied [591]. One of the important challenges lies in how to efficiently allocate resources of physical wireless networks to multiple virtual wireless networks operated by SPs. Recent research has investigated resource allocation problems in wireless network virtualizaton. The work in [592] models the resource allocation problem for wireless network virtualization using a stochastic game, in which SPs bid for the resources to satisfy their service objectives. In [593], a wireless resource allocation problem in terms of transmitting power and wireless spectrum is studied using game theory to maximize the aggregate spectrum efficiency. A dynamic sharing spectrum resource in virtual networks has been proposed in [594]. A resource allocation problem for virtualized full-duplex relaying networks is formulated in [595] to maximize the total utility. In [596], the design and implementation of a network virtualization substrate in cellular networks is described.

## 11.4.1 Wireless Network Virtualization

We consider a wireless network with a set of InPs. Each InP possesses and operates a physical network, also call substrate network. The physical network is composed of physical nodes connected by physical links that form the physical topology. Based on the virtualization frameworks, the physical networks of all InPs are virtualized into a unique physical topology, denoted by a directed graph $G_s = (\mathcal{N}_s, \mathcal{L}_s)$, where $\mathcal{N}_s$ is the set of physical nodes and $\mathcal{L}_s$ is the set of physical links.

Suppose that there is a set $\mathcal{K} \triangleq \{1, 2, \ldots, K\}$ of SPs requesting $K$ different virtual networks,[3] which is composed of a set of virtual nodes and virtual links, each established over the same physical network, denoted by a directed graph $G_k = (\mathcal{N}_k, \mathcal{L}_k)$. $K$ virtual networks coexist and operate over the same physical network as illustrated in Figure 11.10. Here, we assume the virtual network mapping result from each $G_k$ to $G_s$ is already known and focus on resource allocation for virtual networks.

Depend on the resource request from SPs, InPs will allocate bandwidth capacity of each substrate link $l \in \mathcal{L}_s$ to virtual links of SPs. For each substrate link $l \in \mathcal{L}_s$, let $w_{l,k}$ be the bandwidth that substrate link $l$ allocates to a virtual link of virtual network $k$. Then, we have bandwidth allocation vector

$$\mathbf{w}_l \triangleq \{w_{l,1}, w_{l,2}, \ldots, w_{l,K}\}. \tag{11.28}$$

For any virtual network $k$ that does not have virtual link operating on top of the substrate link $l$, the bandwidth allocation must be equal to zero:

$$w_{l,k} = 0, \ l \notin \mathcal{L}_k. \tag{11.29}$$

The total bandwidth allocated to all virtual links must be less than the bandwidth capacity of the physical link, which can be expressed as the following constraint:

---

[3] Since each virtual network is operated by each SP, we use virtual network index and SP index interchangeably

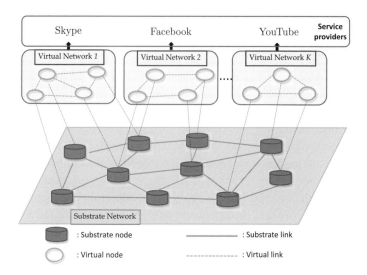

**Figure 11.10** The model of multiple virtual networks operating on top of a single substrate network.

$$\sum_{k=1}^{K} w_{l,k} \leq W_l^{max}, \tag{11.30}$$

where $W_l$ is the maximum capacity of physical link $l$.

## 11.4.2    Routing Model for Virtual Network

Each virtual network $k$, denoted by a directed graph $G_k = (\mathcal{N}_k, \mathcal{L}_k)$, has a collection of $N_k$ virtual nodes that can send, receive, and relay data across virtual communication links. The network topology with respect to the interactions between virtual nodes and virtual links of virtual network $k$ can be compactly represented by a node-link incidence matrix $A_k \in R^{N_k \times L_k}$. An entry $A_k[n_k][l_k]$ of the matrix $A_k$ associated with node $n_k \in \mathcal{N}_k$ and link $l_k \in \mathcal{L}_k$, is given by [597]

$$A_k[n_k][l_k] = \begin{cases} 1 & \text{if node } n_k \text{ is the start node of link } l_k, \\ -1 & \text{if node } n_k \text{ is the end node of link } l_k, \\ 0 & \text{otherwise.} \end{cases} \tag{11.31}$$

We consider a network flow model for routing data to a single destination in each virtual network. The data flows are assumed to be lossless in each virtual link and flow conservation law is assumed to be satisfied at each virtual node in each virtual network. In addition, a virtual source node may need a number of relay nodes to route the data stream to its destination node. We assume each SP uses multipath routing protocol where the traffic from each virtual source node is split into several flows which follow different multi-hop paths to reach the desired destination.

We assume that $d_k$ is the destination node of virtual network $k$. Each source node $n_k \in \mathcal{N}_k$ ($n_k \neq d_k$) generates data with an average rate of $r_{n_k}$ to destination $d_k$. Then the total data rate at the destination node $d_k$ is

$$r_{d_k} = -\sum_{n_k \neq d_k} r_{n_k}. \tag{11.32}$$

For each virtual network $k \in \mathcal{K}$, we also define a source-sink vector $\mathbf{r}_k \in R^{N_k}$ as

$$\mathbf{r}_k \triangleq \{r_1, r_2, \ldots, r_{n_k}, \ldots, r_{N_k}\}, \tag{11.33}$$

whose the $n_k$th ($n_k \neq d_k$) entry $r_{n_k}$ denotes the amount of data that virtual source node $n_k$ injects into the network and destined for virtual destination nodes $d_k$, and the $d_k$th entry is the total data rate at the destination node $d_k$, determined as in (11.32).

On each virtual link $l_k \in \mathcal{L}_k$ of virtual network $k$, we let $f_{k,l_k} \geq 0$ be the aggregate flow for destination node $d_k$. The aggregate on each link may come from different virtual source nodes under the multipath routing model. At each virtual node $n_k \in \mathcal{N}_k$, the total flow going into a virtual node is the same as the total flow going out of that virtual node

$$\sum_{l_k \in \mathcal{O}(n_k)} f_{k,l_k} - \sum_{l_k \in \mathcal{I}(n_k)} f_{k,l_k} = r_{n_k}, \tag{11.34}$$

where $\mathcal{O}(n_k)$ be the set of outgoing links of node $n_k$, and $\mathcal{I}(n_k)$ be the set of incoming links to virtual node $n_k$. The compact expression for the flow conservation law across the whole virtual network $k$ can be expressed as

$$A_k \mathbf{f}_k = \mathbf{r}_k, \tag{11.35}$$

where $\mathbf{f}_k$ be the flow vector in virtual network $k$, which can be defined as

$$\mathbf{f}_k \triangleq \{f_{k,l_k}\}_{\forall l_k \in \mathcal{L}_k}. \tag{11.36}$$

The total amount of traffic on each virtual link must be less than the bandwidth that substrate link allocates to virtual links in virtual network $k$

$$f_{k,l_k} \leq w_{l,k}, \forall l \in \mathcal{L}_k. \tag{11.37}$$

### 11.4.3     Joint Resource and Routing Optimization for Wireless Virtual Networks

Next, we formulate a resource and routing optimization problem for wireless network virtualization.

Each InP has a cost function for operating substrate link $l \in \mathcal{L}_s$, which is assumed to be a strictly convex function on the total bandwidth, motivated by energy consumption cost [598]. Then the total cost for operating the substrate network is $\sum_{l=1}^{L_s} C_l(\mathbf{w}_l)$, where $C_l(.)$ denotes the cost function for operating physical link $l$.

Given $K$ virtual networks operate on top of the substrate network and fixed traffic demand from source nodes of each virtual network, the objective is to find an optimal bandwidth allocation for virtual links such that all traffic demand injected from source nodes is delivered to the desired destination in each virtual network with a minimum operation cost of the substrate network. The operation cost minimization problem can be formulated as

$$\min \quad \sum_{l=1}^{L} C_l(\mathbf{w}_l) \tag{11.38}$$

$$\text{s.t.} \quad A_k \mathbf{f}_k = \mathbf{r}_k, \forall k \in \mathcal{K}, \tag{11.39}$$

$$f_{k,l_k} \leq w_{l,k}, \forall l_k \in \mathcal{L}_k, \forall k \in \mathcal{K}, \tag{11.40}$$

$$\sum_k w_{l,k} \leq W_l^{max}, \forall l \in \mathcal{L}_s, \tag{11.41}$$

$$\text{variables:} \quad \{\mathbf{f}_k\}_{\forall k}, \{\mathbf{w}_l\}_{\forall l}.$$

The constraints in (11.39) represent the flow conservation low for each virtual network. The constraints in (11.40) ensure the amount of traffic on each virtual link to be less than the bandwidth that substrate link allocates for that virtual link, and the constraints in (11.41) represent the maximum amount of bandwidth of each substrate link.

The problem in (11.38)–(11.41) is convex and can be solved using the convex optimization techniques such as ADMM to obtain the optimal solution. The optimal solution will fully satisfy for all traffic demand to be delivered from sources to destination in all virtual networks.

## 11.5    Summary

Utilizing both key mathematical tools and state-of-the-art research results, this chapter explores the principles underpinning large-scale information processing and examines the crucial interaction between big data and its associated communication networks. This book employs two complementary approaches: first, analyzing how the underlying network constrains the upper layer of collaborative big data processing and, second, examining how big data processing may boost performance in wireless networks. This chapter unifies the scope of software-defined networks, cloud computing, and data center and network virtualization so as to provide the reader with the ability to master the fundamental principles for dealing with big data over large communication systems. Other applications can be enlightened and solved in a similar way.

# 12 Big Data Optimization for Smart Grid Systems

## 12.1 Introduction

The development of smart grid, impelled by the increasing demand from industrial and residential customers together with the aging power infrastructure, has become an urgent global priority due to its potential economic, environmental, and societal benefits. Smart grid refers to the next-generation electric power system which aims to provide reliable, efficient, secure, and quality energy generation/distribution/consumption using modern information, communications, and electronics technology. A distributed and user-centric system will be introduced in smart grid, which will incorporate end consumers into its decision processes to provide a cost-effective and reliable energy supply. In smart grid, the modern communication infrastructure [599] will play a vital role in managing, controlling, and optimizing different devices and systems. Information and communication technologies will offer the power grid with the capability of supporting two-way energy and information flow, quick isolating and restoring power outages, facilitating the integration of renewable energy sources into the grid, and empowering the consumer with tools for optimizing their energy consumption.

The inevitable coupling between information/communication technologies and physical operations is expected to present unique challenges as well as opportunities for smart grid. On the one hand, we are observing increasing integration between cyber operations and physical infrastructures for generation, transmission, and distribution control in the electric power grid. Yet the security and reliability of the power grid are not guaranteed at all times and some failures can cause significant problems for the producers and consumers of electricity. For example, the 2003 Northeast power blackout [600] showed that even a small failure in a part of the grid has cascading effects causing billions of dollars in economic losses. Nowadays, the consolidation of physical and cyber components gives rise to security threats in the power grid, which can result in power outages and even system blackouts [601], or substantial economical loss due to its non-optimal operation.

On the other hand, the anticipated smart grid data deluge, generated by the sensing and measurement devices and reinforced by communication and information technologies, provides us with the potential to enhance the security and reliability of smart grid. This "big data," if effectively managed and translated into actionable insights, has the potential to increase revenue and operational efficiency, improve energy conservation, streamline the utilization of sustainable energy sources, and ensure grid resiliency. New difficulties, however, will emerge as unprecedented volume, velocity, and variety of information increases system complexity, introduces network security risks, raises end-user data privacy issues, and increases generation and load uncertainty. Therefore, new computational mathematical models and methodologies must be explored to effectively operate an ever-complicated power grid and achieve the vision of a smart grid.

In this chapter, the applications of big data processing techniques for smart grid security are investigated from two perspectives: how to exploit the inherent structure of the data set, and how to deal with the huge data sets. Two specific applications are included in this chapter: sparse optimization for false data injection detection, and a distributed parallel approach for the security-constrained optimal power flow (SCOPF) problem. The rest of this chapter is organized as follows. Background on the two applications is provided in Section 12.2. Section 12.3 describes sparse optimization for false data injection detection. The distributed parallel approach for the security-constrained optimal power flow problem is developed in Section 12.4. Finally, some conclusions are drawn in Section 12.5.

## 12.2     Backgrounds

### 12.2.1     False Data Injection Attacks against State Estimation

State estimation [602], which estimates the power system operating state based on a real-time electric network model, is a key function of the Energy Management System (EMS). A linearized measurement model is often used to estimate the states in power systems based on measurements from remote meters on buses or transmission lines. Specifically, every several seconds or minutes, the Energy Control Center (ECC) collects active/reactive power flows and injections from transmission lines and buses around the power system as measurement data via a Supervisory Control and Data Acquisition (SCADA) system. The state estimation results reflect the real-time power grid operation state and are essential for operators to make decisions in order to maintain security and stability of the system.

The accuracy of state estimation can be affected by bad measurements in the grid. Bad data could be due to topology errors in the grid, measurement abnormalities caused by meter failures, or malicious attacks. To detect and identify bad measurements in the power grid state, techniques based on the statistical testing of measurement residuals [602] have been developed and are widely used. However, [603] reveals the fact that false data injection attacks are able to circumvent traditional detection methods based on residual testing. By exploiting the configuration of a power system, synchronized data injection attacks on meters can be launched to tamper with their measurements. Moreover, attack vectors can be systematically and efficiently constructed even when the attacker is limited in the resources required to compromise meters, which will mislead the state estimation process, and thus affect power grid control algorithms. Hence, attention should be paid to the vulnerability of state estimation to false data injection attacks, which may cause catastrophic consequences in the power gird.

**State Estimation in Power Systems**

Let $\boldsymbol{\theta} = (\theta_1, \theta_2, \ldots, \theta_n)^\top$ denote the power system state variables, where $\theta_i$ is the phase angle on bus $i$. The measurement at the control center is expressed as $\mathbf{z} = (z_1, z_2, \ldots, z_m)^\top$ and is related to $\boldsymbol{\theta}$ by

$$\mathbf{z} = \mathbf{h}(\boldsymbol{\theta}) + \mathbf{e}, \tag{12.1}$$

where $\mathbf{h}(\boldsymbol{\theta}) = (h_1(\boldsymbol{\theta}), h_2(\boldsymbol{\theta}), \ldots, h_m(\boldsymbol{\theta}))^\top$, and $h_i(\boldsymbol{\theta})$ is a nonlinear function relating the $i$th measurement to the state vector $\boldsymbol{\theta}$. The vector $\mathbf{e}$ denotes independent Gaussian measurement errors with zero mean and known covariance $\mathbf{R}$.

To analyze the efficiency of various state estimation methods solely related to the measurement configuration in a power system, a simplified DC approximation model is utilized. Assuming that the bus voltage magnitudes are already known and normalized, and neglecting all shunt elements and branch resistances, the active power flow from bus $i$ to bus $j$ can be approximated[1] [604] by the first-order Taylor expansion as

$$P_{ij} = \frac{\theta_i - \theta_j}{X_{ij}} + \omega, \tag{12.2}$$

where $X_{ij}$ is the reactance of the transmission line between bus $i$ and bus $j$, and $\omega$ is the measurement error. Similarly, a power injection measurement at bus $i$ can be expressed as

$$P_i = \sum_j P_{ij} + v, \tag{12.3}$$

where $v$ is the measurement error.

Hence, the DC model for the real power measurement can be written in a linear matrix form as

$$\mathbf{z} = \mathbf{H}\boldsymbol{\theta} + \mathbf{e}, \tag{12.4}$$

where $\mathbf{z}$ is the measurement vector including active power flows and injection measurements, and $\mathbf{H} \in \mathbb{R}^{m \times n}$ is the Jacobian matrix of the power system, which is assumed to be known to the independent system operator (ISO).

Suppose the measurement errors $\mathbf{e}$ in (12.4) are not correlated, and thus the covariance matrix $\mathbf{R}$ is a diagonal matrix. The weighted least-squares estimator of the linearized state vector $\boldsymbol{\theta}$ is given by

$$\hat{\boldsymbol{\theta}} = (\mathbf{H}^\top \mathbf{R}^{-1} \mathbf{H})^{-1} \mathbf{H}^\top \mathbf{R}^{-1} \mathbf{z}. \tag{12.5}$$

Let $\mathbf{K} = (\mathbf{H}^\top \mathbf{R}^{-1} \mathbf{H})^{-1} \mathbf{H}^\top \mathbf{R}^{-1}$, and then the measurement residuals can be expressed as

$$\mathbf{r} = \mathbf{z} - \mathbf{H}\hat{\boldsymbol{\theta}} = (\mathbf{I} - \mathbf{K})(\mathbf{H}\boldsymbol{\theta} + \mathbf{e}) = (\mathbf{I} - \mathbf{K})\mathbf{e}, \tag{12.6}$$

where $\mathbf{I}$ is the identity matrix and the matrix $(\mathbf{I} - \mathbf{K})$ is called the residual sensitivity matrix.

The detection and identification of bad data in measurements can be accomplished by processing of the measurement residuals. Specifically, the $\chi^2$-test can be applied on

---

[1] In general, one can approximate the impedance of a transmission line with its reactance due to the high reactance over resistance $(X/R)$ ratio

the measurement residuals to detect bad data. Upon detection of bad data, two kinds of methods, the largest normalized residual test and hypothesis testing identification method, can be used to identify the specific measurement that actually contain bad data [602].

### False Data Injection Attacks

Malicious attack vectors are able to circumvent existing statistical tests for bad data detection if they leave the measurement residuals unchanged. One such example is the false data injection attack, which is defined as follows:

DEFINITION 16. *(False data injection attack) [603] The malicious attack vector* $\mathbf{a} = (a_1, a_2, \ldots, a_m)^\top$ *is called a false data injection attack if* $\mathbf{a}$ *can be expressed as a linear combination of the columns of* $\mathbf{H}$*; that is,* $\mathbf{a} = \mathbf{Hc}$ *for some vector* $\mathbf{c}$*.*

If a false data injection attack is applied to the power system, the collected measurements at the ISO can be expressed as

$$\mathbf{z_a} = \mathbf{z_0} + \mathbf{a} = \mathbf{H}(\boldsymbol{\theta} + \mathbf{c}) + \mathbf{e}. \tag{12.7}$$

Suppose the state estimate using the malicious measurement $\mathbf{z_a}$ is $\boldsymbol{\theta}_\mathbf{a}$, the norm of the measurement residual $||\mathbf{z_a} - \mathbf{H}\boldsymbol{\theta}_\mathbf{a}||_2^2$ in this case is

$$||\mathbf{z_a} - \mathbf{H}\boldsymbol{\theta}_\mathbf{a}||_2^2 = ||\mathbf{z_0} + \mathbf{a} - \mathbf{H}(\boldsymbol{\theta} + \mathbf{c})||_2^2 = ||\mathbf{z_0} - \mathbf{H}\boldsymbol{\theta}||_2^2, \tag{12.8}$$

which means the measurement residuals are unaffected by the injection attack vector $\mathbf{a}$, and the attacker successfully tricks the system into believing that the true state is $\boldsymbol{\theta}_\mathbf{a} = \boldsymbol{\theta} + \mathbf{c}$ instead of $\boldsymbol{\theta}$. Note that $\mathbf{a}$ is the attack vector, which is under the control of the attackers, while $\mathbf{c}$ reflects error induced by $\mathbf{a}$.

Unveiling false data injection attacks is crucial to the security and reliability of power systems. This task is challenging, since attackers may be able to construct false data attack vectors against the protection scheme, and inject attack vectors into the power grid that can bypass the traditional methods for bad measurement detection. Furthermore, the incomplete measurement data due to intended attacks or meter failures complicate the task of malicious attack detection, and thus make state estimation even more difficult.

The effects of false data injection attacks have been studied in [603, 605, 606]. False data injection attacks against state estimation in the electric power grid are presented in [603]. By capitalizing on the configuration of the power system, malicious attacks can be launched to bypass the existing bad measurement detection techniques and manipulate the results of state estimation arbitrarily. [606, 605] demonstrated that false data injection attacks are able to circumvent the bad data identification techniques equipped in an EMS, and could lead to congestion of transmission lines as well as profitable financial misconduct in the power market.

On the other hand, schemes to protect against false data injection attacks are investigated in [607, 608, 609, 610, 611, 612, 613]. [607] proposes an efficient method for computing the security index for sparse attack vectors, and describes a protection scheme to strengthen system security by placing encrypted devices in the electric power grid appropriately. [608] models and analyzes this situation as a zero-sum game between the attacker and the defender. [609] characterizes two kinds of malicious attacks on electric power grids: the strong attack regime, in which false data injection attacks exist, and the weak attack regime, in which the generalized likelihood ratio test can be used to detect attacks. [610] formulates the bad data detection problem as a low-rank matrix recovery problem, which is solved by convex optimization that minimizes a combination of the nuclear norm and the $l1$ norm. In [611], a low-complexity attacking strategy is designed to construct sparse false data injection attack vectors, and strategic protection schemes are also proposed based on greedy approaches. [612] gives a survey of existing detection methods for false data injection attacks, and [613] studies the fundamental limits of cyber-physical security in the presence of false data injection attacks in the system.

## 12.2.2  Security-Constrained Optimal Power Flow

The deregulation of electric power grids offers the opportunity for electricity market participants to exercise least-cost or profit-based operations [614]. Despite the market-driven tendency of the electric power business, security remains a significant concern of sustainable power system operations which cannot be compromised. Security-constrained optimal power flow [615, 616] aims at minimizing the cost of system operation while satisfying a set of postulated contingency constraints. It is an important management task allowing optimal control of power systems securely.

SCOPF is an extension of the conventional optimal power flow (OPF) problem [617], whose objective is to determine a generation schedule that minimizes system operating costs while satisfying the system operation constraints such as hourly load demand, fuel limitations, environmental constraints, and network security requirements. It has been recognized [618] that the optimal control of the normal state may violate the system operation constraints after the occurrence of the event of a major disturbance or contingency, and thus jeopardize the security of power systems. To address this problem, SCOPF is performed by considering both pre-contingency and post-contingency constraints to guarantee sustainable operations of the electric grid. The system security level is improved by taking into account a number of contingencies in a dedicated selected contingency list. The solution to SCOPF should satisfy the so-called $N - 1$ criterion, which requires that the operational limits of the power system should not be violated in case of a single contingency (line and/or generator outage).

SCOPF can be broadly classified as preventive, where control variables are restricted to their pre-contingency condition settings, and corrective, whose control variables are allowed to be rescheduled [619]. We will focus on the corrective model in this example. The seminal paper [618] proposed the generalized Benders decomposition method to solve the corrective SCOPF problem. Since then, an extensive literature has appeared for SCOPF in power systems both for traditional operations and under market environments [620, 621, 622, 623, 616]. The nested Benders decomposition method is utilized in [620] to solve the SCOPF problem for determining the optimal daily generation scheduling in a pool-organized electricity market, and is tested in an actual example of the Spanish power system. [621] embedded SCOPF into the security-constrained unit commitment (SCUC) model, and designed an effective corrective/preventive contingency dispatch over a 24-hour period which balanced the economics and security in the restructured markets. An iterative approach is proposed in [622] to obtain the solution of SCOPF, which aims to efficiently identify an as small as possible superset of the binding contingencies to achieve the SCOPF optimum. [623] applied the Benders decomposition to decompose the traditional SCOPF problem, and the underlying computational complexity is analyzed in this approach. [616] solved the SCOPF problem by a nondecomposed method based on the compression of the post-contingency networks, which can reduce the size of the security constraints and relieve the computational burden in the problem.

Before presenting the distributed parallel approach for this problem, it is useful to recall a general formulation of the conventional SCOPF problem compactly described as follows:

$$\min_{\mathbf{x}^0,\ldots,\mathbf{x}^C;\mathbf{u}^0,\ldots,\mathbf{u}^C} f^0(\mathbf{x}^0,\mathbf{u}^0) \tag{12.9}$$

$$\text{subject to } \mathbf{g}^0(\mathbf{x}^0,\mathbf{u}^0) = 0, \tag{12.10}$$

$$\mathbf{h}^0(\mathbf{x}^0,\mathbf{u}^0) \le 0, \tag{12.11}$$

$$\mathbf{g}^c(\mathbf{x}^c,\mathbf{u}^c) = 0, \tag{12.12}$$

$$\mathbf{h}^c(\mathbf{x}^c,\mathbf{u}^c) \le 0, \tag{12.13}$$

$$|\mathbf{u}^0 - \mathbf{u}^c| \le \Delta_c, \quad c = 1,\ldots,C, \tag{12.14}$$

where $f^0$ is the objective function, through which (12.9) aims to maximize the total social welfare or equivalently minimize the offer-based energy and production cost. $\mathbf{x}^c$ is the vector of state variables, which includes magnitude and voltage angle at all buses, and $\mathbf{u}^c$ is the vector of control variables, which can be generators' real powers or terminal voltages. The superscript $c = 0$ corresponds to the pre-contingency configuration, and $c = 1,\ldots,C$ correspond to different post-contingency configurations. $\Delta_c$ is the maximal allowed adjustment between the normal and contingency states for contingency $c$.

In the conventional SCOPF problem, the equality constraints $\mathbf{g}^c, c = 0,\ldots,C$ represent the system nodal power flow balance over the entire grid, and inequality constraints $\mathbf{h}^c, c = 0,\ldots,C$ represent the physical limits on the equipment, such as the operational limits on the branch currents and bounds on the generators' power output. Constraints (12.10)–(12.11) capture the economic dispatch and enforce the feasibility of the pre-contingency state. Constraints (12.12)–(12.13) incorporate the security-constrained dispatch and enforce the feasibility of the post-contingency state. Constraint (12.14) introduces the security-constrained dispatch with rescheduling, which couples control variables of pre-contingency and post-contingency states and prevents unrealistic post-contingency corrective actions. Note that there are some variations on the objective function and constraints of the SCOPF problem, and we focus on the above conventional formulation in this chapter.

## 12.3 Sparse Optimization for False Data Injection Detection

### 12.3.1 Sparse Optimization Problem Formulation

Assume the measurement of the electric power system observed by the ISO at time $k$ is denoted as $\mathbf{z}_k$. In presence of false data injection attacks, the measurement $\mathbf{z}_k$ is contaminated by the attack vector $\mathbf{a}_k$. Denote $\mathbf{Z}_0 = [\mathbf{z}_1, \mathbf{z}_2, \ldots, \mathbf{z}_t] \in \mathbb{R}^{m \times t}$ as the measurement of the power state for a time period of $t$, and $\mathbf{A} = [\mathbf{a}_1, \mathbf{a}_2, \ldots, \mathbf{a}_t] \in \mathbb{R}^{m \times t}$ as the false data attack matrix. The obtained temporal observations $\mathbf{Z}_a$ can be expressed as

$$\mathbf{Z}_a = \mathbf{Z}_0 + \mathbf{A}. \tag{12.15}$$

Note that gradually changing power system state variables will typically lead to a low-rank measurement matrix $\mathbf{Z}_0$. In addition, due to the capability limitation of the attacker, the attacks are either constrained to some specific measurement meters or unable to compromise measurement meters persistently. Hence, only a small fraction of the observations can be anomalous at a given time instant. This implies that the false data injection matrix $\mathbf{A}$ is sparse across both rows and columns. With a slight abuse of notation, we use $\text{Rank}(\mathbf{Z}_0)$ to denote the rank of the matrix $\mathbf{Z}_0$, and $\|\mathbf{A}\|_0$ to represent the number of nonzero entries of the matrix $\mathbf{A}$. Noticing the intrinsic structures of $\mathbf{Z}_0$ and $\mathbf{A}$, the detection and identification of false data injection attacks can be converted to a matrix separation problem as

$$\min_{\mathbf{Z}_0, \mathbf{A}} \text{Rank}(\mathbf{Z}_0) + \|\mathbf{A}\|_0, \quad s.t. \quad \mathbf{Z}_a = \mathbf{Z}_0 + \mathbf{A}. \tag{12.16}$$

Solving (12.16) extracts the power state measurement matrix $\mathbf{Z}_0$ and sparse attack matrix $\mathbf{A}$ from their sum $\mathbf{Z}_a$. Considering the missing measurements due to meter failures or communication link outages in practical applications, (12.16) can be formulated as

$$\min_{\mathbf{Z}_0, \mathbf{A}} \text{Rank}(\mathbf{Z}_0) + \|\mathbf{A}\|_0, \quad s.t. \quad \mathcal{P}_\Omega(\mathbf{Z}_a) = \mathcal{P}_\Omega(\mathbf{Z}_0 + \mathbf{A}), \tag{12.17}$$

where $\Omega$ is an index subset, and $\mathcal{P}_\Omega(\cdot)$ is the projection operator. Specifically, $\mathcal{P}_\Omega(\mathbf{M})$ is the projection of a matrix $\mathbf{M}$ onto the subspace of matrices whose nonzero entries are restricted to $\Omega$ as

$$[\mathcal{P}_\Omega(\mathbf{M})]_{ij} = 0, \quad \forall (i,j) \notin \Omega. \tag{12.18}$$

In the following, we propose two methods to solve this problem.

### 12.3.2 Nuclear Norm Minimization

The optimization problem in (12.16) characterizes the low-rank property of the power state measurement matrix $\mathbf{Z}_0$ as well as the sparseness of the malicious attack major $\mathbf{A}$. However, it is known to be impractical to directly solve (12.16). One possible approach is to replace $\text{Rank}(\mathbf{Z}_0)$ and $\|\mathbf{A}\|_0$ with their convex relaxations, $\|\mathbf{Z}_0\|_*$ and $\|\mathbf{A}\|_1$, respectively. Here, $\|\mathbf{Z}_0\|_*$ is the nuclear norm of $\mathbf{Z}_0$, which is the sum of its singular values, and $\|\mathbf{A}\|_1$ is the $l_1$ norm of $\mathbf{A}$, which is the sum of absolute values of its entries. Hence, (12.16) can be reformulated as the following convex optimization problem:

$$\min_{\mathbf{Z_0},\mathbf{A}} \|\mathbf{Z_0}\|_* + \lambda\|\mathbf{A}\|_1, \quad s.t. \quad \mathbf{Z_a} = \mathbf{Z_0} + \mathbf{A}, \tag{12.19}$$

where $\lambda$ is a regularization parameter. Correspondingly, (12.17) can be formulated as

$$\min_{\mathbf{Z_0},\mathbf{A}} \|\mathbf{Z_0}\|_* + \lambda\|\mathbf{A}\|_1, \quad s.t. \quad \mathcal{P}_\Omega(\mathbf{Z_a}) = \mathcal{P}_\Omega(\mathbf{Z_0} + \mathbf{A}). \tag{12.20}$$

The optimization problem in (12.20) has been extensively studied in the fields of compressive sensing [624] and matrix completion [413, 625], and can be solved by many off-the-shelf convex optimization algorithms. Motivated by [626], an algorithm that applies the techniques of augmented Lagrange multipliers is utilized here to detect the false data matrix $\mathbf{A}$ as well as to recover the measurement matrix $\mathbf{Z_0}$.

Augmented Lagrange multipliers are used to solve constrained optimization problems as follows:

$$\min f(\mathbf{X}), \quad s.t. \quad \mathbf{h}(\mathbf{X}) = \mathbf{0}, \tag{12.21}$$

where $f : \mathbb{R}^n \to \mathbb{R}$ and $h : \mathbb{R}^n \to \mathbb{R}^m$. The augmented Lagrangian is defined as

$$\mathbf{L}(\mathbf{X},\mathbf{Y},\mu) = \mathbf{f}(\mathbf{X}) + \langle\mathbf{Y},\mathbf{h}(\mathbf{X})\rangle + \frac{\mu}{2}\|\mathbf{h}(\mathbf{X})\|_2^2, \tag{12.22}$$

where $\mu$ is a positive scalar, and $\mathbf{Y}$ contains the Lagrange multipliers. $\langle\mathbf{Y},\mathbf{h}(\mathbf{X})\rangle$ denotes the trace of $\mathbf{Y}^\top\mathbf{h}(\mathbf{X})$. The optimization problem in (12.20) can be solved iteratively via the method of augmented Lagrange multipliers [51]. The Lagrangian for (12.20) is given by

$$L(\mathbf{Z_0},\mathbf{A},\mathbf{Y},\mu) = \|\mathbf{Z_0}\|_* + \lambda\|\mathbf{A}\|_1 + \langle\mathbf{Y},\mathcal{P}_\Omega(\mathbf{Z_a} - \mathbf{Z_0} - \mathbf{A})\rangle$$
$$+ \frac{\mu}{2}\|\mathcal{P}_\Omega(\mathbf{Z_a} - \mathbf{Z_0} - \mathbf{A})\|_2^2. \tag{12.23}$$

The value of $\lambda$ is set to $\frac{1}{\sqrt{\max(m,t)}}$, where $m$ and $t$ are the dimensions of the measurement matrix $\mathbf{Z_a}$. With $k = 1, 2, \ldots$, indexing iterations, $\mathbf{Z_0}$ and $\mathbf{A}$ are optimized according to

$$\mathbf{A}_{[k+1]} = \arg\min_{\mathbf{A}} L(\mathbf{Z_0}_{[k]}, \mathbf{A}, u_{[k]}, \mathbf{Y}_{[k]}), \tag{12.24}$$

$$\mathbf{Z_0}_{[k+1]} = \arg\min_{\mathbf{Z_0}} L(\mathbf{Z_0}, \mathbf{A}_{[k]}, u_{[k]}, \mathbf{Y}_{[k]}), \tag{12.25}$$

where (12.24) can be explicitly computed from the soft shrinkage formula, and (12.25) can be solved via the singular value shrinkage operator [627]. Specifically, we define this operator as $\mathcal{S}_\tau\{x\} = \mathrm{sgn}(x)\max(|x| - \tau, 0)$ for a real variable $x$, where sgn is the sign function. This operator can be extended to vectors and matrices by applying it element-wise. Using this operator, (12.24) can be solved iteratively via

$$\mathbf{A}_{[k+1]} = \mathcal{S}_{\lambda u_{[k]}^{-1}}\{\mathbf{Z_a} - \mathbf{Z_0}_{[k]} + u_{[k]}^{-1}\mathbf{Y}_{[k]}\}. \tag{12.26}$$

To solve (12.25), a singular value decomposition (SVD) is applied to the matrix $\mathbf{Z_a} - \mathbf{A}_{[k+1]} + u_{[k]}^{-1}\mathbf{Y}_{[k]}$:

$$(\mathbf{Z_a} - \mathbf{A}_{[k+1]} + u_{[k]}^{-1}\mathbf{Y}_{[k]}) = \mathbf{U}\mathbf{S}\mathbf{V}^\top, \tag{12.27}$$

where $\mathbf{U} \in \mathbb{R}^{m \times m}$ and $\mathbf{V} \in \mathbb{R}^{t \times t}$ are unitary matrices, and $\mathbf{S} \in \mathbb{R}^{m \times t}$ is a diagonal matrix containing the singular values of $(\mathbf{Z_a} - \mathbf{A}_{[k+1]} + u_{[k]}^{-1} \mathbf{Y}_{[k]})$. The singular values are arranged in decreasing order, and $\mathbf{Z_0}$ is updated via

$$\mathbf{Z}_{0[k+1]} = \mathbf{U} \mathcal{S}_{u_{[k]}^{-1}} \{\mathbf{S}\} \mathbf{V}^\top. \tag{12.28}$$

During each iteration of the optimization, both the Lagrange multipliers $\mathbf{Y}$ and $\mu$ are updated, which improves the performance of the algorithm:

$$\mathbf{Y}_{[k+1]} = \mathbf{Y}_{[k]} + u_{[k]}(\mathbf{Z_a} - \mathbf{Z}_{0[k+1]} - \mathbf{A}_{[k+1]}), \tag{12.29}$$

$$\mu_{[k+1]} = \alpha \mu_{[k]}, \tag{12.30}$$

where $\alpha$ is a positive constant. The algorithm is outlined as Algorithm 11.

---

**Algorithm 11** Nuclear Norm Minimization Approach

Input: $\mathbf{Z_a} \in \mathbb{R}^{m \times t}$; $\lambda = \frac{1}{\sqrt{\max(m,t)}}$;
Initialize: $\mathbf{Y}_{[0]} = 0$; $\mathbf{Z}_{0[0]} = 0$; $\mathbf{A}_{[0]} = 0$; $\mu_{[0]} > 0$; $\alpha > 0$; $k = 0$;
**while** not converge **do**
 $\mathbf{Z}_{0[k+1]} = \mathbf{Z}_{0[k]}$; $\mathbf{A}_{[k+1]} = \mathbf{A}_{[k]}$; $j = 0$;
 **while** not converge **do**
  $\mathbf{A}_{[k+1]}^{[j+1]} = \mathcal{S}_{\lambda u_{[k]}^{-1}} \{\mathbf{Z_a} - \mathbf{Z}_{0[k+1]}^{[j]} + u_{[k]}^{-1} \mathbf{Y}_{[k]}\}$;
  $(\mathbf{Z_a} - \mathbf{A}_{[k+1]} + u_{[k]}^{-1} \mathbf{Y}_{[k]}) = \mathbf{U} \mathbf{S} \mathbf{V}^\top$;
  Obtain $[\mathbf{U}, \mathbf{S}, \mathbf{V}]$;
  $\mathbf{Z}_{0[k+1]}^{[j+1]} = \mathbf{U} \mathcal{S}_{u_{[k]}^{-1}} \{\mathbf{S}\} \mathbf{V}^\top$;
  $j = j + 1$;
 **end while**
 $\mathbf{Y}_{[k+1]} = \mathbf{Y}_{[k]} + u_{[k]}(\mathbf{Z_a} - \mathbf{Z}_{0[k+1]} - \mathbf{A}_{[k+1]})$;
 $\mu_{[k+1]} = \alpha \mu_{[k]}$;
 $k = k + 1$;
**end while**
Return $\mathbf{Z}_{0[k]}$; $\mathbf{A}_{[k]}$;
Output $\mathbf{Z}_{0[k]}$; $\mathbf{A}_{[k]}$;

---

### 12.3.3 Low-Rank Matrix Factorization

The speed and scalability of the nuclear norm minimization approach are limited by the computational complexity of singular value decomposition. When the matrix size and rank increase, the computational operations for singular value decomposition will become quite expensive. To improve the scalability of solving large-scale problems of malicious attack detection in power systems, a low-rank matrix factorization approach is proposed here.

Given the observations $\mathbf{Z_a}$, the measurements $\mathbf{Z_0}$ and false data injection attack matrix $\mathbf{A}$ can be separated by the minimization problem:

$$\min_{\mathbf{U},\mathbf{V},\mathbf{Z_0}} \|\mathbf{Z_a} - \mathbf{Z_0}\|_1, \quad s.t. \quad \mathbf{UV} - \mathbf{Z_0} = \mathbf{0}, \tag{12.31}$$

where the low-rank matrix $\mathbf{Z_0}$ is expressed as a product of $\mathbf{U} \in \mathbb{R}^{m \times r}$ and $\mathbf{V} \in \mathbb{R}^{r \times n}$ for some adjustable rank estimate $r$. Correspondingly, (12.20) can be rewritten as

$$\min_{\mathbf{U},\mathbf{V},\mathbf{Z_0}} \|\mathcal{P}_\Omega(\mathbf{Z_a} - \mathbf{Z_0})\|_1, \quad s.t. \quad \mathbf{UV} - \mathbf{Z_0} = \mathbf{0}. \tag{12.32}$$

Note that a low-rank matrix factorization is explicitly applied to $\mathbf{Z_0}$ instead of minimizing its nuclear norm as in (12.20), which avoids singular value decomposition completely. To solve the minimization problem in (12.32), the augmented Lagrangian can be expressed as

$$L(\mathbf{U}, \mathbf{V}, \mathbf{Z_0}, \mathbf{Y}, \mu) = \|\mathcal{P}_\Omega(\mathbf{Z_a} - \mathbf{Z_0})\|_1 + \langle \mathbf{Y}, \mathbf{UV} - \mathbf{Z_0} \rangle$$
$$+ \frac{\mu}{2} \|\mathbf{UV} - \mathbf{Z_0}\|_2^2, \tag{12.33}$$

where $\mu$ is a penalty parameter and $\mathbf{Y}$ is the vector of Lagrange multipliers corresponding to the constraint $\mathbf{UV} - \mathbf{Z_0} = \mathbf{0}$. Motivated by the idea in the alternating direction method for convex optimization, the augmented Lagrangian can be minimized with respect to the block variables $\mathbf{U}, \mathbf{V}$, and $\mathbf{Z_0}$ individually via the following framework at each iteration $k$ [628]:

$$\mathbf{U}_{[k+1]} = \arg\min_{\mathbf{U}} L(\mathbf{U}, \mathbf{V}_{[k]}, \mathbf{Z_0}_{[k]}, \mathbf{Y}_{[k]}, \mu_{[k]}), \tag{12.34}$$

$$\mathbf{V}_{[k+1]} = \arg\min_{\mathbf{V}} L(\mathbf{U}_{[k+1]}, \mathbf{V}, \mathbf{Z_0}_{[k]}, \mathbf{Y}_{[k]}, \mu_{[k]}), \tag{12.35}$$

$$\mathbf{Z_0}_{[k+1]} = \arg\min_{\mathbf{Z_0}} L(\mathbf{U}_{[k+1]}, \mathbf{V}_{[k+1]}, \mathbf{Z_0}, \mathbf{Y}_{[k]}, \mu_{[k]}), \tag{12.36}$$

where (12.34) and (12.35) are least-squares problems:

$$\mathbf{U}_{[k+1]} = (\mathbf{Z_0} - u_{[k]}^{-1}\mathbf{Y}_{[k]})\mathbf{V}^\top(\mathbf{VV}^\top)^{-1}, \tag{12.37}$$

$$\mathbf{V}_{[k+1]} = (\mathbf{U}^\top\mathbf{U})^{-1}\mathbf{U}^\top(\mathbf{Z_0} - u_{[k]}^{-1}\mathbf{Y}_{[k]}). \tag{12.38}$$

(12.36) can be solved by the shrinkage formula:

$$\mathcal{P}_\Omega(\mathbf{Z_0}_{[k+1]}) = \mathcal{P}_\Omega(\mathcal{S}_{u_{[k]}^{-1}}\{\mathbf{U}_{[k+1]}\mathbf{V}_{[k+1]} - \mathbf{Z_a} + u_{[k]}^{-1}\mathbf{Y}_{[k]}\}). \tag{12.39}$$

The Lagrangian multipliers $\mathbf{Y}$ and $\mu$ are updated during each iteration as follows:

$$\mathbf{Y}_{[k+1]} = \mathbf{Y}_{[k]} + u_{[k]}(\mathbf{U}_{[k+1]}\mathbf{V}_{[k+1]} - \mathbf{Z_0}_{[k+1]}), \tag{12.40}$$

$$\mu_{[k+1]} = \alpha\mu_{[k]}, \tag{12.41}$$

where $\alpha$ is a positive constant. At the end of each iteration, a rank estimation strategy [629] is applied to update $r$ to ensure the success of the algorithm. The proposed algorithm is illustrated in Algorithm 12.

## 12.3.4    Numerical Results

Numerical simulations are presented here to evaluate the performance of the proposed algorithms. Power flow data for IEEE 57 bus, IEEE 118 bus test cases, and the Polish system [630] during winter the peak conditions in 2007–2008 are used to validate the effectiveness of proposed algorithm.

### Receiver Operating Characteristic Analysis

Assume the loads on each bus in the power system are uniformly distributed between 50% and 150% of its base load. When the state estimation measurements are collected, a small portion $\epsilon$ of the measurement data are compromised by malicious attackers with an arbitrary amount of injection data, and $\epsilon$ is defined as the attack ratio in this context. Methods for false data injection attack construction can be found in [609, 611]. Here, we focus on the protection scheme and suppose the locations of the attacks are chosen randomly and are of duration $\Delta t$.[1] Totally a number of $T$ time instance measurements are obtained for analysis. The receiver operating characteristic analysis of the proposed algorithms are first given, and then compared with the principal component analysis (PCA)[2] based detection method. In this analysis, the attack ratio is fixed at $\epsilon = 0.1$ and SNR = 10dB.

The ROC curves for IEEE 57 bus and IEEE 118 bus cases are shown in Figures 12.1 and 12.2, respectively. From the figures, it is apparent that the proposed algorithms can detect the false data accurately at a low false alarm rate. For example in the IEEE 57 bus system, the true positive rate of nuclear norm minimization is 93% and it is 95% for low-rank matrix factorization when the false alarm rate $p_f = 10\%$. The IEEE 118

---

**Algorithm 12** Low-Rank Matrix Factorization

Input: $\mathbf{Z_a} \in \mathbb{R}^{m \times t}$; Initial rank estimate $r$.
Initialize: $\mathbf{U} \in \mathbb{R}^{m \times r}$; $\mathbf{V} \in \mathbb{R}^{r \times t}$; $\mathbf{Z_{0[0]}} = U * V$; $\mathbf{Y_{[0]}} = 0$; $\mu_{[0]} > 0$; $\alpha > 0$; $k = 0$.
**while** not converge **do**
    $\mathbf{U}_{[k+1]} = (\mathbf{Z_0} - u_{[k]}^{-1}\mathbf{Y}_{[k]})\mathbf{V}^{\top}(\mathbf{V}\mathbf{V}^{\top})^{-1}$;
    $\mathbf{V}_{[k+1]} = (\mathbf{U}^{\top}\mathbf{U})^{-1}\mathbf{U}^{\top}(\mathbf{Z_0} - u_{[k]}^{-1}\mathbf{Y}_{[k]})$;
    $\mathbf{Z_{0[k+1]}} = \mathcal{S}_{u_{[k]}^{-1}}\{\mathbf{U}_{[k+1]}\mathbf{V}_{[k+1]} - \mathbf{Z_a} + u_{[k]}^{-1}\mathbf{Y}_{[k]}\}$;
    $\mathbf{Y}_{[k+1]} = \mathbf{Y}_{[k]} + u_{[k]}(\mathbf{U}_{[k+1]}\mathbf{V}_{[k+1]} - \mathbf{Z_{0[k+1]}})$;
    $\mu_{[k+1]} = \alpha\mu_{[k]}$;
    $k = k + 1$;
    Check $r$, possibly re-estimate $r$ and adjust sizes of the iterates;
**end while**
Return $\mathbf{Z_{0[k]}}$;
Output $\mathbf{Z_{0[k]}}$; $\mathbf{Z_a} - \mathbf{Z_{0[k]}}$;

---

[1] Note that the attack vectors used in this chapter are more general compared to those described in [609, 611] and will not affect the efficiency of the proposed algorithms.

[2] For PCA, we retain the largest $K$ singular values of the matrix such that $\frac{\sum_1^K s_i}{\sum_1^N s_i} > 95\%$.

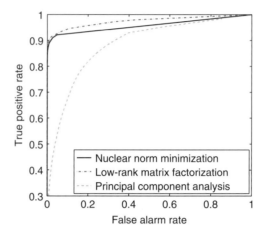

**Figure 12.1** ROC performance for the IEEE 57 bus system. SNR = 10dB.

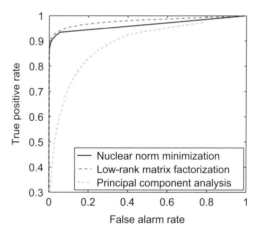

**Figure 12.2** ROC performance for the IEEE 118 bus system. SNR = 10dB.

bus system shows similar results. Moreover, the low-rank matrix factorization approach performs slightly better than the nuclear norm minimization method. The reason for this phenomenon is model-related, and in this case, the sparse attack matrix is not the dominant part in the measurements, which makes the low-rank matrix factorization approach more suitable. Figures 12.1 and 12.2 show that the proposed algorithms outperform the PCA-based approach significantly. The PCA method does not take the corruptions of malicious attacks into consideration. Even though the matrix $\mathbf{Z_0}$ is low rank, the sum of $\mathbf{Z_0}$ and $\mathbf{A}$ will not be low rank any more. Directly applying the PCA method will result in a poor performance. However, proposed algorithms exploit the low-rank structure of the anomaly-free measurement matrix, and the fact that malicious attacks are quite sparse, which renders a better performance.

### Performance vs. Missing Measurement Ratio

Next, we investigate the performance of the proposed algorithms under different missing measurement ratios. In particular, we assume that a portion of the measurements collected at the control center are missing due to meter failures or communication link outage, and evaluate the performance of the proposed algorithms under different missing measurement ratios up to 10% on the IEEE 118 bus system. The attack ratio is fixed at $\epsilon = 0.1$ and SNR = 10dB.

The ROC curves for the IEEE 118 bus case are depicted in Figure 12.3. From the figure we can see that with 10% missing measurements, the proposed algorithms are still able to detect the malicious attacks at acceptable true positive rates, and the low-rank matrix factorization method performs slightly better. By comparing with Figure 12.2, we can see that the missing measurements deteriorate the performance of proposed algorithms as we would expect. Since the PCA-based method is unable to detect the anomalies in this case, we omitted the simulation result for this case. Note that the existence of missing entries will result in an incorrect estimation of the low-dimensional subspace of matrix $\mathbf{Z_0}$, which leads to the failure of PCA.

To investigate the performance under different missing measurement ratios, the percentage of missing measurements is varied from 0% (no missing measurements) to 10%, and the results are shown in Figure 12.4. The true positive rates are calculated for both algorithms when the false alarm rate equals 10%. It is shown that the performance is improved monotonically as more and more measurements are collected. In the worst case when 10% of measurements are missing, the proposed algorithms can still achieve a true positive rate of 85% and 90% for nuclear norm minimization and low-rank matrix factorization, respectively.

A more detailed demonstration of the proposed algorithms' recoverability for power system states is shown in Figure 12.5. Here, we assume 10% of the measurements are missing and SNR = 10dB, and the cumulative distribution functions of the relative

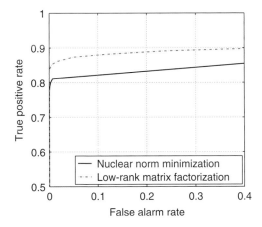

**Figure 12.3** ROC curves of the proposed algorithms for the IEEE 118 bus system. 10% measurements are missing and SNR = 10dB.

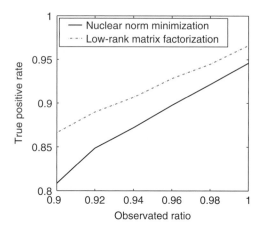

**Figure 12.4** Performance of the proposed algorithms under different missing ratios for the IEEE 118 bus system. The false alarm rate is 10% and SNR = 10dB.

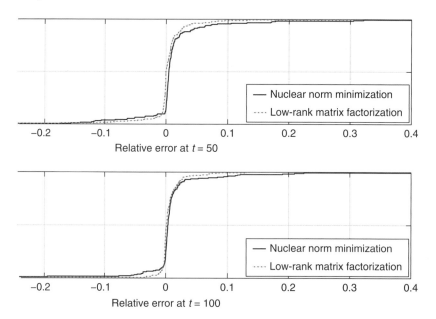

**Figure 12.5** Power state reconstruction performance of the proposed algorithms at specific time instance $t = 50$ and $t = 100$. 10% of the measurements are missing and SNR = 10dB.

reconstruction errors at $t = 50$ and $t = 100$ are calculated. The relative reconstruction error is defined as

$$\varepsilon = (\hat{\boldsymbol{\theta}} - \boldsymbol{\theta})./|\boldsymbol{\theta}|, \qquad (12.42)$$

where $./$ denotes component-wise division, and $|\boldsymbol{\theta}|$ denotes the element-wise absolute value of the vector $\boldsymbol{\theta}$ . $\boldsymbol{\theta}$ (in radian units) is obtained from the recovered $\mathbf{Z_0}$, and the vector $\boldsymbol{\varepsilon}$ represents the relative error of each component in the state vector $\boldsymbol{\theta}$. We calculate the relative error for each bus in the system, and plot the corresponding cumulative distribution functions. From the figures we can see that the proposed algorithms are able to reconstruct the power system states quite accurately. At $t = 50$, the majority of the relative errors focus between interval $[-0.1 \quad 0.1]$, and similar results are shown at $t = 100$. These imply that the proposed algorithms are able to precisely detect the malicious attacks as well as accurately estimate power system states, even under the severe situation of partial measurements.

### Performance vs. Attack Ratio

Third, we investigate the performance of the proposed algorithms under different attack ratios for the IEEE 118 bus system. In particular, $\epsilon$ is varied from 5% to 15%, and SNR = 10dB.

From Figure 12.6, the true positive rate is quite high at low sparsity ratios for both proposed algorithms. Particularly, when the sparsity ratio is 5%, the true positive rates are 93.6% and 94.3% at $f_a = 10\%$ for nuclear norm minimization and low-rank matrix factorization, respectively. Compared with the PCA-based method, the performance of the proposed algorithms are quite stable as the attack ratio increases. When the attack ratio reaches 15%, the true positive rates for both algorithms are still around 90%. The true positive rates of the proposed algorithms will decrease dramatically when attackers attack the power system massively. This is because, when the attack matrix is not sparse enough, the mixed-norm minimization is not able to separate the low-rank anomaly-free matrix and attack matrix.

### Performance on Large-Scale System

Finally, we analyze the scalability and computational efficiency of the proposed algorithms on power flow data for the Polish system during winter peak conditions in 2007–2008. The attack ratio is fixed at $\epsilon = 0.1$ and SNR = 10dB.

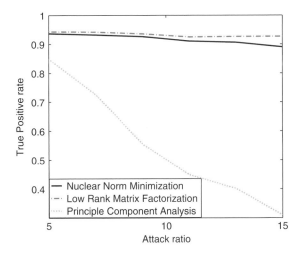

**Figure 12.6** Performance of the proposed algorithms under different attack ratios for the IEEE 118 bus system. The false alarm rate is 10% and SNR = 10dB.

The ROC curve is shown in Figure 12.7. It is shown in the figure that the performance of the proposed algorithms is quite stable on the large-scale system compared
to the IEEE 57 bus system and the IEEE 118 bus system. A comparison of the computational efficiency of the two proposed algorithms is shown in Figure 12.8. The data
matrix row dimension $m$ is varied from 100 to 3,400, in steps of 300 rows. The proposed algorithms are applied to a subset of the measurement matrix each time, and the
CPU computation time is logged. It is shown in Figure 12.8 that as the dimension of the
measurement matrix increases, the CPU time for computation will increase, and the low-
rank matrix factorization approach performs better than the nuclear norm minimization
method, which demonstrates a better scalability to large problems, as expected.

The numerical results validate the effectiveness of the proposed algorithms. According to the simulation results, both the low-rank matrix factorization technique and the

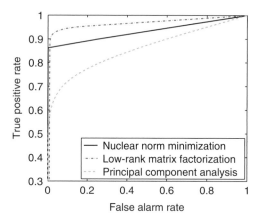

**Figure 12.7** Performance on power flow data for the Polish system during winter peak conditions,
2007–2008. SNR = 10dB.

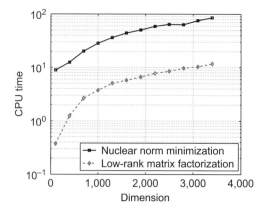

**Figure 12.8** CPU time vs. matrix dimension for the proposed nuclear norm minimization and
low-rank matrix factorization algorithms.

nuclear norm minimization technique can solve the matrix separation problem, and the performance of the low rank-matrix factorization is slightly better than that of the nuclear norm minimization technique. From the perspective of recoverability, since the false data attack matrix $\mathbf{A}$ is not the dominant part compared with $\mathbf{Z_0}$ in this setting, the performance of the low-rank matrix factorization technique is better. From the perspective of computation time, the low-rank matrix factorization technique is much faster than the nuclear norm minimization due to its SVD-free feature. A detailed comparison of the complexity of the two algorithms is beyond the scope of this chapter, and useful discussions can be found in [628].

## 12.4 Distributed Parallel Approach for Security-Constrained Optimal Power Flow

Following the standard approach to formulating the SCOPF problem, the objective here is to minimize the cost of generation while safeguarding power system sustainability. For the sake of simplicity and computational tractability, constraints (12.10)–(12.13) are modeled with the linear DC load flow, and we assume the list of contingencies is given. Thus, assuming a DC power network modeling and neglecting all shunt elements, the standard SCOPF problem can be simplified to the following optimization problem:

$$\min_{\theta^0,\dots,\theta^C;\mathbf{P}^{g,0},\dots,\mathbf{P}^{g,C}} \sum_{i\in\mathcal{G}} f_i^g(\mathbf{P}_i^{g,0}) \tag{12.43}$$

$$\text{subject to} \quad \mathbf{B}_{bus}^0\theta^0 + \mathbf{P}^{d,0} - \mathbf{A}^{g,0}\mathbf{P}^{g,0} = 0, \tag{12.44}$$

$$\mathbf{B}_{bus}^c\theta^c + \mathbf{P}^{d,c} - \mathbf{A}^{g,c}\mathbf{P}^{g,c} = 0, \tag{12.45}$$

$$|\mathbf{B}_f^0\theta^0| - \mathbf{F}_{max} \le 0, \tag{12.46}$$

$$|\mathbf{B}_f^c\theta^c| - \mathbf{F}_{max} \le 0, \tag{12.47}$$

$$\underline{\mathbf{P}^{g,0}} \le \mathbf{P}^{g,0} \le \overline{\mathbf{P}^{g,0}}, \tag{12.48}$$

$$\underline{\mathbf{P}^{g,c}} \le \mathbf{P}^{g,c} \le \overline{\mathbf{P}^{g,c}}, \tag{12.49}$$

$$|\mathbf{P}^{g,0} - \mathbf{P}^{g,c}| \le \Delta_c, \tag{12.50}$$

$$i \in \mathcal{G}, \quad c = 1,\dots,C, \tag{12.51}$$

where the notation is given in Table 12.1.

The solution to (12.43) ensures economic dispatch while guaranteeing power system security, by taking into account a set of postulated contingencies. The major challenge of SCOPF is the size of the problem, especially for large systems with numerous contingency cases to be considered. Directly solving the SCOPF problem by simultaneously imposing all post-contingency constraints will result in prohibitive memory requirements and a substantial CPU burden. To achieve efficient and secure operations of the

**Table 12.1** Notation definitions

| | |
|---|---|
| $\mathcal{G}$ | Set of generators |
| $\mathcal{N}$ | Set of buses |
| $\mathcal{B}$ | Set of branches |
| $\boldsymbol{\theta}^c \in \mathbb{R}^{|\mathcal{N}|}$ | Vector of voltage angles |
| $\mathbf{P}^{g,c} \in \mathbb{R}^{|\mathcal{G}|}$ | Vector of real power flows |
| $f_i^g$ | Generation cost function |
| $\mathbf{P}_i^{g,0}$ | Displaceable real power of each individual generation unit $i$ for the pre-contingency configuration |
| $\mathbf{B}_{bus}^c \in \mathbb{R}^{|\mathcal{N}| \times |\mathcal{N}|}$ | Power network system admittance matrix |
| $\mathbf{B}_f^c \in \mathbb{R}^{|\mathcal{B}| \times |\mathcal{N}|}$ | Branch admittance matrix |
| $\mathbf{P}^{d,c} \in \mathbb{R}^{|\mathcal{N}|}$ | Real power demand |
| $\mathbf{A}^{g,c} \in \mathbb{R}^{|\mathcal{N}| \times |\mathcal{G}|}$ | Sparse generator connection matrix, whose $(i,j)$th element is 1 if generator $j$ is located at bus $i$ and 0 otherwise |
| $\mathbf{F}_{max}$ | Vector for the maximum power flow |
| $\overline{\mathbf{P}^{g,c}}$ | Upper bound on real power generation |
| $\underline{\mathbf{P}^{g,c}}$ | Lower bound on real power generation |
| $\Delta_c$ | Predefined maximal allowed variation of power outputs |

entire electrical grid, a distributed and parallel optimization algorithm is designed to solve the SCOPF problem in large-scale power systems in the following.

## 12.4.1    An Introduction to ADMM

In this section, a distributed algorithm is proposed to solve the SCOPF problem by decomposing it into a set of simpler and parallel subproblems corresponding to the base case and each contingency case, respectively. The approach is based on the alternating direction method of multipliers (ADMM) [548, 518], whose general form is described as follows:

$$\min_{\mathbf{x,z}} \quad f(\mathbf{x}) + g(\mathbf{z}) \tag{12.52}$$

$$\text{subject to} \quad \mathbf{Ax} + \mathbf{Bz} = \mathbf{c}, \tag{12.53}$$

where $\mathbf{x} \in \mathbb{R}^n$, $\mathbf{z} \in \mathbb{R}^m$ and $\mathbf{c} \in \mathbb{R}^p$, and $\mathbf{A} \in \mathbb{R}^{p \times n}$ and $\mathbf{B} \in \mathbb{R}^{p \times m}$. The functions $f$ and $g$ are closed, convex, and proper. The scaled augmented Lagrangian can be expressed as

$$\mathcal{L}_\rho(\mathbf{x}, \mathbf{z}, \boldsymbol{\mu}) = f(\mathbf{x}) + g(\mathbf{z}) + \frac{\rho}{2} \|\mathbf{Ax} + \mathbf{Bz} - \mathbf{c} + \boldsymbol{\mu}\|_2^2, \tag{12.54}$$

where $\rho > 0$ is the penalty parameter and $\boldsymbol{\mu}$ is the vector of scaled dual variables. Using the scaled dual variables, $\mathbf{x}$ and $\mathbf{z}$ are updated in a Gauss–Seidel fashion. At each iteration $k$, the update process can be expressed as

$$x[k+1] = \arg\min_{\mathbf{x}} f(\mathbf{x}) + \frac{\rho}{2}\|\mathbf{Ax} + \mathbf{Bz}[k] - \mathbf{c} + \boldsymbol{\mu}[k]\|_2^2,$$

$$z[k+1] = \arg\min_{\mathbf{z}} g(\mathbf{z}) + \frac{\rho}{2}\|\mathbf{Ax}[k+1] + \mathbf{Bz} - \mathbf{c} + \boldsymbol{\mu}[k]\|_2^2.$$

Finally, the scaled dual variable vector is updated via

$$\boldsymbol{\mu}[k+1] = \boldsymbol{\mu}[k] + \mathbf{Ax}[k+1] + \mathbf{Bz}[k+1] - \mathbf{c}.$$

The use of ADMM for optimization in power systems has been considered in [631] and [632]. ADMM-based methods offer a general framework for distributed optimization, and a corresponding distributed and parallel approach to SCOPF is introduced below.

## 12.4.2   Distributed and Parallel Approach for SCOPF

The optimization problem (12.43) cannot be readily solved using ADMM immediately, since the constraint (12.50) couples the pre-contingency and post-contingency variables, and the inequalities make the problem even more complicated. To address these challenges, the optimization problem (12.43) can be reformulated by introducing a slack variable $\mathbf{p}^c \in \mathbb{R}^{|\mathcal{G}|}$:

$$\begin{aligned}
&\text{minimize} &&\text{(12.43)} &&\text{(12.55)}\\
&\text{subject to} &&\text{Constraints (12.44)–(12.49),} &&\text{(12.56)}\\
&&&\mathbf{P}^{g,0} - \mathbf{P}^{g,c} + \mathbf{p}^c = \Delta_c, &&\text{(12.57)}\\
&&&0 \le \mathbf{p}^c \le 2\Delta_c, \quad c = 1,\dots,C. &&\text{(12.58)}
\end{aligned}$$

The above optimization problem can be solved distributively using ADMM. The scaled augmented Lagrangian can be calculated as

$$\mathcal{L}_\rho(\mathbf{P}^{g,0},\dots,\mathbf{P}^{g,C};\mathbf{p}^1,\dots,\mathbf{p}^C;\boldsymbol{\mu}^1,\dots,\boldsymbol{\mu}^C)$$

$$= \sum_{i\in\mathcal{G}} f_i^g(\mathbf{P}_i^{g,0}) + \sum_{c=1}^{C} \frac{\rho^c}{2}\|\mathbf{P}^{g,0} - \mathbf{P}^{g,c} + \mathbf{p}^c - \Delta_c + \boldsymbol{\mu}^c\|_2^2. \qquad (12.59)$$

The optimization variables $\mathbf{P}^{g,0}$, $\mathbf{P}^{g,c}$, and $\mathbf{p}^c$ are arranged into two groups, $\{\mathbf{P}^{g,0}\}$ and $\{\mathbf{P}^{g,c},\mathbf{p}^c\}$, and updated iteratively. The variables in each group are optimized in parallel on distributed computing nodes, and coordinated by the dual variable vector $\boldsymbol{\mu}^c$ during each iteration.

At the $k$th iteration, the $\mathbf{P}^{g,0}$-update solves the base scenario with square regularization terms enforced by the coupling constraints and expressed as

$$\mathbf{P}^{g,0}[k+1] = \arg\min_{\mathbf{P}^{g,0}} \sum_{i\in\mathcal{G}} f_i^g(\mathbf{P}_i^{g,0}) +$$

$$\sum_{c=1}^{C} \frac{\rho^c}{2}\|\mathbf{P}^{g,0} - \mathbf{P}^{g,c}[k] + \mathbf{p}^c[k] - \Delta_c + \boldsymbol{\mu}^c[k]\|_2^2$$

$$\text{subject to} \quad \text{Constraints (12.44), (12.46), and (12.48).} \qquad (12.60)$$

---

**Algorithm 13** Distributed SCOPF

---

Input: $\mathbf{B}_{bus}^c$, $\mathbf{B}_f^c$, $\mathbf{A}^{g,c}$, $\mathbf{P}^{d,c}$, $\overline{\mathbf{P}^{g,c}}$, $\underline{\mathbf{P}^{g,c}}$, $\Delta_c$;

Initialize: $\boldsymbol{\theta}^c$, $\mathbf{P}^{g,c}$, $\mathbf{p}^c$, $\boldsymbol{\mu}^c$, $\rho^c$, $k = 0$;

**while** not converge **do**

  $\mathbf{P}^{g,0}$-update:

  $\mathbf{P}^{g,0}[k+1] = \arg\min_{\mathbf{P}^{g,0}} \sum_{i \in \mathcal{G}} f_i^g(\mathbf{P}_i^{g,0})$

  $+ \sum_{c=1}^{C} \frac{\rho^c}{2} \|\mathbf{P}^{g,0} - \mathbf{P}^{g,c}[k] + \mathbf{p}^c[k] - \Delta_c + \boldsymbol{\mu}^c[k]\|_2^2$

  subject to Constraints (12.44), (12.46), and (12.48).

  $\mathbf{P}^{g,c}$-update, distributively at each computing node:

  $\mathbf{P}^{g,c}[k+1] = \arg\min_{\mathbf{P}^{g,c},\mathbf{p}^c} \frac{\rho^c}{2} \|\mathbf{P}^{g,0}[k+1] - \mathbf{P}^{g,c} + \mathbf{p}^c - \Delta_c + \boldsymbol{\mu}^c[k]\|_2^2$

  subject to Constraints (12.45), (12.47), (12.49), and (12.58),

  $\boldsymbol{\mu}^c[k+1] = \boldsymbol{\mu}^c[k] + \mathbf{P}^{g,0}[k+1] - \mathbf{P}^{g,c}[k+1] + \mathbf{p}^c[k+1] - \Delta_c$.

  Adjust penalty parameter $\rho^c$ is necessary;

  $k = k + 1$;

**end while**

Return $\boldsymbol{\theta}^c$, $\mathbf{P}^{g,c}$;

Output $\boldsymbol{\theta}^c$, $\mathbf{P}^{g,c}$;

---

The $\mathbf{P}^{g,c}$-update solves a number of independent optimization subproblems correspond to post-contingency scenarios and can be calculated distributively at the $c$th computing nodes via

$$\mathbf{P}^{g,c}[k+1] = \arg\min_{\mathbf{P}^{g,c},\mathbf{p}^c} \frac{\rho^c}{2} \|\mathbf{P}^{g,0}[k+1] - \mathbf{P}^{g,c}$$

$$+ \mathbf{p}^c - \Delta_c + \boldsymbol{\mu}^c[k]\|_2^2$$

$$\text{subject to} \quad \text{Constraints (12.45), (12.47), (12.49), and (12.58),} \qquad (12.61)$$

where the scaled dual variable vector is also updated locally at the $c$th computing utility as

$$\boldsymbol{\mu}^c[k+1] = \boldsymbol{\mu}^c[k] + \mathbf{P}^{g,0}[k+1] - \mathbf{P}^{g,c}[k+1] + \mathbf{p}^c[k+1] - \Delta_c. \qquad (12.62)$$

At the $k$th iteration, the original problem is divided into $C + 1$ subproblems of approximately the same size. The computing node handling $\mathbf{P}^{g,0}$ needs to communicate with all computing nodes solving (12.61) during the iterations. The results of the $\mathbf{P}^{g,0}$-update $\{\mathbf{P}^{g,0}\}$ will be scattered among the computing nodes for the $\mathbf{P}^{g,c}$-update. After the $\mathbf{P}^{g,c}$-update, the computed $\{\mathbf{P}^{g,c}, \mathbf{p}^c, \boldsymbol{\mu}^c\}$ will be collected to calculate the pre-contingency control variables. The subproblem data are iteratively updated so that in the end the block-coupling constraints (12.57) are satisfied. Note that since each of the subproblems is a smaller-scale OPF problem, existing techniques for OPF can be applied with minor modifications. The proposed algorithm is illustrated in Algorithm 13.

**Table 12.2** Characteristics of test cases

| Case | $|\mathcal{N}|$ | $|\mathcal{G}|$ | $|\mathcal{B}|$ | Number of contingency cases |
|---|---|---|---|---|
| IEEE 57 bus | 57 | 7 | 80 | 50 |
| IEEE 118 bus | 118 | 54 | 186 | 100 |
| IEEE 300 bus | 300 | 69 | 411 | 100 |

ADMM is a primal-dual algorithm in which each computing node $c$ solves its own subproblem (12.61), and variations to constraint (12.57) are systematically penalized at certain prices through the scaled dual variable to each individual subproblem. Note that in the ADMM framework for distributed computing the dual variables, or prices, are not uniformly set for all nodes, which will require costly synchronization. For convex optimization problems, ADMM converges to the optimum geometrically [633], and the convergence rate can be improved by using the warm start technique [634].

### 12.4.3 Numerical Results

In this section, numerical studies are examined to evaluate the performance of the proposed algorithm. Three classical test systems are used to study the SCOPF problem: the IEEE 57 bus, IEEE 118 bus, and IEEE 300 bus [630], whose structures and characteristics are summarized in Table 12.2.

Two kinds of contingencies are considered in the numerical tests: branch outage and generator failure. The contingencies are artificially generated and the numbers of contingencies considered are listed in Table 12.2. We follow the physical limits on the equipment of the test systems and assume every active generator is able to reschedule up to 50% of its maximum real power capacity. The numerical tests are implemented via MATLAB 7.10 on a PC with an Intel Q8200 2.33GHz processor and 8GB memory. The basic OPF problem solver is the same for all test systems. The performance of the convergence and computing time of the proposed algorithm are investigated in the following. The results are averaged over a total of 500 Monte Carlo implementations.

**Convergence Performance**

We first consider the convergence of the proposed algorithm. Since the number of contingencies and the optimal value for each test system differ, the relative error is used here to present the results. Suppose $r[k]$ is the result of the value of the objective function at the $k$th iteration, and $r^*$ is the optimal solution. The relative error $e$ is defined as $e = \left| \frac{r[k]-r^*}{r[0]-r^*} \right|$. The convergence performance is shown in Figure 12.9. It can be seen that after a moderate number of iterations, the proposed algorithm converges to the optimal values in the cases considered. From Figure 12.9, we can see that the IEEE 57 system gives the fastest convergence rate. This is mainly due to the small scale of the test system as well as the number of contingencies in the system. A large system leads to a large-scale optimization problem, and the large number of contingencies considered will make the problem scale even larger. Note that, after very few iterations, the algorithm

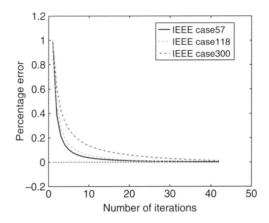

**Figure 12.9**  Convergence performance of the proposed algorithm on the test systems.

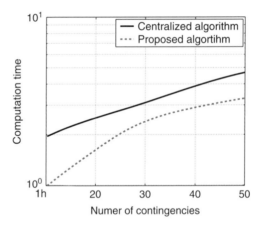

**Figure 12.10**  Computing time for the IEEE 57 bus system with different numbers of contingency cases.

gets very close to the optimal value, which means that the proposed algorithm is able to yield a good approximation to the optimal value in a short time.

### Computing Time Performance

The computing time for the test systems with different numbers of contingency cases is investigated and the results are given in Figures 12.10, 12.11, and 12.12. The number of contingencies is increased by 20% each time and the computing time is recorded. It can be seen from these figures that with an increase in the number of contingency cases in the SCOPF problem, the computing time of the centralized algorithm increases much faster than that of the proposed algorithm. Thus, the proposed distributed algorithm is more scalable and stable than the centralized approach.

The computing time to achieve an approximate solution with a relative error of $e = 1\%$ is also considered for the distributed case. To better illustrate the numerical

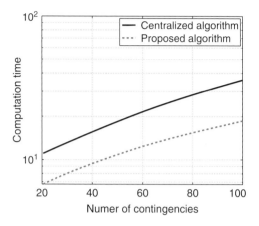

**Figure 12.11** Computing time for the IEEE 118 bus system with different numbers of contingency cases.

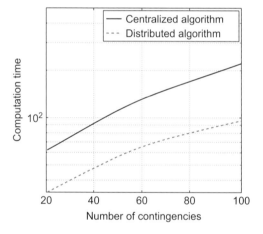

**Figure 12.12** Computing time for the IEEE 300 bus system with different numbers of contingency cases.

results, a speed-up factor is defined as $S_p = T_c/T_p$, where $T_c$ is the computing time of the centralized approach, and $T_p$ is the computing time of the distributed approach. The results of the computing time performance are presented in Table 12.3. It is shown in Table 12.3 that the proposed distributed approach obtains the same optimum as the centralized approach, and can achieve a speed-up factor $S_p$ of $1.4 \sim 2.4$. Note that if only an approximate result is needed, the speed-up factor can even be improved to $S_p$ of $4.4 \sim 4.8$ by using the proposed distributed algorithm. The speed-up factor for the smallest test system, IEEE 57 bus, is the smallest. This is due to the communication overhead between different computing nodes during the simulations. A larger $S_p$ can be achieved on a large-scale test system because the communication overhead is negligible compared with the computing time of the optimization subproblem handled by each computing node.

**Table 12.3**  Computing time performance of the proposed algorithm on different test systems

| Cases | Centralized | | Distributed ADMM | | | |
|---|---|---|---|---|---|---|
| | Cost | Time | Cost | Time | Cost (e = 1%) | Time |
| IEEE 57 bus | 487.53 | 5.22 | 487.53 | 3.55 | 492.40 | 1.18 |
| IEEE 118 bus | 1606.73 | 36.03 | 1606.73 | 18.92 | 1622.79 | 7.93 |
| IEEE 300 bus | 9567.12 | 221.67 | 9567.12 | 95.74 | 9662.80 | 52.87 |

## 12.5    Concluding Remarks

In this chapter, we have investigated the applications of big data processing techniques to security in smart grid. We have introduced two security concerns: false data injection attacks against state estimation and security-constrained optimal power flow in power system. We have explored the possibilities of exploiting the inherent structure of the data sets and effectively processing the large data sets to enhance power system security. We have designed a sparse optimization approach for false data injection detection, and a distributed parallel approach for security-constrained optimal power flow problem. We have performed numerical studies to validate the effectiveness of the proposed approaches. We have shown that effective management and processing of "big data" has the potential to significantly improve smart grid security.

# 13 Processing Large Data Sets in MapReduce

## 13.1 Introduction

While the performance of a data parallelism structure is primarily determined by the amount of data that need to be processed, since data are processed by different machines in parallel, the completion time of an application job is determined by the last finished machine. It is a grand challenge to distribute data load evenly to different machines. In the default configuration of MapReduce Hadoop, this is done by developing a hash function in the shuffling phase so that the intermediate results outputted by the mapper phase will be distributed evenly to different machines in the reduce phase. Such default configuration emphasizes the "common" case. Clearly, it is not optimal for various scenarios, and in many situations such configuration performs poorly.

There are many research studies on this grand problem. There are the reactive approach and the proactive approach. In the reactive approach, the workloads of different machines are monitored and workloads can be migrated from one machine to another in case there is a large skew of the workloads. SkewTune is one good example representing reactive approach [22]. In the proactive approach, systems try to understand the possible workloads the data to be processed can generate on each machine, so that when the data are dispatched to different machines in a load aware manner.

In this chapter, we study a proactive approach to improve performance of processing large data in MapReduce using sublinear algorithms.

### 13.1.1 The Data Skew Problem of MapReduce Jobs

When running MapReduce jobs, most of the studies have assumed that the input data are of uniform distribution, which, often being hashed to reduce worker nodes, naturally leads to a desirable balanced load in the later stages. The real-world data, however, are not necessarily uniform, and often exhibit a remarkable degree of skewness. For example, in PageRank, the graph commonly includes nodes with much higher degrees of incoming edges than others [21], and in Inverted Index, certain content can appear in many more documents than others [22]. Such a skewed distribution of the input or intermediate data can result in a small number of mappers or reducers taking a significantly longer time to complete than others [19]. Recent experimental studies [21, 19, 22] have shown that, in the CloudBurst application with a biology data set of a bimodal distribution [635], the slowest map task takes five times as long to complete as the fastest.

PageRank with Cloud9 data [636] is even worse, as the slowest map task takes twice as long to complete as the second slowest, and the latter is already five times slower than the average. Our experiments with the WordCount application produced similar results. Given that the overall finishing time is bounded by the slowest task, it can be dramatically prolonged with such skewed data.

In distributed databases, data skew is a known common phenomenon. There are mature solutions such as joining, grouping, aggregation, and others [637, 638]. Unfortunately, these can hardly be applied in the MapReduce context. The map function transfers the raw input data into (key, value) pairs, and the reduce function merges all intermediate values associated with the same intermediate key. In the database case, the pairs that share the same key do not need to be processed in a single machine. MapReduce, on the other hand, must guarantee that these pairs belong to the same partition – in other words, that they are distributed to the same reducer.

There are pioneering works dealing with the data skew in MapReduce [20, 22, 639]. Most of them are on offline heuristics, where the solution is to wait for all of the mappers to finish so as to obtain the key frequencies, or to engage in sampling before the map tasks to estimate the data distribution and then partition in advance, or to repartition the reduce tasks to balance the load among the servers. These approaches can be time-consuming with excessive I/O costs or network overheads. The solutions also lack theoretical bounds, given that most of them are heuristics.

## 13.1.2     Chapter Outline

In this chapter we examine the problem of accommodating data skew in MapReduce with online operations. In contrast with solutions that address the problem in the very late reduce stage [22] or after seeing all of the data [26], we address the skew from the very beginning of the inputting of data, and make no assumptions about *a priori* knowledge of the distribution of the data, nor require synchronized operations. We examine the keys in a continuous fashion and adaptively assign the tasks with a load-balanced strategy. We show that the optimal strategy is a constrained version of the *online minimum makespan problem* [640], and demonstrate that, in the MapReduce context where tasks with identical keys must be scheduled to the same machine, there is an online algorithm with a provable 2-competitive ratio. We further suggest that the online solution can be enhanced by a sample-based algorithm, which identifies the most frequent keys and assigns associated tasks in advance. We show that, probabilistically, it achieves a 3/2-competitive ratio with a bounded error.

Note that in the development of our algorithm, there is an embedded sublinear algorithm. More specifically, our first online algorithm provides a baseline competitive ratio where there is no requirement to "peek" at the data in advance. Clearly, the ratio provided by this algorithm is loose. To tighten such a competitive ratio, we "peek" at a sample set of data and develop an advanced algorithm based on this information. Intuitively, such information helps us to obtain some knowledge of which data are more important, so that when handling the skewness of these data, it is possible to apply a more refined scheme. We develop the algorithm, study how much data (sublinear to the whole data set) we should "peek" at, and analyze a tightened competitive ratio.

We evaluate our algorithms on both synthetic data and a real public data set. Our simulation results show that, in practice, the maximum loads of our online and sample-based algorithms are close to those of the offline solutions, and are significantly lower than those with the naive hash function in MapReduce. They enjoy comparable computation times as those with the hash function, which are much shorter than those of the offline solutions.

In Section 13.2, we analyze in detail the data skew problem in MapReduce. We then formulate the problem. In Section 13.3, we present the first online algorithm, which achieves a baseline 2-competitive ratio. We further develop our enhanced online algorithm in Section 13.4, which has a sublinear algorithm embedded within. We evaluate the performance of our algorithms in Section 13.5. In Section 13.6, we summarize the chapter.

## 13.2    Server Load Balancing: Analysis and Problem Formulation

### 13.2.1    Background and Motivation

The MapReduce libraries have been written in different programming languages. We take Apache Hadoop (high-availability distributed object-oriented platform), one of the most popular free implementations, as an example. Hadoop is based on a master-worker architecture, where a master node makes scheduling decisions and multiple worker nodes run tasks dispatched from the master.

In the map phase, the master node divides the large data set into small blocks and distributes them to the map workers. The map workers generate a large amount of intermediate (key, value) pairs and report the locations of these pairs on the local disk to the master, who is responsible for forwarding these locations to the reduce workers.

A hash function then assigns the values to different worker nodes for processing in the reduce phase. In Hadoop, the default hash function is

$$Hash(HashCode(intermediate\ key)\ mod\ ReducerNumber)$$

This is a simple hash function. It is highly efficient and naturally achieves load balance if the keys are uniformly distributed. This, however, can fail with skewed inputs. We look at an example of WordCount. WordCount is a classic MapReduce application for counting the number of words in a big document. It is also a commonly used benchmark application to evaluate the performance of MapReduce whenever an improvement is developed.

In a document, popular words such as "the," "a," and "of" appear much more frequently than other words. After hashing, they impose heavier workloads on the corresponding reduce workers. Consider a toy example shown in Figure 13.1 with a skewed input. The naive hash function will assign keys $a$, $d$, and $g$ to the first machine, keys $b$

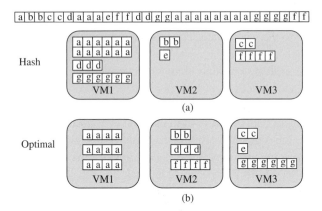

**Figure 13.1** An illustrative example for key assignment in MapReduce. There are three machines in this example. The "Hash" and "Optimal" rows represent the result load distribution of each scheduling, respectively.

and $e$ to the second machine, and keys $c$ and $f$ to the third machine. As a result, the first machine will achieve a maximum load with 19, six times more than that of the least load, while the maximum load of the balanced solution will be 12, as shown in the "optimal" row. Since the overall finishing time is bounded by the slowest, such simple hash-based scheduling is simply not satisfactory.

It is also worth noting that if an algorithm is designed to handle such load balancing problem, the algorithm must be online. This is because Hadoop starts to execute the reduce phase before every corresponding partition is available (i.e., it is activated when only part of the map phase has been completed (5% by default) [22]). The rationale behind this synchronous operation is to overlap the map and the reduce phases and consequently reduce the maximum finishing time; yet it can prevent making a partition of the map phase and the reduce phase in advance. In fact, the reduce phase further consists of three subphases: *shuffle*, in which the task pulls the map outputs; *sort*, in which the map outputs are sorted by keys; and *reduce function execution*, in which a user-defined function takes the map outputs with each key and, after all of the mappers have finished working, starts to run and generates the final outputs.

In Figure 13.2, we show a detailed measurement of the processing times of all of the phases for a WordCount application running on Amazon EC2. The *map* phase starts at time zero. The *reduce* phase, and the *shuffle* subphase of the *reduce* then starts at about 200s. We can see that the shuffle finishing time is much longer than the reduce function executing time, because the reduce workers need to wait for the map workers to generate intermediate pairs while using remote procedure calls to read the buffered data from the local disks of the map workers. Also note that the maximum map finishing time is quite close to that of shuffle subphase. Therefore, if we wait until all of the keys are generated, in other words, if we start the *shuffle* subphase after all of the map workers have finished, the overall job finishing time will double. This is unfortunately what the state-of-the-art

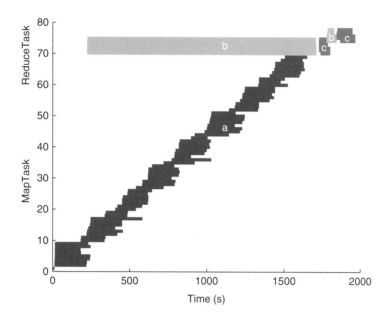

**Figure 13.2** We ran WordCount application on Amazon EC2 with four instances and we set 70 map tasks and seven reduce tasks. This figure describes the timing flow of each *Map* task and *Reduce* task. Region *a* represents the actual map function executing time. Region *b* represents the shuffling time and region *c* represents the actual reduce function executing time. The regions between both *b* and *c* represent the sorting time.

offline algorithms do for load balancing. It is what have motivated us to design an online solution to start the *shuffle* subphase as soon as possible, while making the maximum load of the reduce workers as low as possible.

## 13.2.2   Problem Formulation

We consider a general scenario where, during the map phase, the mapper nodes generate many intermediate values (data) with associated keys. Each of these, which we denote as $(k_i, l_i)$, forms a (key, location) pair, where location $l_i$ refers to where the (key, value) pair is stored, and the worker nodes report the pair to the master node.[1] In the rest of this chapter, we will be mainly interested in the issue of processing the *key* attribute of such a pair; for the sake of simplicity, we will usually refrain from discussing the location attribute.

The master node then assigns these pairs to different machines based on the key values. Each such pair must be assigned to one machine for processing, with the additional restriction that pairs with the same key must be assigned to the same machine. The number of pairs assigned to a machine makes up the *load* of the machine. Here, we assume that each machine will have a finishing time that is directly proportional to its

---

[1] Note that the value is *not* being reported, and thus, the information received by the master node for each item will only require a small amount of space.

load, and that the finishing time of the machine with the highest load (which we call the *makespan*) will be the overall finishing time.

The objective then is to minimize the overall finishing time by minimizing the maximum load of all of the machines.[2]

Formally, for the master node, the input is a stream $S = (b_1, b_2, \cdots, b_N)$ of length $N$, where each $b_i$ denotes a (key, location) pair. Let $N'$ denote the number of different keys; we denote $C = \{c_1, c_2, \cdots, c_{N'}\}$ as the universal set of the different keys, with $b_i \in C$ for every $i \in N$. We assume that there are $m$ identical machines numbered $1, \ldots, m$. We denote the load of machine $i$ by $M_i$ (i.e., the number of pairs assigned to machine $i$). Initially, all loads are 0. Our goal is to assign each $b_i$ in $S$ to a machine, so as to obtain

$$\min_{} \underset{i \in \{1, 2, \cdots, m\}}{\mathrm{Max}} M_i$$

such that any two pairs $(k_1, l_1)$ and $(k_2, l_2)$ will be assigned to the same machine if $k_1 = k_2$.

### 13.2.3     Input Models

Basically, we see that this is a streaming problem, where we need to make online decisions to balance the loads. We consider two input models, leading to two computational models. In our first model, we allow arbitrary (possibly adversarial) input, and stipulate that the master node will assign the pairs in a purely online fashion. In our second model, we assume that the input comes from a probability distribution, and, in order to exploit this fact, the master node is allowed to store and process *samples* of its input before starting to assign the samples and the rest of the pairs in an online fashion. In the next two sections, we develop algorithms that follow each of these two models respectively.

### 13.3     A 2-Competitive Fully Online Algorithm

In order to minimize the overall finishing time, it makes sense to start the *shuffle* subphase as soon as possible, with as much of an overlap with the *Map* phase as possible. In this section, we give an *online* algorithm, List-Based Online Scheduling, for assigning the keys to the machines. Our algorithm decides, upon receiving a (key, location) pair, which machine to assign that item to without any knowledge of what other items may be received in the future. We assume that the stream of items can be arbitrary; that is, after our algorithm makes a particular assignment, it can possibly receive the "worst" stream of items for that assignment. Our algorithm will be analyzed for the worst-case scenario: we will compare its effectiveness to that of the best offline algorithm (i.e., one that makes its decisions with knowledge of the entire input).

For assigning items to machines based on their keys, we adopt a Greedy-Balance load balancing approach [641] of assigning unassigned keys to the machine with the smallest load once they come in.

---

[2] Here we make an implicit assumption that each pair represents a workload of unit size, but our algorithm can easily work also for variable integer workload weights.

---

**Algorithm 14** List-Based Online Scheduling

---
1: Read pair $(k_i, l_i)$ from $S$
2: **if** $k_i$ has been assigned to machine $j$ **then**
3:     Assign the pair to the machine $j$
4: **else**
5:     Assign the pair to the machine with the least load
6: **end if**

---

We now show that this algorithm yields an overall finishing time that is at most twice that of the best offline algorithm.

THEOREM 23.  *List-based Online Scheduling has a competitive ratio of 2.*

*Proof.* Let OPT denote the offline optimum makespan, which is the maximum finishing time. Assume that machine $j$ is the machine with the longest finishing time in the optimal offline solution and that $T'$ is the number of pairs read just before the last new key, say $c_j$, is assigned to machine $j$. Obviously, $T'$ must be less than $N$, the total length of the input. Then, at the time that $c_j$ is assigned to machine $j$, $j$ must have had the smallest load, say $L_j$. Thus we have:

$$L_j \leq \frac{T'}{m} < \frac{N}{m} \leq OPT$$

Let $|c_j|$ denote the number of pairs with key $c_j$ in $S$. Then, the finishing time of machine $j$, denoted by $T$, which is also the makespan, is

$$T = L_j + |c_j| \leq OPT + OPT = 2OPT$$

Thus, with our list-based online algorithm, the makespan can achieve a 2-competitive ratio to the offline optimal makespan.  □

## 13.4    A Sampling-Based Semi-Online Algorithm

Our previous algorithm made no assumptions about our advance knowledge of the key frequencies. Clearly, if we had some *a priori* knowledge about these frequencies, we could make the key assignments more efficiently.

In this section, we assume that the pairs are such that their keys are drawn independently from an unknown distribution. In order to exploit this, we compromise on the online nature of our algorithm and start by collecting a small number of input pairs into a *sample* before making any assignments. We then use this sample to estimate the frequencies of the $K$ most frequent keys in this distribution, and use this information later to process our stream in an online fashion. In order to observe the advantages of such a scheme, consider another toy example shown in Figure 13.3. If we can wait for a short period before making any assignments, for instance, collect the first nine keys in the example and assign the frequent keys to the machine with the least load in order of frequency, the maximum load is reduced to 12 from 15.

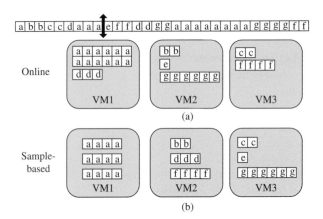

**Figure 13.3**  An illustrative example showing benefits of sampling. The setting is the same as in Figure 13.1. The "Online" and "Sample-based" rows represent the respective result load of each machine in the online and sample-based schedules.

Our algorithm classifies keys into two distinct groups: the $K$ most frequent, called *heavy* keys, and the remaining, less frequent, keys. The intuition is that the heavy keys contribute much more strongly to the finishing time than the other keys, and thus need to be handled more carefully. As a result, our algorithm performs assignments of the keys to the machines differently for the two groups.

We first consider how to identify the heavy keys. Clearly, if one could collect all of the stream $S$, the problem can be solved exactly and easily. However, this would use too much space and delay the assignment process. Instead, we would like to trade-off the size of our samples (and wait before we start making the assignments) with the accuracy of our estimate of the key frequencies. We explore the parameters of this trade-off in the rest of this section.

Our first goal is to show that we can identify the most frequent (heavy) $K$ keys reliably. We first analyze the sample size necessary for this task using techniques from probability and sampling theory. We then move on to an algorithm for assigning the heavy keys as well as the remaining, less frequent, ones.

## 13.4.1   Sample Size

In this section we analyze what our sample size needs to be in order to obtain a reliable estimate of the key frequencies. Estimating probabilities from a given sample is well understood in probability and statistics; our proof below follows standard lines [642].

Let $S'$ denote our sample of size $n$ and $n'$ denote the number of distinct keys in $S'$. For simplicity, we will ignore the fact that $S'$ consists of (key, location) pairs, instead considering it as a stream (or set) of keys.

Let $p_i$ denote the proportion of key $c_i$ in the stream $S$, and let $X_i$ denote the number of occurrences of $c_i$ in the sample $S'$. (It is possible to treat $p_i$ as a probability as well, without any changes to our algorithm.) Then $X_i$ can be regarded as a binomial random variable with $E(X_i) = np_i$ and $\sigma_{X_i} = \sqrt{p_i(1 - p_i)n}$. Provided that $n$ is large (i.e., $X_i \geq 5$ and $n - X_i \geq 5$), the Central Limit Theorem implies that $X_i$ has an approximately normal distribution regardless of the nature of the item distribution.

In order to select the heavy keys, we need an estimate of key probabilities. To this end, we estimate $p_i$ as $\hat{p}_i = X_i/n$, which is the sample fraction of key $i$ in $S'$. Since $\hat{p}_i$ is just $X_i$ multiplied by the constant $1/n$, $\hat{p}_i$ also has an approximately normal distribution. Thus, $E(\hat{p}_i) = p_i$ and $\sigma_{\hat{p}_i} = \sqrt{p_i(1 - p_i)/n}$, and we have the following theorem bounding the size of the sample that we need in order to have a good estimate of the key frequencies.

THEOREM 24. *Given a sample $S'$ of the size of $n = (z_{\alpha/2}/\epsilon)^2$, consider any key $c_i$ with proportion $p_i$, satisfying $X_i \geq 5$ and $n - X_i \geq 5$, and let $\hat{p}_i = X_i/n$. Then, $|p_i - \hat{p}_i| \leq \epsilon\sqrt{(\hat{p}_i)(1 - \hat{p}_i)}$ with a probability of $1 - \alpha$.*

Note that $z_{\alpha/2}$ is a parameter of the normal distribution whose numeric value depends on $\alpha$ and can be obtained from the normal distribution table.

## 13.4.2 Heavy Keys

We first state our notion of the more frequent (i.e., heavy) keys. The following guarantees that we will explore all keys with a length of at least $OPT/2$.

DEFINITION 17. *(Heavy key) A key $i$ is said to be heavy if $\hat{p}_i \geq 1/2m + \epsilon$.*

Note, then, that a key whose length (i.e., the number of times that it occurs in $S$) is greater than $N/2m$ is very likely to be heavy. Then, it is easy to see that there could be up to $2m$ heavy keys.

It is worth noting that one might need to see $O(m)$ samples to sample a particular heavy key. Thus we will need to increase our sample size by an $O(m \log m)$ factor to ensure that we sample the heavy keys and estimate the lengths of each of the heavy keys reliably, resulting in a sample size of $n = O((z_{\alpha/2}/\epsilon)^2 m \log m)$.

## 13.4.3 A Sample-Based Algorithm

We are now ready to present an algorithm for assigning the heavy keys, similar to the sorted-balance algorithm [641] for load balancing. Our sample-based algorithm first collects samples. It then sorts the keys in the sample in a non-increasing order of observed key frequencies, and selects the $K$ most frequent keys. Then, going through this list, it assigns each type of key to the machine with the least current load. For assigning all other keys, we use Algorithm 14.

The following lemma bounds the size of the last key assigned to the machine that ends up with the longest finishing time.

LEMMA 11. *If the makespan obtained by the sample-based algorithm is larger than $OPT + \epsilon N$, with a probability of at least $1 - 2\alpha$, the last key added to the machine has a frequency of at most $(OPT/2N + \epsilon)$.*

---

**Algorithm 15** Sample-Based Algorithm

---

    wait until $n = O((z_{\alpha/2}/\epsilon)^2 m \log m)$ pairs are collected to form a sample
    sort the $K$ most frequent keys in the sample in non-increasing order, say $\hat{p}_1 \geq \hat{p}_2 \geq$
    $\cdots \geq \hat{p}_K$
    going over the sorted list, assign each key $i$ to the machine with the smallest load
    **while** a new pair is received with key $i$ **do**
        **if** the $i$ was previously assigned to machine $j$ **then**
            assign $i$ to machine $j$
        **else**
            assign $i$ to the machine with the smallest load
        **end if**
    **end while**

---

*Proof.* Let $OPT$ be the optimal makespan of the given instance. Divide the keys into two groups: $C_L = \{j \in C : \hat{p}_j N > OPT/2 + \epsilon N\}$ and $C_S = C - C_L$, called large and small keys, respectively. With probability $1 - \alpha$, we have $p_j N > \hat{p}_j N - \epsilon N > OPT/2$ for all keys $j$. Note that there can be at most $m$ large keys, otherwise one could not obtain a finishing time of $OPT$ with two such keys scheduled on the same machine. Since the length of a large key is greater than $OPT/2$, this contradicts the view that $OPT$ is the optimal makespan. It is also obvious that we cannot have any keys with a length of greater than $OPT$ (i.e., no $j$ exists such that $p_j N > OPT$). Thus, if the makespan obtained by the algorithm is greater than $OPT + \epsilon N$, with probability $1 - \alpha$, the last new key that is assigned to the makespan machine must be a small key. Using the union bound, with probability $1 - 2\alpha$, the last type of key assigned to the machine with the longest processing time must have a frequency of at most $OPT/2N + \epsilon$.     □

THEOREM 25. *With a probability of at least $1 - 2\alpha$, $0 < \alpha < 1/2$, the sample-based algorithm will obtain an overall finishing time of at most $\frac{3}{2}OPT + N\epsilon$.*

*Proof.* Assume that machine $j$ has the longest finishing time when the Sample-based algorithm is used for the key assignments. Consider the last key $k$ assigned to $j$. Before this assignment, the load of $j$ was $L_j$, and it must be the least load at that point in time among all of the machines. Thus,

$$L_j \leq \frac{N}{m} \leq OPT.$$

Then, after adding the last key $k$, its finishing time becomes at most

$$L_j + Np_k \leq OPT + OPT/2 + \epsilon N \leq \frac{3}{2}OPT + \epsilon N.$$

Note that $L_j \leq OPT$ is deterministically true. Therefore, the probability of the above can be shown to be at least $1 - 2\alpha$, $0 < \alpha < 1/2$. Thus, with at least $1 - 2\alpha$, $0 < \alpha < 1/2$, our sample-based algorithm can achieve $\frac{3}{2}OPT + \epsilon N$.     □

## 13.5    Performance Evaluation

### 13.5.1    Simulation Setup

We evaluate our algorithms on both a real data trace and synthetic data. The real trace is a public data set [643], which contains the Wikipedia page-to-page link for each term. This trace has a data size of 1 Gigabytes. We generate the synthetic data according to a Zipf distribution with a varying parameter $s$, by which we can control the skew of the data distribution.

In our performance evaluation, we not only simulate the data assignment process, but also the procedures by which the reduce workers pull data from a specific place.

We evaluate both of our two algorithms: the list-based online scheduling algorithm (online) and the sample-based algorithm (sample-based). Recall that the former is faster, while the latter has better accuracy. We compare our algorithms to the current MapReduce algorithm with the default hash function (default). To set a benchmark for our online algorithm, we also compare it to the offline version (offline), which sorts the keys by their frequencies and then assigns them to the machine with the least load so far. The primary evaluation criteria for these algorithms are the maximum load and the shuffle finishing time.

The default values in our evaluation are $z_{\frac{\alpha}{2}} = 1.96$, $\epsilon = 0.005$; the number of records is 1,000,000; and the number of identical machines is 20. The parameter $s$ is set to 1 by default, and we also vary it to examine its impact. Note that we scale the $y$ axis to make the figure visually clean.

### 13.5.2    Results on Synthetic Data

Figure 13.4 shows the maximum load as a function of the number of data records on the synthetic data. Our data record is the key, as mentioned earlier. We increase the number of data records from $0.5 \times 10^6$ to $2.3 \times 10^6$. We compare all four algorithms. We can see that the performance of the default algorithm is much worse than that of the other three. When the number of data records is $2.1 \times 10^6$, the maximum load of the default algorithm is $3.78 \times 10^5$ and our online algorithm has a maximum load of only $2.3 \times 10^5$, an improvement of 39.15%. In addition, we see when the number of data records increases, the maximum loads of all of the algorithms will also increase. This is not surprising, as we need to process more data. However, the loads in our algorithms increase at a much slower pace as compared to the baseline algorithm. Further, we can see that the performance of our online algorithm is almost identical to that of the offline algorithm. This indicates our algorithm not only bounds the worse-case scenario theoretically, but also in practice performs much better than the theoretical bound.

Figure 13.5 shows the maximum load as a function of the reducer number on the synthetic data of all of the four algorithms. We increase the reducer number from 10 to 100. We can see that the default algorithm performs much worse than the other three. In particular, when the reducer number is 20, the maximum load of the default algorithm is $2.2 \times 10^5$, while the maximum load of our online algorithm is only $1.3 \times 10^5$, a

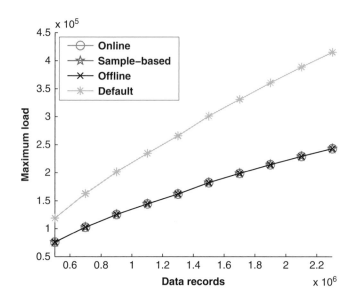

**Figure 13.4**  Maximum load as a function of data record number (synthetic data).

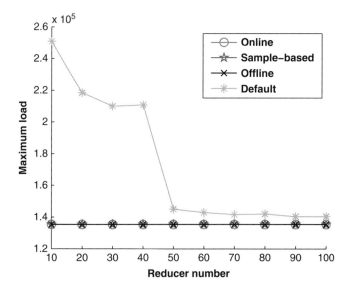

**Figure 13.5**  Maximum load as a function of reducer number (synthetic data).

reduction of 40.90%. It is natural for the maximum load to decrease as the reducer number increases, as the default algorithm shows. However, it is interesting that the other three algorithms do not change much as the reducer number increases. We have checked the data distribution and found that there is one key that is extremely frequent. Our algorithms have indeed identified this key so that the performance of our online algorithm is almost identical to that of the offline algorithm.

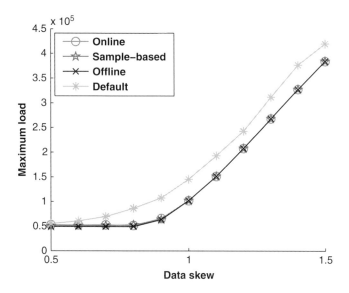

**Figure 13.6** Maximum load as a function of data skew (synthetic data).

Figure 13.6 compares the maximum load of all the four algorithms as a function of the skew on the synthetic data. We set the skew by adjusting parameter $s$ from 0.5 to 1.5 in the Zipf distribution function. The larger $s$ is, the more skew the data has. We can see that when the data are even, the parameter $s$ ranges from 0.5 to 0.7; and the maximum loads of all of the four algorithms are almost the same. When the data distribution becomes increasingly skewed, it is easy to recognize that the default algorithm behaves much worse than all of the other three. Not surprisingly, when the data are highly skewed, the balancing strategies are always better than that of the default algorithm. This is because the maximum load is the frequency of the most frequent key. For the online algorithm and the sample-based algorithm, it is easy to identify the most frequent key without having to see all of the keys.

Figure 13.7 shows the shuffle finishing time as a function of the number of data records on the synthetic data of all of the four algorithms. We still vary the number of records from $0.5 \times 10^6$ to $2.3 \times 10^6$. We can see that the offline algorithm behaves much worse than all of the other three algorithms. In particular, when the number of data records is $2.1 \times 10^6$, the shuffle finishing time of the offline algorithm is 14,000 ms and our online algorithm has a shuffle finishing time of 1,000 ms, an improvement of 14 times. In comparison, the shuffle finishing time of the sample-based algorithm is 5,000 ms, an improvement of almost three times. It is not surprising to see that, as the number of data records grows, the shuffle finishing time increases. However, the loads in our algorithms increase at a much slower pace than the offline algorithm. This shows that the shuffle finishing time of our online algorithm is almost identical to that of the default algorithm, which takes the least amount of time to finish the *shuffle* subphase.

Figure 13.8 shows the shuffle finishing time as a function of the reducer number on the synthetic data with all four algorithms. The result is similar to that depicted in

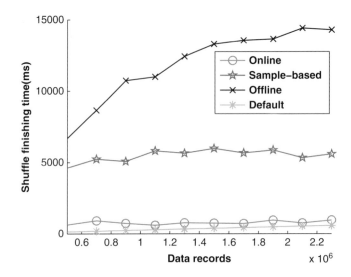

**Figure 13.7** Shuffle finishing time as a function of data record number (synthetic data).

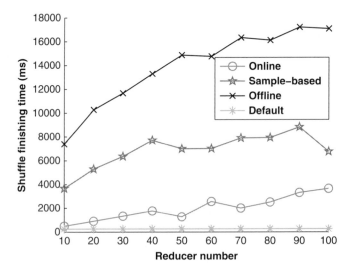

**Figure 13.8** Shuffle finishing time as a function of reducer number (synthetic data).

Figure 13.7. We tested the performance of the online algorithm by increasing the reducer number from 10 to 100. We found the shuffle finishing time of the online algorithm to be good as expected, since the decision to assign a newly incoming key to a specific machine is made earlier in the online algorithm than in the sample-based algorithm, which is earlier than in the offline algorithm. We should note that as the reducer number increases, the shuffle finishing time also increases for all of the algorithms expect for the default algorithm. This is because we need to check whether the reducer machine contains the incoming keys or gets the least load machine in these algorithms.

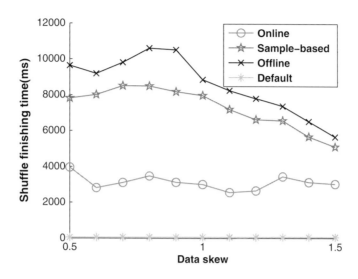

**Figure 13.9**  Shuffle finishing time as a function of data skew (synthetic data).

Figure 13.9 shows the shuffle finishing time of all four algorithms as a function of data skew on synthetic data. We still set the skew parameter from 0.5 to 1.5. Note that the data sets are generated independently for the different skew parameters, with the result that the trend of the shuffle finishing time in each algorithm is not as monotonic as expected. This result is similar to that depicted in Figure 13.8.

### 13.5.3  Results on Real Data

Figure 13.10 shows the maximum load as a function of the number of (key, value) pairs in the real trace data set. We tested the performance by setting the number of (key, value) pairs from $0.5 \times 10^6$ to $2.0 \times 10^6$. Our algorithm always performs better than the baseline algorithm. In particular, when the number of (key, value) pairs is 900,000, the maximum load of the baseline algorithm is $16,993$ unit sizes, while the maximum load of our algorithm is $9,002$ unit size, an improvement of 47.03%. A similar result is found in Figure 13.4 on the synthetic data.

Figure 13.11 shows the maximum load as a function of the number of reducers on both algorithms in the real trace data set. We evaluated the performances by increasing the reducer number from 100 to 190. We can see the performance of the baseline algorithm is much worse than that of our algorithm. It is natural for the maximum loads of all of the algorithms to decrease when the reducer number increases. Further, we can see that the performance of our sample-based algorithm is almost identical to that of the offline algorithm.

Figure 13.12 compares the shuffle finishing time of all four algorithms as a function of the number of data records in the real trace data set, which is similar to the result as shown in Figure 13.8. Our online algorithm, the sample-based algorithm, and the default algorithm achieved better results than that of the offline algorithm. In addition, we found

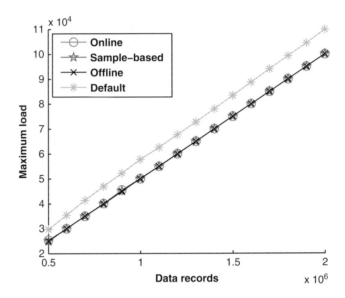

**Figure 13.10**  Maximum load as a function of data record number (real data trace).

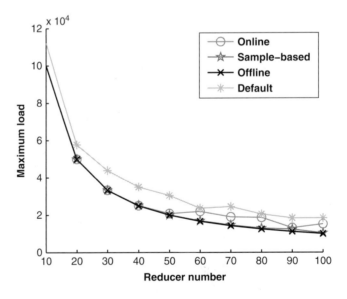

**Figure 13.11**  Maximum load as a function of reducer number (real data trace).

that when the number of data records increases, the maximum loads of all of the algorithms increase. This is because more time is needed to process more data. However, the loads in our algorithms increased in a much slower pace as compared to the offline algorithm. This is because when the data records become larger, a longer processing time is required in the map phase, making the offline algorithm wait much longer.

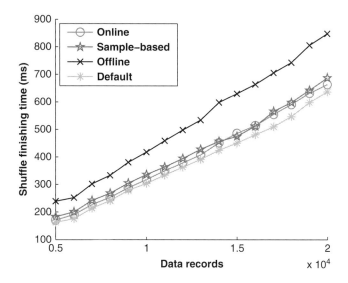

**Figure 13.12** Shuffle finishing time as a function of data record number (real data trace).

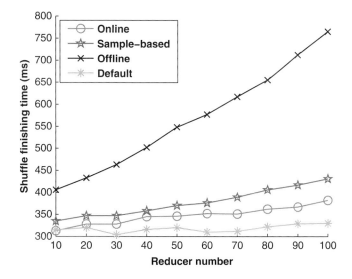

**Figure 13.13** Shuffle finishing time as a function of reducer number (real data trace).

Figure 13.13 shows the shuffle finishing time as a function of the reducer number of all four algorithms. We also found that the shuffle finishing time grows as the number of reduce workers increases. This is illustrative, since it will cost much more time to check whether or not the incoming key was assigned and to find machine with the least load. Interestingly, the shuffle finishing time of the offline algorithm increases much faster than the other three. This gap probably appears because the overall waiting time and the increased cost brought by the increasing number of reducers is very large.

In summary, the performance of our online and sample-based algorithm perform close to that of the offline algorithm in terms of finishing time. The two algorithms consistently perform better than the MapReduce default algorithm from a maximum load point of view. Our algorithms also show a comparable shuffle finishing time to that of the default algorithm, and are better than the offline algorithm in that regard.

## 13.6     Summary

In this chapter, we presented one problem related to big data processing. More specifically, we faced a data skew problem in the shuffling phase of MapReduce. We observed that the data skew problem can be translated into a heavy hitter problem. This is where the sublinear algorithm is applied. Such a linkage can be common when applying sublinear algorithms to real-world applications.

The original sorted-balance algorithm can achieve $\frac{4}{3}OPT$ [641] in the maximum finishing time. Our semi-online algorithm achieved $\frac{3}{2}OPT$ plus some additive error for the $K$ most frequent keys. A further shrinking of this gap could be achieved through advanced algorithm design, or the error could be further reduced. The keys could also be finely classified into different groups according to certain weights, so as to refine the results. In addition, if the distribution of data follows a known distribution (e.g., the Zipf distribution), the parameters can be better estimated, making the process of the identifying the $K$ initial keys much easier and more accurate. It should also be possible to make the additive error smaller as well.

# 14 Massive Data Collection Using Wireless Sensor Networks

## 14.1 Introduction

Wireless sensor networks provide a model in which sensors are deployed in large numbers where traditional wired or wireless networks are not available/appropriate. Their intended uses include terrain monitoring, surveillance, and discovery [644] with applications to geological tasks such as tsunami and earthquake detection, military surveillance, search and rescue operations, building safety surveillance (e.g., for fire detection), and biological systems.

The major difference between sensor networks and traditional networks is that unlike a host computer or a router, a sensor is typically a tightly constrained device. Sensors not only lack long life spans due to their limited battery power, but also possess little computational power and memory storage [645]. As a result of the limited capabilities of individual sensors, one sensor usually can only collect a small amount of data from its environment and carry out a small number of computations. Therefore, a single sensor is generally expected to work in cooperation with other sensors in the network. As a result of this unique structure, a sensor network is typically data-centric and query-based [646]. When a query is made, the network is expected to distribute the query, gather values from individual sensors, and compute a final value. This final value typically represents key properties of the area where the network is deployed; examples of such values are MAXIMUM, MINIMUM, QUANTILE, AVERAGE, and SUM [647, 648] over the individual parameters of the sensors, such as temperature, air or water composition, and so on. As an example, consider a sensor network monitoring the average vibration level around a volcano. Each sensor lying in the crater area submits its own value representing the level of activity in a small area around it. Then the data values are relayed through the network; in this process, they are aggregated so that fewer messages need to be sent. Ultimately, the base station obtains the aggregated information about the area being monitored.

In addition to their distributed nature, most sensor networks are highly redundant to compensate for the low reliability of the sensors and environmental conditions. Since the data from a sensor network is the aggregation of data from individual sensors, the number of sensors in a network has a direct influence on the delay incurred in answering a query. In addition, significant delay is introduced by in-network aggregation [649, 650, 647], since intermediate parent nodes have to wait for the data values collected from their children before they can aggregate them with their own data.

While most of the techniques for fast data gathering focus on delay-energy efficiencies, they lack provable guarantees for the accuracy of the result. In this chapter, we focus on a new approach to address the delay and accuracy challenges. We propose a simple distributed architecture which consists of layers, where each layer contains a subset of the sensor nodes. Each sensor randomly promotes itself into different layers, where large layers contain a superset of the sensors on smaller layers. The key difference between our layered architecture and hierarchical architectures is that each sensor in our network only represents itself and submits its own data for each query, without the need to act as a "head" of a cluster of sensors. In this model a query will be made to a particular layer, resulting in an aggregation tree with fewer hops, and thus smaller delay. Unfortunately, the reduction in delay comes with a price tag; since only a subset of the sensors submit their data, the accuracy of the answer to the query is compromised.

In this chapter, we study the trade-off between the delay and the accuracy with proving bounds. We implement this study in the context of five key properties of the network, MAX, MIN, QUANTILE, AVERAGE, and SUM. Given a user-defined accuracy level, we analyze what the layer of the network should be queried for these properties. We show that different queries do show distinct characteristics which affect the delay/accuracy trade-off. Meanwhile, we present that for certain types of queries such as AVERAGE and SUM, additional statistical information obtained from the history of the environment can help further reduce the number of sensors involved in answering a query. We then investigate the new trade-offs given the additional information.

The algorithm that we propose for our architecture is fully distributed; there is no need for the sensors to keep information about other sensors. Using the fact that each sensor is independent of others, we show how to balance the power consumption at each node by reconstructing the layered structure periodically, which results in an increase in the life expectancy of the whole network.

## 14.1.1     Background and Related Work

Wireless sensor networks have gained tremendous attention from the very beginning of its proposal. There is a wide range of applications; initially it started in the wild and battlefield and recently has moved to urban applications. The key advantages of a wireless sensor network is the wireless communication making it cheap and readily deployable; its self-organizing nature; and its deep penetration to the physical environments. Some surveys on the challenges, techniques, and protocols of wireless sensor networks can be found in [646, 645, 651].

One key objective of wireless sensor network is data collection. Different from data transmission of traditional networking, which is address-based and end-to-end, wireless sensor data collection is data-centric, commonly integrated with in-network aggregation. More specifically, each individual sensor contributes its own data and the sensors of the whole network collectively achieve a certain task. There are many research issues related to sensor data collection; in particular, many focus on trade-off between key parameters, such as query accuracy, delay, and energy usage (or load balancing).

SPIN [652] is the first data-centric protocol which uses flooding; Directed Diffusion [653] is proposed to select more efficient paths. Several related protocols with similar concepts can be found in [654, 655, 656]. As an alternative to flat routing, hierarchical architectures have been proposed for sensor networks; in LEACH [644], heads are selected for clusters of sensors; they periodically obtain data from their clusters. When a query is received, a head reports its most recent data value. In [657], energy is studied in a more refined way in which a secondary parameter such as node proximity or node degree is included. Clustering techniques are studied in a different fashion in several papers, where [658] focuses on non-homogeneously dispersed nodes and [659] considers spanning tree structures. In-network data aggregation is a widely used technique in sensor networks [647, 648, 660]. Ordered properties, for example QUANTILE, are studied in [661]. A recent result in [662] considers power-aware routing and aggregation query processing together, building energy-efficient routing trees explicitly for aggregation queries.

Delay issues in sensor networks are mentioned in [650, 647] where the aggregation introduces high delay since each intermediate node and the source have to wait for the data values from the leaves of the tree, as confirmed by [663]. In [649], where a modified direct diffusion is proposed, a timer is set up for intermediate nodes to flush data back to the source if the data from their children have not been received within a time threshold. In case of energy-delay trade-offs, [663] formulates delay-constraint trees. A new protocol is proposed in [664] for delay-critical applications, in which energy consumption is of secondary importance. In these algorithms, all of the sensors in the network are queried, resulting in $\Theta(N)$ processing time, where $N$ denotes the number of sensors in the network, which incurs long delay. Embedding hierarchical architectures into the network, where a small set of "head" sensors collect data periodically from their children/clusters and submit the results that queried [644, 665, 657] provides a very useful abstraction, where the length of the period is crucial for the trade-off between the freshness of the data and the overhead.

### 14.1.2 Chapter Outline

We present the system architecture in Section 14.2. Section 14.3 contains the theoretical analysis of the trade-off between the accuracy of query answers and the latency of the system. In Section 14.4, we address the energy consumption of our system. Section 14.5 evaluates the performance of our system using simulations. We further present some variations of the architecture in Section 14.6. In Section 14.7, we summarize this application and how the sublinear algorithms are used in this application.

## 14.2 System Architecture

### 14.2.1 Preliminaries

We assume our network has $N$ sensors denoted by $s_1, s_2, \ldots, s_N$ and deployed uniformly in a square area with side length $D$. We assume that a base station acts as an interface between the sensor network and the users, receiving queries which follow a Poisson distribution with the mean interval length $\lambda$.

We embed a layered structure in our network, with $L$ layers, numbered 0, 1, 2, ..., $L-1$. We use $r(l)$ to denote the transmission range used on layer $l$: during a transmission taking place on layer $l$, all sensors on layer $l$ communicate by using $r(l)$ and can reach one another, in one or multiple hops. Let $e(l)$ be the energy needed to transmit for layer $l$. The energy spent by each sensor for a transmission is $e(l) = r(l)^\alpha$ where $2 \leq \alpha \leq 4$ [666]. Initially, each sensor is at energy level $B$, which decreases with each transmission. $R$ denotes the maximum transmission range of the sensors.

## 14.2.2    Network Construction

We would like to impose a layered structure on our sensor network where each sensor will belong to one or more layers. The properties of this structure are as follows:

(a) The *base layer* contains all sensors $s_1, \ldots, s_N$.
(b) The layers are numbered 0 through $L-1$, with the base layer labeled 0.
(c) The sensors on layer $l$ form a subset of those on layer $l-1$, for $1 \leq l \leq L-1$.
(d) The expected number of sensors on each layer drops exponentially with the layer number.

We now expound on how this structure is constructed. In our scheme, each sensor decides, without requiring any communication with the outside world, to which layer(s) it will belong. We assume that all the sensors have access to a value $0 < p < 1$ (this value may be hardwired into the sensors). Let us consider the decision process that a generic sensor $s_i$ undergoes. All sensors, including $s_i$, exist in the base layer 0. Inductively, if $s_i$ exists on some layer $l$, it will, with probability $p$, *promote* itself to layer $l+1$, which means that $s_i$ will exist on layer $l+1$ *in addition to* all the lower layers $l, l-1, \ldots, 0$. If on some layer $l'$, $s_i$ makes the decision not to promote itself to layer $l'+1$, $s_i$ stops the randomized procedure and does not exist on any higher layers. If $s_i$ promotes itself to the highest layer $L-1$, it stops the promotion procedure since no sensor is allowed to exist beyond layer $L-1$. Thus, any sensor will exist on layers $0, 1, \ldots, k$ for some $0 \leq k \leq L-1$. Figure 14.1 shows the architecture of a sensor network with three layers.

Since our construction does not assume the existence of any mechanism of synchronization, it is possible that some sensors may be late in completing their procedure for promoting themselves up the layers. Since the construction scheme works in a distributed fashion, this is not a problem – the late sensor can simply promote itself using probability $p$ and join its related layers in its own time.

Whenever the base station has a query, the query is sent to a specific layer. Those and only those sensors existing on this layer are expected to take place in the communication. This can be achieved by reserving a small field (of $\log(\log N)$ bits) in the transmission packet for the layer number. Once $l$ is specified by the base station (the method for which will be explained later), all of the sensors on layer $l$ communicate using transmission range $r(l)$. The transmission range can be determined by the expected distance of two

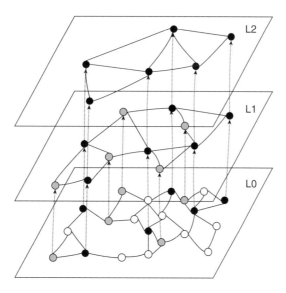

**Figure 14.1** A layered sensor network; a link is presented whenever the sensor nodes in a certain layer are within transmission range.

neighboring sensors on layer $l$, that is, $r(l) = \dfrac{D}{\sqrt{N/2^l}}$, and can be enlarged a little further to ensure higher chances of connectivity.

### 14.2.3 Specifying the Structure of the Layers

Note that in the construction of the layers, the sensors do not promote themselves indefinitely; this is because if there are too few sensors on a layer, the inter-sensor distance will exceed the maximum transmission range $R$. Rather, we "cut off" the top of the layered structure, not allowing more than $L$ layers where $L = \Theta \left( \log \dfrac{N}{(\frac{D}{R}+1)^2} \right)$.

In what follows, we assume that the promotion probability $p = \frac{1}{2}$. We analyze the effect of varying $p$ when appropriate and in our simulations.

### 14.2.4 Data Collection and Aggregation

Given a layered sensor network constructed as above, we now focus on how a query is injected into the network and an answer is returned. We simplify the situation by assuming the same as [667] that the base station is a special node where a query will be initiated. Thus the base station acts as an interface between the sensor network and the user.

When the base station has a query to make, it first determines which layer is to be used for this query. Let this layer be $l$. The base station then broadcasts the query using communication range $r(l)$ for this layer. In this message, the base station specifies the layer number $l$ and the query type (in this chapter, we study MAX, MIN, QUANTILE,

AVERAGE, and SUM). Any sensor on layer $l$ that hears this message will relay information using communication range $r(l)$; those sensors not on layer $l$ will simply ignore this message.

After the query is received by all the sensors on layer $l$, a routing tree rooted at the base station is formed. Each leaf node then collects its data and sends it to its parent, which then aggregates its own data with the data from its children, relaying it up to its parent. Once the root has the aggregated information, it can obtain the answer to the query.

Note that our schemes are independent of the routing and aggregation algorithms used in the network. Our goal is to specify the layer number $l$ which will reduce the number of sensors, as well as the number of messages, used in responding to a query. Once $l$ is determined, the distribution of the query and the collection of the data can be performed in a number of ways, such as that proposed in [662]. In fact, once the layer to be used for a particular query has been identified, the particular routing/aggregation algorithm to be used is transparent to our algorithm.

## 14.3     Evaluation of the Accuracy and the Number of Sensors Queried

In this section, we explore how the accuracy of the answers to queries and the latency relate to the layer which is being queried. In general, we would like to obtain the answers to the queries with as little delay as possible. This delay is a function of the number of sensors whose data are being utilized for a particular query. Thus, the delay is reflected by the layer to which the query is sent. We would also like to get as accurate answers to our queries as possible. When a query utilizes data from all the sensors, the answer is accurate; however, when readings from only a subset of the sensors are used, errors are introduced. We now analyze how these concerns of delay and accuracy relate to the number of sensors queried, and thus to the layer used.

To explore the relation between the accuracy of the answer to a query and the layer $l$ to which the query has been sent, we recall that the current configuration of the layers have been reached by each sensor locally, which decides how many layers it will exist. Due to the randomized nature of this process, the number of sensors on each layer is a random variable. In the next lemma, we investigate which layer must be queried if one would like to have input from at least $k$ sensors.

LEMMA 12. *Let* $l < \log N - \log\left(k + \ln\frac{1}{\delta} + \sqrt{\ln\frac{1}{\delta}(2k + \ln\frac{1}{\delta})}\right)$, *where $k \le$ the expected number of sensors on layer l. Then, the probability that there are fewer than k sensors on layer l is less than $\delta$.*

*Proof.* Define random variable $Y_i$ for $i = 1, \ldots, N$ as follows. $Y_i = 1$ if $s_i$ is promoted to layer $l$; and $Y_i = 0$ otherwise. Clearly, $Y_1, \ldots, Y_N$ are independent. $Pr[Y_i = 1] = 1/2^l$, and $Pr[Y_i = 0] = 1 - 1/2^l$. On layer $l$ there are $Y = \sum_{i=1}^{N} Y_i$ sensors. Therefore, $Pr[Y < k] = Pr[Y < \frac{k}{E[Y]}E[Y]] < e^{-(1-\frac{k}{E[Y]})^2 E[Y]/2}$ by Chernoff's inequality. Since $E[Y] = N/2^l$, to have $e^{-(1-\frac{k}{E[Y]})^2 E[Y]/2} < \delta$, we must have $l < \log N - \log \left( k + \ln\frac{1}{\delta} + \sqrt{\ln\frac{1}{\delta}(2k + \ln\frac{1}{\delta})} \right)$        □

In what follows, we analyze the accuracy and the layer in the context of certain types of queries.

### 14.3.1    MAX and MIN Queries

In general, exact answers to maximum or minimum queries cannot be obtained unless all sensors in the network contribute to the answer, since any missed sensors might contain an arbitrarily high or low data value. The following theorem is immediate.

THEOREM 26. *The queries for MAX and MIN must be sent to the base layer to avoid arbitrarily high error.*

### 14.3.2    QUANTILE Queries

As we cannot obtain an exact quantile by querying a proper subset of the sensors in the network we first introduce the notion of *an approximate of quantile*.

DEFINITION 18. *The $\phi$-quantile ($\phi \in (0, 1]$) of an ordered sequence S is the element whose rank in S is $\phi|S|$.*

DEFINITION 19. *An element of an ordered sequence S is the $\epsilon$-approximation $\phi$-quantile of S if its rank in S is between $(\phi - \epsilon)|S|$ and $(\phi + \epsilon)|S|$.*

The following lemma shows that a large enough subset of $S$ has similar quantiles to $S$.

LEMMA 13. *Let $Q \subseteq S$ be picked at random from the set of subsets of size k of S. Given error bound $\epsilon$ and confidence parameter $\delta$, if $k \geq \frac{\ln\frac{2}{\delta}}{2\epsilon^2}$, with probability at least $1 - \delta$, the $\phi$-quantile of Q is an $\epsilon$-approximation $\phi$-quantile of S.*

*Proof.* The element with rank $\phi|Q|$ in $Q$ [1] does not have rank within $(\phi \pm \epsilon)|S|$ in $S$ if and only if one of the following holds: (a) More than $\phi|Q|$ elements in $Q$ have rank less than $(\phi - \epsilon)|S|$ in $S$, or (b) more than $(1 - \phi)|Q|$ elements in $Q$ have rank greater than $(\phi + \epsilon)|S|$ in $S$.

Since $|Q| = k$, the distribution of elements in $Q$ is identical to the distribution where $k$ elements are picked uniformly at random without replacement from $S$. This is due to the fact that any element of $S$ is as likely to be included in $Q$ as any other element in either scheme, and both schemes include $k$ elements in $Q$.

Since the two distributions mentioned above are identical, we can think of the construction of $Q$ as $k$ random draws without replacement from a *0-1 box* that contains $|S|$ items, of which those with rank less than $(\phi - \epsilon)|S|$ are labeled "1" and the rest are

---

[1]  Wherever rank in a set is mentioned, it should be understood that this rank is over a sequence obtained by sorting the elements of the set.

labeled "0." For $i = 1, \ldots, k$, let $X_i$ be the random variable for the label of the $i$th element in $Q$. Then $X = \sum_{i=1}^{k} X_i$ is the number of elements in $Q$ that have rank less than $(\phi - \epsilon)|S|$ in $S$. Clearly, $E[X] = (\phi - \epsilon)k$. Hence, $Pr[X \geq \phi k] = Pr[X - E[X] \geq \phi k - (\phi - \epsilon)k] = Pr[X - E[X] \geq \epsilon k] = Pr[\frac{X}{k} - E[\frac{X}{k}] \geq \epsilon]$. This is at most $e^{-2\epsilon^2 k}$, by Hoeffding's inequality. Note that Hoeffding's inequality applies to random samples chosen without replacement from a finite population, as shown in Section 6 of Hoeffding's original paper [668], without the need for independence of the samples.

Similarly, it can be shown that the probability that more than $(1 - \phi)|Q|$ elements in $Q$ have rank greater than $(\phi + \epsilon)|S|$ in $S$ is also at most $e^{-2\epsilon^2 k}$. Setting $2e^{-2\epsilon^2 k} \leq \delta$, we have $k \geq \frac{\ln\frac{2}{\delta}}{2\epsilon^2}$. $\square$

We now show which layer we must use for given error and confidence bounds.

THEOREM 27. *If a $\phi$-quantile query is sent to layer $l < \log N - \log \left( \frac{\ln\frac{4}{\delta}}{2\epsilon^2} + \ln\frac{2}{\delta} + \sqrt{\ln\frac{2}{\delta}(2\frac{\ln\frac{4}{\delta}}{2\epsilon^2} + \ln\frac{2}{\delta})} \right)$, then the answer will be the $\epsilon$-approximation $\phi$-quantile of the whole network with probability greater than $(1 - \delta)$.*

*Proof.* By Lemma 12, the probability that layer $l < \log N - \log \left( k + \ln\frac{2}{\delta} + \sqrt{\ln\frac{2}{\delta}(2k + \ln\frac{2}{\delta})} \right)$ has fewer than $k$ sensors is less than $\frac{\delta}{2}$. By Lemma 13, if the number of sensor nodes on layer $l$ is at least $\frac{\ln 2(\frac{2}{\delta})}{2\epsilon^2} = \frac{\ln\frac{4}{\delta}}{2\epsilon^2}$, the probability that the $\phi$-quantile on layer $l$ is $\epsilon$-approximation $\phi$-quantile of the sensor network is at least $1 - \frac{\delta}{2}$. Hence, the answer returned by layer $l < \log N - \log \left( \frac{\ln\frac{4}{\delta}}{2\epsilon^2} + \ln\frac{2}{\delta} + \sqrt{\ln\frac{2}{\delta}(2\frac{\ln\frac{4}{\delta}}{2\epsilon^2} + \ln\frac{2}{\delta})} \right)$ is $\epsilon$-approximation $\phi$-quantile of the sensor network with probability greater than $(1 - \delta)$. $\square$

### 14.3.3    AVERAGE and SUM Queries

#### The Initial Algorithm

AVERAGE queries and SUM queries are correlated queries where the AVERAGE is just SUM/$N$. Since we know the number of the sensors in advance, we just analyze the AVERAGE queries in this section and do not explicitly explain the SUM queries.

We now consider approximating the average data value over the whole sensor network by querying a particular layer. The below lemma indicates that the expectation of the average data value of an arbitrary layer is the same as the average of the base layer, which is the exact average of the sensor network.

LEMMA 14. *Let $a_1, a_2, \ldots, a_N$ be the data values collected by the nodes $s_1, s_2, \ldots, s_N$ of the sensor network. Let $k$ be the number of sensors on layer $l$. Let $X_1, X_2, \ldots, X_k$ be the random variables describing the $k$ data values on layer $l$. Let $\overline{X} = \frac{1}{k}\sum_{i=1}^{k} X_i$. Then $E[\overline{X}] = \frac{1}{N}\sum_{i=1}^{N} a_i$.*

*Proof.* Since each sensor independently promotes itself to layer $l$ with the same probability, $Pr[X_i = a_1] = Pr[X_i = a_2] = \ldots = Pr[X_i = a_N] = \frac{1}{N}$, for $i = 1, 2, \cdots, k$. Then $E[X_i] = \frac{1}{N}(a_1 + a_2 + \cdots + a_N)$. Hence, $E[\overline{X}] = E[\frac{1}{k} \sum_{i=1}^{k} X_i] = \frac{1}{k} \sum_{i=1}^{k} E[X_i] = \frac{1}{k} \frac{k}{N}(a_1 + a_2 + \cdots + a_N) = \frac{1}{N} \sum_{i=1}^{N} a_i$. □

Thus, we propose that the average returned by the queried layer be output as the average of the whole network. The next theorem shows that, given the appropriate layer, this constitutes an $\epsilon$-approximation to the actual average with probability greater than $1 - \delta$.

THEOREM 28. *Let the data value at each sensor come from the interval $[a, b]$, and let $l$ be such that $l < \log N - \log \left( \frac{(b-a)^2 \ln \frac{4}{\delta}}{2\epsilon^2} + \ln \frac{2}{\delta} + \sqrt{\ln \frac{2}{\delta}(2\frac{(b-a)^2 \ln \frac{4}{\delta}}{2\epsilon^2} + \ln \frac{2}{\delta})} \right)$, then the probability that the average of the data values on layer $l$ deviates from the exact average by more than $\epsilon$ is less than $\delta$.*

*Proof.* Let $k$ be the number of sensors on layer $l$. As we have explained in Lemma 3, these $k$ sensors can be considered to be random samples without replacement from all of the $N$ sensors. Let $X_1, X_2, \ldots, X_k$ be the random variables describing the $k$ sensor values on layer $l$, as in Lemma 14. Then $a \leq X_i \leq b$ for $i = 1, 2, \cdots, k$. Let $\overline{X} = \frac{1}{k} \sum_{i=1}^{k} X_i$. By Lemma 14, $E[\overline{X}]$ is the exact average of the sensor network. For any $\epsilon > 0$, $Pr[|\overline{X} - E[\overline{X}]| \geq \epsilon] \leq 2e^{\frac{-2k\epsilon^2}{(b-a)^2}}$, by Hoeffding's inequality. Setting $2e^{\frac{-2k\epsilon^2}{(b-a)^2}} \leq \delta/2$, we have $k \geq \frac{(b-a)^2 \ln \frac{4}{\delta}}{2\epsilon^2}$. By Lemma 12, the probability that layer $l < \log N - \log \left( k + \ln \frac{2}{\delta} + \sqrt{\ln \frac{2}{\delta}(2k + \ln \frac{2}{\delta})} \right)$ has fewer than $k$ sensors is less than $\frac{\delta}{2}$. Thus, if we send an AVERAGE query to layer $l < \log N - \log \left( \frac{(b-a)^2 \ln \frac{4}{\delta}}{2\epsilon^2} + \ln \frac{2}{\delta} + \sqrt{\ln \frac{2}{\delta}(2\frac{(b-a)^2 \ln \frac{4}{\delta}}{2\epsilon^2} + \ln \frac{2}{\delta})} \right)$, the probability that the estimated average deviates from the exact average more than $\epsilon$ is less than $\frac{\delta}{2} + \frac{\delta}{2} = \delta$. □

## Utilizing Statistical Information about the Behavior of Data

If we have access to additional information regarding the characteristics of the objects that the sensor network is monitoring, we can reduce the latency even further. In what follows, we show that the knowledge that the change in data values over time respects a certain distribution (such as the normal distribution) can be used to improve the quality of our estimates.

Assume the change of the data value in one time unit for each single sensor follows a normal distribution with mean $\mu$. For instance, we might know that the electricity consumption is likely to rise around 10 degrees from 2 am to 12 pm, and fall around 6 degrees from 12 pm to 8 pm. Small variations might happen but substantial changes are less likely (see Figures 14.2 and 14.3). The change in the average value also follows a normal distribution since the sum of normal distributions is still a normal distribution with mean and variance equal to the sum of the individual means and variances.

Temperature change from 2 am to 12 pm

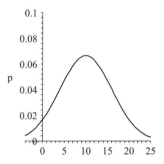

**Figure 14.2**  Electricity consumption changes from 2 am to 12 pm.

Temperature change from 12 pm to 8 pm

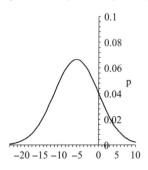

**Figure 14.3**  Electricity consumption changes from 12 pm to 8 pm.

To make use of the statistical information regarding the change in the value, we adopt a history-based approach, where we assume that we know the distribution of the change of the environment to be a normal distribution with mean $\mu$.

The intuition of our strategy is as follows. First we obtain an initial estimate *avg* of the average data value in the network, which, by our computations above, is likely to be close to the true average. After one unit of time, the true average is likely to have changed by some value close to $\mu$. Thus, $avg + \mu$ is likely to be a good estimate for the average for that point in time. However, errors have been introduced into our estimate. One cause for possible error is the fact that only a subset of the sensors have been queried. The other contribution to the error comes from our inability to know the exact change in the data value; we only know from the normal distribution that the change is "likely" to be "around" $\mu$. Since the quantity of the error, as well as its likelihood, increases with each step of this procedure, we need to make sure that our error and confidence bounds remain at acceptable levels.

To ensure low error, we adopt a multi-stage approach to our estimation of the average. In the first stage, which we call the "Query Average" (QueryAvg), we query a relatively large subset of the sensors – more precisely, we query a low enough layer to obtain an error of $\epsilon_1 < \epsilon$ with confidence level $\delta_1 < \delta$. This high guarantee will leave some room for extra error to be incurred in the following stages.

In the following stage (after one time unit has elapsed), which we call the "Test Average" (TestAvg), and subsequent ones, we will query higher layers, thus involving a smaller number of sensors, to see whether the expected change pattern is followed. The result of doing this is that either (a) we will boost the confidence to an acceptable level or (b) we will observe an "anomaly," that is, a deviation from expected behavior, which we will attempt to resolve by querying a lower layer with a larger number of sensors. In case of (a), we will have obtained a fast and acceptable answer by querying only a very small number of sensors. Case (b) on the other hand is, by definition of the normal distribution, an anomaly that will not often happen. In the unlikely event of an "accident" near one of the nodes, in the form of an atypical value, our system will experience a longer query time for the sake of accuracy. In the long run, we will see more "expected" cases and will observe a lower average query time.

Before we go into the specifics of the algorithm, we present an example (Figure 14.4). Suppose $\epsilon = 8$ and $\delta = 0.2$. In the first stage, we see that we get the average data value $avg_1 = 60°F$ by using error bound, say $\epsilon_1 = 2$ and a confidence level, say 0.9, (i.e., error probability $\delta_1 = 0.1$). After ten hours, we expect that the electricity consumption changes to $avg_1 + \mu = 70°F$. Let $\hat{\epsilon}$ and $\hat{\delta}$ denote the confidence interval and confidence level of the normal distribution. To ensure an error within the user-specified bound of 8, then $\hat{\epsilon} \leq 6$, as shown in Figure 14.4, resulting in a confidence level of $\hat{\delta} = 0.8$. Therefore, after this period of time, the overall confidence level is $0.8 \times 0.9 = 0.72$ (i.e., the error probability is larger than the 0.2 specified as acceptable). To boost the confidence level to 0.8, we need to query a few more sensors with an error bound of $\epsilon_2 = 8$ and a much looser error probability of $\delta_2 = \frac{0.2}{1-0.72} = 0.714$. If the returned value of TestAvg is 70, we will return this value. Otherwise, if the returned value falls outside of $70 \pm 8$, this indicates that an anomaly might be present,[2] in which

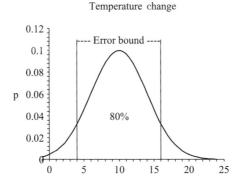

Temperature change

**Figure 14.4** The possible change for electricity consumption after a time unit follows a normal distribution with $\mu = 10$ and $\sigma = 4$. To ensure the ultimate error bound of $\epsilon = 8$, the error bound $\epsilon_n = 6$.

---

[2] Here we use the word anomaly to indicate a situation whose likelihood is small according to the given normal distribution.

---

**Algorithm** QueryAvg $(\epsilon, \delta)$
1 Select $\epsilon_1 < \epsilon$ and $\delta_1 < \delta$.
2 $l_1 = \log N - \log \left( \dfrac{(b-a)^2 \ln \frac{4}{\delta_1}}{2\epsilon_1^2} + \ln \frac{2}{\delta_1} + \sqrt{\ln \frac{2}{\delta_1} (2 \dfrac{(b-a)^2 \ln \frac{4}{\delta_1}}{2\epsilon_1^2} + \ln \frac{2}{\delta_1})} \right)$
3 $avg_1 =$ Query$(l_1)$.
4 **return** $avg_1$

**Figure 14.5**  Algorithm QueryAvg.

---

**Algorithm** TestAvg $(i, \epsilon, \delta, \mu, \sigma, avg_1, \epsilon_1, \delta_1)$
1 $\hat{\epsilon} = \epsilon - \epsilon_1$.
2 Calculate $\hat{\delta} = Pr(\mu - \epsilon_n < X < \mu + \epsilon_n)$ by Normal Distribution.
3 **if** $1 - \hat{\delta} \times (1 - \delta_1) < \delta$,
4     $avg_i = avg_1 + \mu$, **return** $avg_i$
5 **else**
6     $k = 0, \delta_i = \dfrac{\delta}{1 - \hat{\delta} \times (1 - \delta_1)}$
    **repeat**
7       $\epsilon_k = \epsilon - k, \epsilon_i = \epsilon_k$
8       $l_i = \log N - \log \left( \dfrac{(b-a)^2 \ln \frac{4}{\delta_i}}{2\epsilon_i^2} + \ln \frac{2}{\delta_i} + \sqrt{\ln \frac{2}{\delta_i} (2 \dfrac{(b-a)^2 \ln \frac{4}{\delta_i}}{2\epsilon_i^2} + \ln \frac{2}{\delta_i})} \right)$
9       **if** (Query$(l_i) \leq avg_1 + \mu + k$)
              **and** (Query$(l_i) \geq avg_1 + \mu - k$)
10      **then** $avg_i = avg_1 + \mu$, **return** $avg_i$
11      **else** increase $k$
12      **if** $k \geq \epsilon$, Goto QueryAvg

**Figure 14.6**  Algorithm TestAvg.

---

case we perform QueryAvg to determine the new electricity consumption value. If the returned value falls within $70 \pm 8$, to ensure the error bound, we perform TestAvg with a more stringent error bound until an anomaly is found or the new average value is confirmed.

Below we explain our algorithm in higher detail and analyze its properties mathematically. Figure 14.5 shows algorithm QueryAvg. It takes as input the error and confidence parameters $\epsilon, \delta$. We assume that Query$(l)$ returns the average data value for sensors on layer $l$.

Our next algorithm (Figure 14.6) shows how to perform TestAvg given QueryAvg. It takes as input the error and confidence parameters $\epsilon, \delta$, as well as the mean $\mu$ and standard deviation $\sigma$ of the distribution of the change of the data value. Also it takes $avg_1$ which is the average obtained from QueryAvg and the round number $i$.

In line 2 of algorithm TestAvg, we calculate the probability that the change will fall within interval $\hat{\epsilon}$. In lines 1–5, if QueryAvg has already guaranteed the error probability, we do not perform any further queries. This might occur when the number of sensors queried in QueryAvg is large enough. Line 12 displays the threshold that we should perform the query again.

THEOREM 29. *Assume the data value collected by each sensor is bounded by $[a, b]$ and the change in the average of the values at all sensors follows a normal distribution with mean $\mu$ and standard deviation $\sigma$. The probability that algorithm QueryAvg and TestAvg will deviate from the exact average by more than $\epsilon$ is less than $\delta$.*

*Proof.* Let us first consider QueryAvg. Choose any $\epsilon_1, \delta_1$ such that $\epsilon_1 < \epsilon$ and $\delta_1 < \delta$. By Theorem 28, we can obtain the desired accuracy by sending queries to any layer $l_1$ where $l_1 < \log N - \log \left( \frac{(b-a)^2 \ln \frac{4}{\delta_1}}{2\epsilon_1^2} + \ln \frac{2}{\delta_1} + \sqrt{\ln \frac{2}{\delta_1} (2 \frac{(b-a)^2 \ln \frac{4}{\delta_1}}{2\epsilon_1^2} + \ln \frac{2}{\delta_1})} \right)$.

For TestAvg, let $\overline{Y}_i$ denote the average value over all the sensors at round $i$. Define $\Delta_i = \overline{Y}_i - \overline{Y}_1$. We know *a priori* that the probability distribution for each $\Delta_i$ is a normal distribution with mean $\mu_i$ and variance $\sigma_i^2$.

In round $i$, $\hat{\epsilon} = \epsilon - \epsilon_1$ is the confidence interval for normal distribution, hence, $\hat{\delta}$ is probability that the change in the value of the average will not exceed $\hat{\epsilon}$.

Therefore, with probability $1 - \hat{\delta} \times (1 - \delta_1)$, we can guarantee an error bound of $\epsilon_1 + \hat{\epsilon} < \epsilon$. If $1 - \hat{\delta} \times (1 - \delta_1) < \delta$, then our query satisfies both bounds $\epsilon$ and $\delta$, and we can compute the value to be returned from the value in the previous round and the expected change. Otherwise, we choose $\delta_i = \frac{\delta}{1 - \hat{\delta} \times (1 - \delta_1)}$ which ensures that the confidence error $\delta$ will be bounded in the $i$th round. For error bound $\epsilon_i$ in the $i$th round, since we do not know the returned value, we can use all $\epsilon_i = \epsilon_k$ where $\epsilon \geq \epsilon_k \geq 0$ as long as the returned value $avg_i \pm (\epsilon_k)$ will be bounded by $(avg_1 + \mu) \pm \epsilon$. If that happens, the change is confirmed. Otherwise, to bound $\epsilon$ and $\delta$, a new QueryAvg must be performed. In our algorithm, we reduce $\epsilon_i$ iteratively from $\epsilon$ to 0, and use $\epsilon_i$ and $\delta_i$ to query layer $l_i < \log N - \log \left( \frac{(b-a)^2 \ln \frac{4}{\delta_i}}{2\epsilon_i^2} + \ln \frac{2}{\delta_i} + \sqrt{\ln \frac{2}{\delta_i} (2 \frac{(b-a)^2 \ln \frac{4}{\delta_i}}{2\epsilon_i^2} + \ln \frac{2}{\delta_i})} \right)$ so that the number of sensors in TestAvg will increase little by little and stop as early as possible. □

## 14.3.4   Effect of the Promotion Probability $p$

The promotion probability $p$ will only affect the logarithmic base of the system. Therefore, we only give out comparable theorems without detailing the proofs.

THEOREM 30. *(w.r.t. Theorem 27) To attain the $\phi$-quantile of the sensor readings with error bound $\epsilon$ and confidence level $\delta$, the query must be sent to layer $l < \log_{\frac{1}{p}} N - \log_{\frac{1}{p}} \left( \frac{\ln \frac{4}{\delta}}{2\epsilon^2} + \ln \frac{2}{\delta} + \sqrt{\ln \frac{2}{\delta} (2 \frac{\ln \frac{4}{\delta}}{2\epsilon^2} + \ln \frac{2}{\delta})} \right)$, then the answer will be the $\epsilon$-approximation $\phi$-quantile of the whole network with probability greater than $(1 - \delta)$.*

THEOREM 31. *(w.r.t. Theorem 28) Let the data value at each sensor come from the interval $[a, b]$, and let $l$ be such that $l < \log_{\frac{1}{p}} N - \log_{\frac{1}{p}} \left( \frac{(b-a)^2 \ln \frac{4}{\delta}}{2\epsilon^2} + \ln \frac{2}{\delta} + \sqrt{\ln \frac{2}{\delta} (2 \frac{(b-a)^2 \ln \frac{4}{\delta}}{2\epsilon^2} + \ln \frac{2}{\delta})} \right)$, then the probability that the average of the data values on layer $l$ deviates from the exact average by more than $\epsilon$ is less than $\delta$.*

The algorithms of QueryAvg and TestAvg can be adjusted accordingly and the corresponding theorems also follow.

## 14.4    Energy Consumption

It can be readily observed that in our system higher-layer sensors will be transmitting at longer ranges than their lower-layer counterparts. Given that any high-layer sensor is also presented in all the lower layers, if nothing is done to balance out the energy consumption, the higher-layer sensors may get depleted much faster than the lower-layer ones. To balance out the energy consumption, our system reconstructs the layered network periodically by deciding each layer from scratch, so that the top-layer sensors change over time. An appropriate timing scheme for the reconstruction will lead to relatively uniform energy consumption across the sensors in the network. Note that the frequency of reconstructions has no expected effect on accuracy, since we are as likely to be stuck with a "good" sample of sensors (in which case reconstruction is likely to give us a worse sample) as with a "bad" one. Given the above and the overhead of building a new aggregation tree for each new construction, we are interested in infrequently repeating this procedure for making the energy consumption more even across sensors.

Let the lifetime of the network be the time between its initial construction and the first time that a sensor runs out of power [657]. We investigate the trade-off between the timing of the reconstructions and the expected lifetime of our system in our simulations. In this section, we analyze our system assuming that each sensor has sufficient power to let it undergo several reconstructions, and that we run reconstructions enough times. Ideally, we have a totally symmetric scenario where the service that each sensor has performed on each layer is identical across sensors. Since the layers are chosen in a randomized fashion, given a large enough number of reconstructions, one will see that most sensors have served on different layers.

The energy spent by each sensor for a query directly depends on the distance between the sensor and its neighbors. Recall that since there are an expected $(N/2^l)$ nodes on layer $l$, the transmission range is set to be $r(l) = \dfrac{D}{\sqrt{N/2^l}}$. Therefore, the energy spent by each sensor for each query on layer $l$ is $e(l) = \left(\dfrac{D}{\sqrt{N/2^l}}\right)^{\alpha}$, which is what we will use below to estimate the overall system lifetime.

### 14.4.1    Overall Lifetime of the System

In this section, we assume that the queries are uniformly distributed across different layers due to the error bounds and confidence levels coming independently from the users.

We now present a theorem which estimates the expected lifetime of our system depending on the network parameters.

THEOREM 32. *In a setting where each level is equally likely to be queried, the expected lifetime of our system is* $E(t) = \dfrac{BL(\sqrt{N})^{\alpha}(1-(\sqrt{2})^{\alpha-2})}{\lambda D^{\alpha}(1-(\sqrt{2})^{L(\alpha-2)})}$

*Proof.* We assume that each layer has the same probability $\frac{1}{L}$ of being queried. The probability that a sensor exists on layer $l$ is $\frac{1}{2^l}$, therefore the energy consumption for this sensor is $\sum_{l=0}^{L-1} \frac{1}{2^l} e(l) \frac{1}{L}$. Let the life expectancy of this sensor be $t$. Recall that $B$ is the battery power, $\lambda$ is the incoming query interval following (assumed) Poisson distribution, and $e(l) = \left(\frac{D}{\sqrt{N/2^l}}\right)^\alpha$. We have $\sum_{l=0}^{L-1} \frac{1}{2^l} e(l) \frac{1}{L} \lambda t = B$. Therefore, the expected lifetime of the system is $E(t) = \frac{BL(\sqrt{N})^\alpha (1-(\sqrt{2})^{\alpha-2})}{\lambda D^\alpha (1-(\sqrt{2})^{L(\alpha-2)})}$ □

## 14.5 Evaluation Results

We use numerical simulations to test the performance of our system, as well as to observe the effects of the parameters of the algorithm and the re-election time on the performance.

### 14.5.1 System Settings

We set the default number of sensors to be $N = 10,000$. The default promotion probability is $p = \frac{1}{2}$. We focus on QUANTILE and AVERAGE queries in our simulations.

### 14.5.2 Layers vs. Accuracy

We first evaluate the trade-off between the layer answering a query and parameters relating to the quality of the answer to the query.

**QUANTILE Queries**

We first study QUANTILE queries in Figure 14.7. It can be seen clearly that, as $\epsilon$ and $\delta$ increase, the layer that the query should be sent to also increases, as confirmed by our computations. Here it can be observed that even though the layer number monotonically increases with both parameters, $\epsilon$ has more impact on it than $\delta$ as can be seen from Figure 14.7 (a) and (b). This is because the confidence parameter $\delta$ can easily be improved using standard boosting techniques from probability theory and randomized algorithms. In fact, repeating the algorithm $O(\log k)$ times and returning the median answer will improve $\delta$ to $\delta/k$, since the probability of getting an incorrect answer $(\log k)/2$ times is at most $\delta^{\log k}$, which is $O(\delta/k)$. On the other hand, to reduce $\epsilon$ by a constant factor $k$, $O(k)$ repetitions of the experiment are needed. Figure 14.7 (c) shows the trade-off between $p$ and the layer number. As $p$ increases, the variation in the layer number for the same query is more obvious. This is because there are fewer layers for smaller $p$ and the choice of layer is more coarse-grained than for larger $p$.

**AVERAGE Queries**

We show the trade-off of the layer number with $\delta$, $\epsilon$ and $p$ for average queries in Figure 14.8. We observe the same effect as with the QUANTILE queries (i.e., $\epsilon$ has a larger

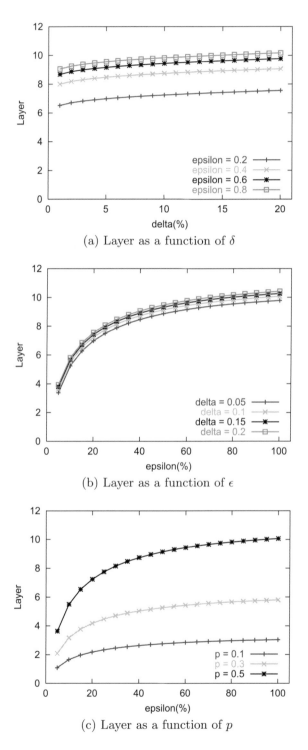

(a) Layer as a function of $\delta$

(b) Layer as a function of $\epsilon$

(c) Layer as a function of $p$

**Figure 14.7** QUANTILE queries, the relation among the layer as a function of the confident parameter $\delta$, the error parameter $\epsilon$, and the promotion probability $p$.

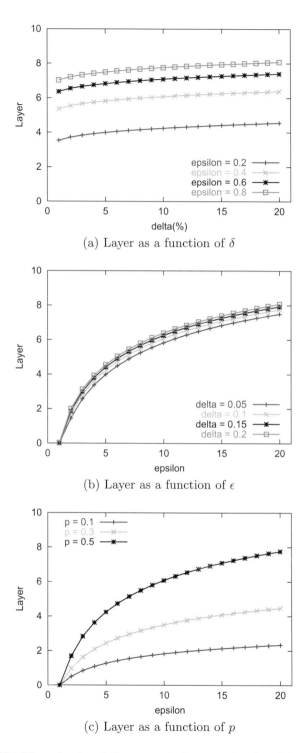

(a) Layer as a function of $\delta$

(b) Layer as a function of $\epsilon$

(c) Layer as a function of $p$

**Figure 14.8** AVERAGE queries, the relation among the layer as a function of the confident parameter $\delta$, the error parameter $\epsilon$, and the promotion probability $p$.

impact than $\delta$), for the same reason. This gives us hints for building test queries as we have additional statistical information.

To investigate the question of how to choose the parameters introduced in our algorithms, QueryAvg and TestAvg, we fix the following parameters and vary the others. The upper and lower bounds of the data values are $a = 20$ and $b = 100$. We set the user-defined error bound and confidence parameter to be $\epsilon = 15.1$ and $\delta = 25\%$, respectively. The mean and standard deviation for the normal distribution are set to $\mu = 20$ and $\sigma = 8$. We choose $\epsilon_1$ in a range of [6, 15], and two different $\delta_1$, 0.05 and 0.2, respectively, to compare this two-phase algorithm with no test at all (i.e., using $\epsilon$ and $\delta$ directly).

Note that, when $p = 0.5$, the expected number of sensors in each layer increases by 2 as we go down each layer. To reduce overall delay, every query performs on some layer $i - 1$ rather than layer $i$ must be compensated by two or more runs of TestAvg performed on layer $i + 1$ rather than layer $i$, or one or more on $i + 2$ rather than $i + 1$, etc. Thus, the combination of QueryAvg and TestAvg is more profitable when the change in the data is highly predictable. Thus, QueryAvg and TestAvg can be used for emergency monitoring applications in stable environments whereas QueryAvg alone can be used in applications of data acquisition in changing environments.

Figure 14.9 (a) shows the effect of $\epsilon_1$ and $\delta_1$ for QueryAvg. Figure 14.9 (b) shows the effect of $\epsilon_1$ and $\delta_1$ for TestAvg. In our simulations, we see that regardless of the choice of $\epsilon_1$ and $\delta_1$, the QueryAvg procedure will query a larger number of sensors and the TestAvg procedure will query a smaller number of sensors, comparing to using no test at all. In QueryAvg, however, varying $\delta_1$ has relatively larger effect. In addition, we note that $\epsilon_1$ changes more for QueryAvg than for TestAvg. As a result, we suggest choosing a larger $\epsilon_1$ and $\delta_1$ for QueryAvg, so that we can save more for QueryAvg and pay less in TestAvg.

The effect of TestAvg also depends on the readings of the second phase, (or $i$th query of TestAvg from the QueryAvg). If the $i$th query is far from the expected value, then we have to narrow down $\epsilon_i$ to investigate the sensor network with higher accuracy. For example, let $r$ denote the output of TestAvg, in our algorithm, if the answer of $r \pm \epsilon_i$ is out of the range of $avg_1 + \mu \pm \epsilon$, then we set $\epsilon_i$ more stringently. This, however, leads to possibly involving a larger number of sensors for this query. In theory, we stop at $\epsilon_i = 0$; however, in practice, we can stop earlier and move to QueryAvg for efficiency reasons. We observe this effect in Figure 14.9 (c) where we set $\epsilon_1 = 13$ and $\delta_1 = 0.2$. We see that in this setting, there are reasonable chances that we will stop with savings since when the results from TestAvg fall within $avg_1 + \mu \pm 5$, fewer sensors are queried. This also confirms that $\epsilon$ has a greater impact than $\delta$ for our architecture.

Next, we study the effect of $\sigma$, the standard deviation of the normal distribution. $\sigma$ represents the rate of change in the data. One can see in Figure 14.10 that $\sigma$ has a big influence on efficiency. The discontinuity of the lines in Figure 14.10 (a) and (b) indicate that QueryAvg has tested more sensors than required, making some of the following rounds of TestAvg unnecessary.

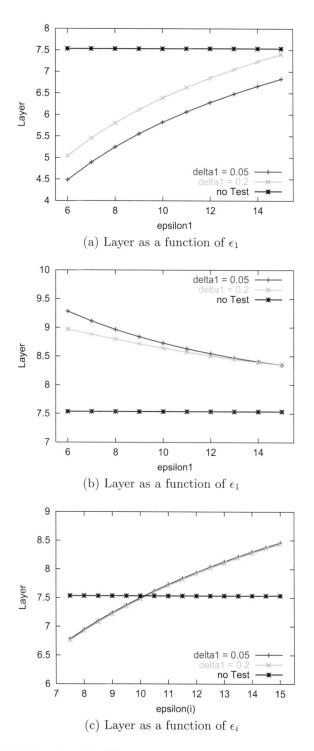

(a) Layer as a function of $\epsilon_1$

(b) Layer as a function of $\epsilon_1$

(c) Layer as a function of $\epsilon_i$

**Figure 14.9** AVERAGE queries with different $\delta_1$ values; no test denotes using the original $\delta$ and $\epsilon$, (a) QueryAvg, (b) TestAvg, and (c) TestAvg based on different readings.

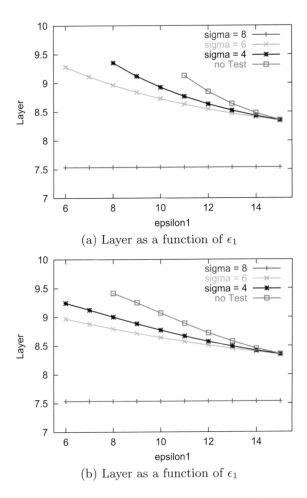

(a) Layer as a function of $\epsilon_1$

(b) Layer as a function of $\epsilon_1$

**Figure 14.10**  Effect of the standard deviation ($\sigma$) of the normal distribution.

## 14.6     Practical Variations of the Architecture

We have given the theoretical analysis of a layer architecture. We now discuss some practical concerns in this section.

In our numerical results, we observe that different $\epsilon$ and $\delta$ values may match to same layers. This is because, in our architecture, the promotion probability $p$ is predefined and the layer architecture is constructed with no consideration for different queries. This makes the structure inflexible to different user parameters (e.g., $\epsilon$ and $\delta$). We call this architecture pro-activate architecture. To solve this problem within this architecture, larger $p$ can be selected, resulting in more layers. Thus, the architecture can deal with the queries in a more refined way. The promotion may also have determined reactively where the sensors determined the $p$ and construct a layer after receiving the query. The query involves two phases, broadcasting the query and collecting the data value. In this situation, the base station determines the transmission range using $\epsilon$ and $\delta$, calculated in the same way as in Section 14.3. We do not need to use the floor function, however. The transmission is sent to the root sensor and the root sensor will use this transmission range to broadcast the message. Every sensor receiving this message will relay this information using the same transmission range. Notice that by using this larger transmission range (compared to the base-layer range), the delay of the broadcasting phase will be controlled. All sensors, after receiving this query, will calculate the promotion probability $p$ which is inversely proportion to the transmission range. Sensors that succeed in promotion will then send the data value back to the root.

In the above settings, all nodes have to relay the message using a large transmission range in the broadcasting session. The energy consumption is thus much higher than the pro-activate architecture. Several optimizations can be applied for saving energy consumption. For example, since, when a sensor sends a message, all neighbor sensors within the transmission range will hear this message [651], only the neighbors that are "farther" away to this sensor will relay the message. We do not elaborate on the broadcasting and collecting techniques as these are not the focus of this chapter. Detailed studies in this area can be found in [653, 647, 657, 663].

## 14.7 Summary

In this chapter, we presented one problem related to wireless sensor data collection. We faced a trade-off between delay-sensitive requirement and data accuracy. We observed that the delay issue can be translated into reducing the number of data collected. We developed a layered architecture, and we discussed in details about how the layered architecture should be constructed and how the queries should be handled. We analyzed the architecture in terms of accuracy and latency. We applied sublinear algorithms to derive provable bounds. In addition, we used side information to further enhance the sublinear algorithms.

# Bibliography

[1] https://en.wikipedia.org/wiki/Andrew_File_System.

[2] https://en.wikipedia.org/wiki/Network_File_System.

[3] S. Ghemawat, H. Gobioff, and S.-T. Leung, "The google file system," in *Proceedings of USENIX Symposium on Operating System Principles*, Lake George, NY, USA, October 2003.

[4] G. Malewicz, M. H. Austern, A. J. C. Bik, J. C. Dehnert, I. Horn, N. Leiser, and G. Czajkowski, "Pregel: A system for large-scale graph processing," in *Proceedings of ACM Special Interest Group on Management of Data Conference*, Indianapolis, IN, USA, June 2010.

[5] Y. Low, J. Gonzalez, A. Kyrola, D. Bickson, C. Guestrin, and J. M. Hellerstein, "Graphlab: A new framework for parallel machine learning," in *Proceedings of Conference on Uncertainty in Artificial Intelligence*, Catalina Island, USA, July 2010.

[6] ——, "Distributed graphlab: A framework for machine learning and data mining in the cloud," in *Proceedings of International Conference on Very Large Data Bases*, Istanbul, Turkey, August 2012.

[7] X. Wang, M. Hong, S. Ma, and Z.-Q. Luo, "Solving multiple-block separable convex minimization problems using two-block alternating direction method of multipliers," *Pacific Journal on Optimization*, vol. 11, no. 4, pp. 645–667, 2015.

[8] J. Tan, S. Meng, X. Meng, and L. Zhang, "Improving reduce task data locality for sequential MapReduce jobs," in *Proceedings of IEEE International Conference on Computer Communications*, Turin, Italy, April 2013.

[9] M. Zaharia, K. Elmeleegy, D. Borthakur, S. Shenker, J. S. Sarma, and I. Stoica, "Delay scheduling: A simple technique for achieving locality and fairness in cluster scheduling," in *Proceedings of Eurosys Conference*, Paris, France, April 2010.

[10] B. W. Lampson, "A scheduling philosophy for multiprocessing systems," *Communication of the ACM*, vol. 11, no. 5, pp. 346–360, May 1968.

[11] A. S. Schulz, "Polytopes and scheduling," PhD dissertation, Technical University of Berlin, 1996.

[12] H. Kasahara and S. Narita, "Practical multiprocessor scheduling algorithms for efficient parallel processing," *IEEE Transactions on Computers*, vol. 33, no. 11, pp. 1023–1029, November 1984.

[13] T. L. Adam, K. M. Chandy, and J. R. Dickson, "A comparison of list schedules for parallel processing systems," *Communications of the ACM*, vol. 17, no. 12, pp. 685–690, December 1974.

[14] M. Queyranne and A. S. Schulz, "Approximation bounds for a general class of precedence constrained parallel machine scheduling problems," *SIAM Journal on Computing*, vol. 35, no. 5, pp. 1241–1253, March 2006.

[15] H. Chang, M. Kodialam, R. R. Kompella, T. V. Lakshman, M. Lee, and S. Mukherjee, "Scheduling in mapreduce-like systems for fast completion time," in *Proceedings of IEEE International Conference on Communications*, Shanghai, China, April 2011.

[16] F. Chen, M. Kodialam, and T. V. Lakshman, "Joint scheduling of processing and shuffle phases in mapreduce systems," in *Proceedings of IEEE International Conference on Communications*, Orlando, FL, USA, March 2012.

[17] Y. Yuan, D. Wang, and J. Liu, "Joint scheduling of mapreduce jobs with servers: Performance bounds and experiments," in *Proceedings of IEEE International Conference on Computer Communications*, Toronto, Canada, April 2014.

[18] J. Dean and S. Ghemawat, "Mapreduce: Simplified data processing on large clusters," in *Proceedings of USENIX Operating System Design and Implementation*, San Francisco, CA, USA, December 2004.

[19] J. Lin, "The curse of zipf and limits to parallelization: A look at the stragglers problem in mapreduce," in *The 7th Workshop on Large-Scale Distributed Systems for Information Retrieval*, Boston, MA, USA, July 2009.

[20] B. Gufler, N. Augsten, A. Reiser, and A. Kemper, "Load balancing in MapReduce based on scalable cardinality estimates," in *Proceedings of IEEE International Conference on Data Engineering*, Washington, DC, USA, April 2012.

[21] Y. Kwon, M. Balazinska, B. Howe, and J. Rolia, "A study of skew in MapReduce applications," in *Open Cirrus Summit 2011*, Atlanta, GA, USA, October 2011.

[22] Y. Kwon, M. Balazinska, B. Howe, and J. Rolia, "Skewtune: Mitigating skew in MapReduce applications," in *Proceedings of ACM Special Interest Group on Management of Data Conference*, Scottsdale, AZ, USA, May 2012.

[23] B. Wang, J. Jiang, and G. Yang, "Actcap: Accelerating MapReduce on heterogeneous clusters with capability-aware data placement," in *Proceedings of IEEE International Conference on Computer Communications*, Hong Kong, China, April 2015.

[24] Y. Le, J. Liu, F. Ergun, and D. Wang, "Online load balancing for MapReduce with skewed data input," in *Proceedings of IEEE International Conference on Computer Communications*, Toronto, Canada, April 2014.

[25] S. Ibrahim, H. Jin, L. Lu, S. Wu, B. He, and L. Qi, "Leen: Locality/fairness-aware key partitioning for mapreduce in the cloud," in *IEEE Second International Conference on Cloud Computing Technology and Science*, Washington, DC, USA, November 2010.

[26] B. Gufler, N. Augsten, A. Reiser, and A. Kemper, "Handling data skew in MapReduce," in *Proceedings of International Conference on Cloud Computing and Services Science*, Noordwijkerhout, the Netherlands, May 2011.

[27] K. Ousterhout, A. Panda, J. Rosen, S. Venkataraman, R. Xin, S. Ratnasamy, S. Shenker, and I. Stoica, "The case for tiny tasks in compute clusters," in *Proceedings of USENIX Hot Topics in Operating Systems*, Santa Ana Pueblo, NM, USA, May 2013.

[28] K. Ousterhout, P. Wendell, M. Zaharia, and I. Stoica, "Sparrow: Distributed, low latency scheduling," in *Proceedings of USENIX Symposium on Operating System Principles*, Farmington, PA, USA, November 2013.

[29] P. Delgado, F. Dinu, A.-M. Kermarrec, and W. Zwaenepoel, "Hawk: Hybrid datacenter scheduling," in *Proceedings of USENIX Annual Technical Conference*, Santa Clara, CA, USA, July 2015.

[30] Y. Yuan, H. Wang, D. Wang, and J. Liu, "On interference-aware provisioning for cloud-based big data processing," in *Proceedings of IEEE/ACM Symposium on Quality of Services*, Montreal, Canada, June 2013.

[31]  D. Xie, N. Ding, Y. C. Hu, and R. Kompella, "The only constant is change: Incorporating time-varying network reservations in data centers," in *Proceedings of the ACM Special Interest Group on Data Communications Annual Conference*, Helsinki, Finland, August 2012.

[32]  R. Shea, F. Wang, H. Wang, and J. Liu, "A deep investigation into network performance in virtual machine based cloud environments," in *Proceedings of IEEE International Conference on Computer Communications*, Toronto, Canada, April 2014.

[33]  C. Reiss, A. Tumanov, G. R. Ganger, R. H. Katz, and M. A. Kozuch, "Heterogeneity and dynamicity of clouds at scale: Google trace analysis," in *Proceedings of ACM Symposium on Cloud Computing*, San Jose, CA, USA, October 2012.

[34]  X. Ling, Y. Yuan, D. Wang, and J. Yang, "Tetris: Optimizing cloud resource usage unbalance with elastic VM," in *Proceedings of IEEE/ACM International Symposium on Quality of Services*, Beijing, China, June 2016.

[35]  B. Hindman, A. Konwinski, M. Zaharia, A. Ghodsi, A. D. Joseph, R. Katz, S. Shenker, and I. Stoica, "Mesos: A platform for fine-grained resource sharing in the data center," in *Proceedings of USENIX Networking System Design and Implementation*, Boston, MA, USA, March 2011.

[36]  M. Schwarzkopf, A. Konwinski, M. Abd-El-Malek, and J. Wilkes, "Omega: Flexible, scalable schedulers for large compute clusters," in *Proceedings of ACM Eurosys Conference*, Prague, Czech Republic, April 2013.

[37]  A. Verma, L. Pedrosa, M. Korupolu, D. Oppenheimer, E. Tune, and J. Wilkes, "Large-scale cluster management at Google with Borg," in *Proceedings of ACM Eurosys Conference*, Bordeaux, France, April 2015.

[38]  E. Boutin, J. Ekanayake, W. Lin, B. Shi, J. Zhou, Z. Qian, M. Wu, and L. Zhou, "Apollo: Scalable and coordinated scheduling for cloud-scale computing," in *Proceedings of USENIX Annual Technical Conference*, Broomfield, CO, USA, October 2014.

[39]  A. Goder, A. Spiridonov, and Y. Wang, "Bistro: Scheduling data-parallel jobs against live production systems," in *Proceedings of USENIX Annual Technical Conference*, Santa Clara, CA, USA, July 2015.

[40]  A. Ghodsi, M. Zaharia, B. Hindman, A. Konwinski, S. Shenker, and I. Stoica, "Dominant resource fairness: Fair allocation of multiple resources types," in *Proceedings of USENIX Networking System Design and Implementation*, Boston, MA, USA, March 2011.

[41]  G. Ananthanarayanan, C. Douglas, R. Ramakrishnan, S. Rao, and I. Stoica, "True elasticity in multi-tenant data-intensive compute clusters," in *Proceedings of ACM Symposium on Cloud Computing*, San Jose, CA, USA, October 2012.

[42]  M. Isard, V. Prabhakaran, J. Currey, U. Wieder, K. Talwar, and A. Goldberg, "Quincy: Fair scheduling for distributed computing clusters," in *Proceedings of USENIX Symposium on Operating System Principles*, Big Sky, MT, USA, October 2009.

[43]  K. Kambatla, A. Pathak, and H. Pucha, "Towards optimizing hadoop provisioning in the cloud," in *Proceedings of USENIX Hot Topics in Cloud Computing*, San Diego, CA, USA, June 2009.

[44]  F. Tian and K. Chen, "Towards optimal resource provisioning for running mapreduce programs in public clouds," in *Proceedings of IEEE International Conference on Cloud Computing*, Washington, DC, USA, July 2011.

[45]  H. Herodotou, F. Dong, and S. Babu, "No one (cluster) size fits all: Automatic cluster sizing for data-intensive analytics," in *Proceedings of ACM Symposium on Cloud Computing*, Cascais, Portugal, October 2011.

[46] L. Zhang, C. Wu, Z. Li, C. Guo, M. Chen, and F. C. Lau, "Moving big data to the cloud: An online cost-minimizing approach," *IEEE Journal on Selected Areas in Communications Special Issue on Networking Challenges in Cloud Computing Systems and Applications*, vol. 31, no. 12, pp. 2710–2721, December 2013.

[47] S.-H. Park, O. Simeone, O. Sahin, and S. Shamai, "Robust and efficient distributed compression for cloud radio access networks," *IEEE Transactions on Vehicular Technology*, vol. 62, no. 2, pp. 692–703, February 2013.

[48] Y. Huangfu, J. Cao, H. Lu, and G. Liang, "Matrixmap: Programming abstraction and implementation of matrix computation for big data applications," in *Proceedings of IEEE International Conference on Parallel and Distributed Systems*, Melbourne, Australia, December 2015.

[49] X. Meng, J. Bradley, B. Yavuz, E. Sparks, S. Venkataraman, D. Liu, J. Freeman, D. Tsai, M. Amde, S. Owen, D. Xin, R. Xin, M. J. Franklin, R. Zadeh, M. Zaharia, and A. Talwalkar, "Mllib: Machine learning in apache spark," *Journal of Machine Learning Research*, vol. 17, no. 34, pp. 1–7, 2016.

[50] http://spark.apache.org/docs/latest/mllib-guide.html.

[51] D. P. Bertsekas, *Nonlinear Programming*, 2nd ed., Belmont, MA, USA: Athena Scientific, 1999.

[52] Y. Nesterov, *Introductory lectures on Convex Optimization: A Basic Course*, Springer, 2004.

[53] Z.-Q. Luo and P. Tseng, "On the convergence of the coordinate descent method for convex differentiable minimization," *Journal of Optimization Theory and Application*, vol. 72, no. 1, pp. 7–35, 1992.

[54] ——, "On the linear convergence of descent methods for convex essentially smooth minimization," *SIAM Journal on Control and Optimization*, vol. 30, no. 2, pp. 408–425, 1992.

[55] M. Hong, M. Razaviyayn, Z.-Q. Luo, and J.-S. Pang, "A unified algorithmic framework for block-structured optimization involving big data," *IEEE Signal Processing Magazine*, vol. 33, no. 1, pp. 57–77, 2016.

[56] D. P. Bertsekas and J. N. Tsitsiklis, *Neuro-Dynamic Programming*, Belmont, MA, USA: Athena Scientific, 1996.

[57] F. Facchinei, S. Sagratella, and G. Scutari, "Flexible parallel algorithms for big data optimization," in *2014 IEEE International Conference on Acoustics, Speech and Signal Processing*, May 2014, pp. 7208–7212.

[58] G. Scutari, F. Facchinei, P. Song, D. P. Palomar, and J.-S. Pang, "Decomposition by partial linearization: Parallel optimization of multi-agent systems," *IEEE Transactions on Signal Processing*, vol. 63, no. 3, pp. 641–656, February 2014.

[59] M. Razaviyayn, "Successive convex approximation: Analysis and applications," PhD thesis, University of Minnesota, 2014.

[60] M. Razaviyayn, M. Hong, Z.-Q. Luo, and J. S. Pang, "Parallel successive convex approximation for nonsmooth nonconvex optimization," in *Proceedings of the Neural Information Processing Systems*, Montreal, Canada, December 2014, pp. 1440–1448.

[61] J. Ortega and W. Rheinboldt, *Iterative Solution of Nonlinear Equations in Several Variables*, Academic Press, 1970.

[62] C. Hildreth, "A quadratic programming procedure," *Naval Research Logistics Quarterly*, vol. 4, no. 1, pp. 79–85, March 1957.

[63] J. Warga, "Minimizing certain convex functions," *Journal of the Society for Industrial and Applied Mathematics*, vol. 11, no. 3, pp. 588–593, September 1963.

[64] A. Auslender, *Optimisation. Méthodes numériques*, Masson, 1976.

[65] R. Sargent and D. Sebastian, "On the convergence of sequential minimization algorithms," *Journal of Optimization Theory and Applications*, vol. 12, no. 6, pp. 567–575, December 1973.

[66] M. J. D. Powell, "On search directions for minimization algorithms," *Mathematical Programming*, vol. 4, no. 1, pp. 193–201, December 1973.

[67] M. Razaviyayn, M. Hong, and Z.-Q. Luo, "A unified convergence analysis of block successive minimization methods for nonsmooth optimization," *SIAM Journal on Optimization*, vol. 23, no. 2, pp. 1126–1153, June 2013.

[68] I. Necoara and A. Patrascu, "A random coordinate descent algorithm for optimization problems with composite objective function and linear coupled constraints," *Computational Optimization and Applications*, vol. 57, no. 2, pp. 307–377, March 2014.

[69] P. Tseng, "Convergence of a block coordinate descent method for nondifferentiable minimization," *Journal of Optimization Theory and Applications*, vol. 103, no. 9, pp. 475–494, June 2001.

[70] B. Chen, S. He, Z. Li, and S. Zhang, "Maximum block improvement and polynomial optimization," *SIAM Journal on Optimization*, vol. 22, no. 1, pp. 87–107, January 2012.

[71] Z.-Q. Luo and P. Tseng, "Error bounds and convergence analysis of feasible descent methods: A general approach," *Annals of Operations Research*, vol. 46, no. 1, pp. 157–178, February 1993.

[72] Y. Nesterov, "Efficiency of coordinate descent methods on huge-scale optimization problems," *SIAM Journal on Optimization*, vol. 22, no. 2, pp. 341–362, April 2012.

[73] A. Beck and L. Tetruashvili, "On the convergence of block coordinate descent type methods," *SIAM Journal on Optimization*, vol. 23, no. 4, pp. 2037–2060, 2013.

[74] M. Hong, X. Wang, M. Razaviyayn, and Z.-Q. Luo, "Iteration complexity analysis of block coordinate descent methods," preprint, 2013, http://arXiv:1310.6957.

[75] Z. Lu and L. Xiao, "Randomized block coordinate non-monotone gradient method for a class of nonlinear programming," preprint, 2013, http://arXiv:1306.5918.

[76] P. Richtárik and M. Takáč, "Iteration complexity of randomized block-coordinate descent methods for minimizing a composite function," *Mathematical Programming*, vol. 144, no. 1–2, pp. 1–38, 2014.

[77] A. P. Dempster, N. M. Laird, and D. B. Rubin, "Maximum likelihood from incomplete data via the EM algorithm," *Journal of the Royal Statistical Society Series B*, vol. 39, no. 1, pp. 1–38, 1977.

[78] A. L. Yuille and A. Rangarajan, "The concave-convex procedure," *Neural Computation*, vol. 15, no. 4, pp. 915–936, April 2003.

[79] D. D. Lee and H. S. Seung, "Algorithms for non-negative matrix factorization," in *Advances in Neural Information Processing Systems*, ed. T. K. Leen, T. G. Dietterich, and V. Tresp, MIT Press, 2001, pp. 556–562.

[80] Q. Shi, M. Razaviyayn, Z.-Q. Luo, and C. He, "An iteratively weighted MMSE approach to distributed sum-utility maximization for a MIMO interfering broadcast channel," *IEEE Transactions on Signal Processing*, vol. 59, no. 9, pp. 4331–4340, September 2011.

[81] M. R. Hestenes, "Multiplier and gradient methods," *Journal of Optimization Theory and Applications*, vol. 4, no. 5, pp. 303–320, 1969.

[82] M. J. D. Powell, "A method for nonlinear constraints in minimization problems," in *Optimization*, ed. R. Fletcher, New York: Academic Press, 1972, pp. 283–298.

[83]  R. Glowinski and P. Le Tallec, *Augmented Lagrangian and Operator-Splitting Methods in Nonlinear Mechanics*, Philadelphia, PA, USA: SIAM, 1989.

[84]  G. Chen and M. Teboulle, "Convergence analysis of a proximal-like minimization algorithm using Bregman functions," *SIAM Journal on Optimization*, vol. 3, no. 3, pp. 538–543, 1993.

[85]  T.-H. Chang, M. Hong, and X. Wang, "Multi-agent distributed optimization via inexact consensus ADMM," *IEEE Transactions on Signal Processing*, vol. 63, no. 2, pp. 482–497, January 2015.

[86]  S. Boyd, N. Parikh, E. Chu, B. Peleato, and J. Eckstein, "Distributed optimization and statistical learning via the alternating direction method of multipliers," *Foundations and Trends in Machine Learning*, vol. 3, no. 1, pp. 1–122, 2011.

[87]  R. Glowinski, *Numerical Methods for Nonlinear Variational Problems*, New York: Springer-Verlag, 1984.

[88]  B. He and X. Yuan, "On the O(1/n) convergence rate of the Douglas-Rachford alternating direction method," *SIAM Journal on Numerical Analysis*, vol. 50, no. 2, pp. 700–709, 2012.

[89]  ——, "On nonergodic convergence rate of Douglas-Rachford alternating direction method of multipliers," *Numerische Mathematik*, vol. 130, no. 3, pp. 567–577, 2015.

[90]  D. Goldfarb, S. Ma, and K. Scheinberg, "Fast alternating linearization methods for minimizing the sum of two convex functions," *Mathematical Programming*, vol. 141, no. 1–2, pp. 349–382, 2012.

[91]  T. Goldstein, B. Oonoghue, and S. Setzer, "Fast alternating direction optimization methods," *UCLA CAM Report 12–35*, 2012.

[92]  W. Deng and W. Yin, "On the global linear convergence of the alternating direction method of multipliers," *Rice University CAAM Technical Report TR12-14*, 2012.

[93]  J. Eckstein and D. Bertsekas, "An alternating direction method for linear programming," *Laboratory for Information and Decision Systems, MIT*, http://hdl.handle.net/1721.1/3197, 1990.

[94]  W. Deng and W. Yin, "On the global and linear convergence of the generalized alternating direction method of multipliers," *Journal of Scientific Computing*, vol. 66, no. 3, pp. 889–916, 2016.

[95]  M. Hong and Z.-Q. Luo, "On the linear convergence of the alternating direction method of multipliers," accepted by Mathematical Programming Series A, 2016, http://arXiv:1208.3922.

[96]  D. Boley, "Local linear convergence of the alternating direction method of multipliers on quadratic or linear programs," *SIAM Journal on Optimization*, vol. 23, no. 4, pp. 2183–2207, 2013.

[97]  C. Chen, B. He, X. Yuan, and Y. Ye, "The direct extension of ADMM for multi-block convex minimization problems is not necessarily convergent," *Mathematical Programming*, vol. 155, no. 1–2, pp. 57–98, 2016.

[98]  B. He, M. Tao, and X. Yuan, "Alternating direction method with Gaussian back substitution for separable convex programming," *SIAM Journal on Optimization*, vol. 22, no. 2, pp. 313–340, 2012.

[99]  B. He, H. Xu, and X. Yuan, "On the proximal Jacobian decomposition of ALM for multipleblock separable convex minimization problems and its relationship to ADMM," *Journal of Scientific Computing*, vol. 66, no. 3, pp. 1204–1217, 2016.

[100]  W. Deng, M. Lai, Z. Peng, and W. Yin, "Parallel multi-block ADMM with o (1/k) convergence," preprint, 2014, http://arXiv:1312.3040.

[101] M. Hong, T.-H. Chang, X. Wang, M. Razaviyayn, S. Ma, and Z.-Q. Luo, "A block successive upper bound minimization method of multipliers for linearly constrained convex optimization," preprint, 2013, http://arXiv:1401.7079.

[102] X. Gao and S. Zhang, "First-order algorithms for convex optimization with nonseparate objective and coupled constraints," *Optimization online*, vol. 3, p. 5, 2015.

[103] Y. Zhang, "An alternating direction algorithm for nonnegative matrix factorization," preprint, 2010.

[104] D. L. Sun and C. Fevotte, "Alternating direction method of multipliers for non-negative matrix factorization with the beta-divergence," in *IEEE International Conference on Acoustics, Speech and Signal Processing*, May 2014, pp. 6201–6205.

[105] B. Ames and M. Hong, "Alternating directions method of multipliers for l1-penalized zero variance discriminant analysis and principal component analysis," *Computational Optimization and Applications*, vol. 64, no. 3, pp. 725–754, 2016.

[106] B. Jiang, S. Ma, and S. Zhang, "Alternating direction method of multipliers for real and complex polynomial optimization models," *Optimization*, vol. 63, no. 6, pp. 883–898, 2014.

[107] R. Zhang and J. T. Kwok, "Asynchronous distributed ADMM for consensus optimization," in *Proceedings of International Conference on Machine Learning*, Beijing, China, June 2014, pp. 1701–1709.

[108] P. A. Forero, A. Cano, and G. B. Giannakis, "Distributed clustering using wireless sensor networks," *IEEE Journal of Selected Topics in Signal Processing*, vol. 5, no. 4, pp. 707–724, August 2011.

[109] Z. Wen, C. Yang, X. Liu, and S. Marchesini, "Alternating direction methods for classical and ptychographic phase retrieval," *Inverse Problems*, vol. 28, no. 11, pp. 1–18, 2012.

[110] M. Hong, Z.-Q. Luo, and M. Razaviyayn, "Convergence analysis of alternating direction method of multipliers for a family of nonconvex problems," *SIAM Journal on Optimization*, vol. 26, no. 1, pp. 337–364, 2016.

[111] W. P. Ziemer, *Weakly Differentiable Functions: Sobolev Spaces and Functions of Bounded Variation*, Graduate Texts in Mathematics, New York, USA: Springer, 1989.

[112] S. Kullback, "The Kullback-Leibler distance," *The American Statistician*, vol. 41, no. 4, pp. 340–341, 1987.

[113] ——, *Information Theory and Statistics*, Mineola, NY, USA: Dover Pubns, 1997.

[114] D. Hosmer and S. Lemeshow, *Applied Logistic Regression*, vol. 354, Hoboken, NJ, USA: Wiley-Interscience, 2000, vol. 354.

[115] J. Duchi, S. Shalev-Shwartz, Y. Singer, and T. Chandra, "Efficient projections onto the l1-ball for learning in high dimensions," in *Proceedings of the International Conference on Machine Learning*, New York, NY, USA: ACM, July 2008, pp. 272–279.

[116] A. Quattoni, X. Carreras, M. Collins, and T. Darrell, "An efficient projection for $\ell_{1,\infty}$ regularization," in *Proceedings of the Annual International Conference on Machine Learning*, Montreal, QC, Canada, 2009, pp. 857–864.

[117] J. Liu and J. Ye, "Efficient Euclidean projections in linear time," in *Proceedings of Annual International Conference on Machine Learning*, Montreal, QC, Canada, 2009, pp. 657–664.

[118] E. van den Berg and M. Friedlander, "Probing the Pareto frontier for basis pursuit solutions," *SIAM Journal on Scientific Computing*, vol. 31, no. 2, pp. 890–912, 2008.

[119] E. van den Berg, M. Schmidt, M. Friedlander, and K. Murphy, "Group sparsity via linear-time projection," *Optimization Online*, 2008.

[120] S. Boyd and L. Vandenberghe, *Convex Optimization*, Cambridge University Press, 2004.

[121] P. Tseng and S. Yun, "A coordinate gradient descent method for nonsmooth separable minimization," *Mathematical Programming*, vol. 117, no. 1, pp. 387–423, 2009.

[122] M. Razaviyayn, M. Hong, and Z. Luo, "A unified convergence analysis of coordinatewise successive minimization methods for nonsmooth optimization," Report of University of Minnesota, Twin Cites 2012.

[123] X. Wei, Y. Yuan, and Q. Ling, "DOA estimation using a greedy block coordinate descent algorithm," *IEEE Transactions on Signal Processing*, vol. 60, no. 12, pp. 6382–6394, December 2012.

[124] D. Donoho, "De-noising by soft-thresholding," *IEEE Transactions on Information Theory*, vol. 41, no. 3, pp. 613–627, May 1995.

[125] J. F. Cai, E. J. Candes, and Z. Shen, "A singular value thresholding algorithm for matrix completion," *SIAM Journal on Optimization*, vol. 20, no. 4, pp. 1956–1982, 2008.

[126] S. Ma, D. Goldfarb, and L. Chen, "Fixed point and Bregman iterative methods for matrix rank minimization," *Mathematical Programming Series A*, vol. 128, no. 1–2, pp. 321–353, 2011.

[127] J. Cai and S. Osher, "Fast singular value thresholding without singular value decomposition," *UCLA CAM Report 10–24*, vol. 5, 2010.

[128] L. Rudin, S. Osher, and E. Fatemi, "Nonlinear total variation based noise removal algorithms," *Physica D: Nonlinear Phenomena*, vol. 60, no. 1–4, pp. 259–268, 1992.

[129] J. Darbon and M. Sigelle, "Image restoration with discrete constrained total variation, Part I: Fast and exact optimization," *Journal of Mathematical Imaging and Vision*, vol. 26, no. 3, pp. 261–276, 2006.

[130] D. Goldfarb and W. Yin, "Parametric maximum flow algorithms for fast total variation minimization," *SIAM Journal on Scientific Computing*, vol. 31, no. 5, pp. 3712–3743, 2009.

[131] T. F. Chan, H. M. Zhou, and R. H. Chan, "Continuation method for total variation denoising problems," *Advanced Signal Processing Algorithms*, vol. 2563, no. 1, pp. 314–325, 1995.

[132] A. Chambolle, "An algorithm for total variation minimization and applications," *Journal of Mathematical Imaging and Vision*, vol. 20, no. 1–2, pp. 89–97, 2004.

[133] ——, "Total variation minimization and a class of binary MRF models," *Energy Minimization Methods in Computer Vision and Pattern Recognition, Lecture Notes in Computer Science 3757*, pp. 136–152, 2005.

[134] B. Wohlberg and P. Rodriguez, "An iteratively reweighted norm algorithm for minimization of total variation functionals," *IEEE Signal Processing Letters*, vol. 14, no. 12, pp. 948–951, December 2007.

[135] Y. Wang, J. Yang, W. Yin, and Y. Zhang, "A new alternating minimization algorithm for total variation image reconstruction," *SIAM Journal on Imaging Sciences*, vol. 1, no. 3, pp. 248–272, 2008.

[136] T. Goldstein and S. Osher, "The split Bregman method for l1 regularized problems," *SIAM Journal on Imaging Sciences*, vol. 2, no. 2, pp. 323–343, 2009.

[137] Y.-L. Yu, "Better approximation and faster algorithm using the proximal average," in *Proceedings of the Neural Information Processing Sytems*, Lake Tohoe, NV, USA, December 2013, pp. 458–466.

[138] ——, "On decomposing the proximal map," in *Proceedings of the Neural Information Processing Sytems*, 2013, pp. 91–99.

[139] M. Figueiredo and R. Nowak, "An EM algorithm for wavelet-based image restoration," *IEEE Transactions on Image Processing*, vol. 12, no. 8, pp. 906–916, August 2003.

[140] C. De Mol and M. Defrise, "A note on wavelet-based inversion algorithms," *Contemporary Mathematics*, vol. 313, pp. 85–96, 2002.

[141] J. Bect, L. Blanc-Feraud, G. Aubert, and A. Chambolle, "A $\ell_1$-unified variational framework for image restoration," *European Conference on Computer Vision, Prague, Lecture Notes in Computer Sciences 3024*, pp. 1–13, 2004.

[142] J. Douglas and H. H. Rachford, "On the numerical solution of the heat conduction problem in 2 and 3 space variables," *Transactions of the American Mathematical Society*, vol. 82, pp. 421–439, 1956.

[143] D. H. Peaceman and H. H. Rachford, "The numerical solution of parabolic elliptic differential equations," *SIAM Journal on Applied Mathematics*, vol. 3, no. 1, pp. 28–41, 1955.

[144] D. Gabay and B. Mercier, "A dual algorithm for the solution of nonlinear variational problems via finite-element approximations," *Computers and Mathematics with Applications*, vol. 2, no. 1, pp. 17–40, 1976.

[145] J. Eckstein and D. P. Bertsekas, "On the Douglas-Rachford splitting method and the proximal point algorithm for maximal monotone operators," *Mathematical Programming*, vol. 55, no. 1–3, pp. 293–318, 1992.

[146] M. Fortin and R. Glowinski, *Augmented Lagrangian Methods*, Ambsterdam, New York, USA: North-Holland, 1983.

[147] P. L. Combettes and V. R. Wajs, "Signal recovery by proximal forward-backward splitting," *Multiscale Modeling and Simulation*, vol. 4, no. 4, pp. 1168–1200, 2005.

[148] R. Rockafellar, *Convex Analysis*, Princeton, USA: Princeton University Press, 1970.

[149] S. Ma, W. Yin, Y. Zhang, and A. Chakraborty, "An efficient algorithm for compressed MR imaging using total variation and wavelets," *IEEE International Conference on Computer Vision and Pattern Recognition*, pp. 1–8, September 2008.

[150] R. Tibshirani, M. Saunders, S. Rosset, J. Zhu, and K. Knight, "Sparsity and smoothness via the fused lasso," *Journal of the Royal Statistical Society: Series B (Statistical Methodology)*, vol. 67, no. 1, pp. 91–108, 2005.

[151] Z.-Q. Luo and P. Tseng, "On the linear convergence of descent methods for convex essentially smooth minimization," *SIAM Journal on Control and Optimization*, vol. 30, no. 2, pp. 408–425, 1990.

[152] E. T. Hale, W. Yin, and Y. Zhang, "Fixed-point continuation for l1-minimization: Methodology and convergence," *SIAM Journal on Optimization*, vol. 19, no. 3, pp. 1107–1130, 2008.

[153] Y. Nesterov, "A method of solving a convex programming problem with convergence rate $O(1/k^2)$," *Soviet Mathematics Doklady*, vol. 27, no. 2, pp. 372–376, 1983.

[154] ——, "Gradient methods for minimizing composite objective function," *www.optimization-online.org, CORE Discussion Paper 2007/76*, 2007.

[155] P. Tseng, "On accelerated proximal gradient methods for convex-concave optimization," submitted to *SIAM Journal on Optimization*, 2008.

[156] A. Beck and M. Teboulle, "A fast iterative shrinkage-thresholding algorithm for linear inverse problems," *SIAM Journal on Imaging Sciences*, vol. 2, no. 1, pp. 183–202, 2009.

[157] S. Becker, J. Bobin, and E. Candès, "NESTA: A fast and accurate first-order method for sparse recovery," *SIAM Journal on Imaging Sciences*, vol. 4, no. 1, pp. 1–39, 2011.

[158] W. Deng, W. Yin, and Y. Zhang, "Group sparse optimization by alternating direction method," *Rice University CAAM Technical Report TR11-06*, 2011.

[159] B. Recht, M. Fazel, and P. Parrilo, "Guaranteed minimum-rank solutions of linear matrix equations via nuclear norm minimization," *SIAM Review*, vol. 52, no. 3, pp. 471–501, 2010.

[160] L. Bregman, "The relaxation method of finding the common points of convex sets and its application to the solution of problems in convex programming," *USSR Computational Mathematics and Mathematical Physics*, vol. 7, no. 3, pp. 200–217, 1967.

[161] W. Yin and S. Osher, "Error forgetting of Bregman iteration," *Journal of Scientific Computing*, vol. 54, no. 2–3, pp. 684–95, 2013.

[162] S. Osher, M. Burger, D. Goldfarb, J. Xu, and W. Yin, "An iterative regularization method for total variation-based image restoration," *SIAM Journal on Multiscale Modeling and Simulation*, vol. 4, no. 2, pp. 460–489, 2005.

[163] J. Yang and Y. Zhang, "Alternating direction algorithms for $\ell_1$-problems in compressive sensing," *SIAM Journal on Scientific Computing*, vol. 33, no. 1, pp. 250–278, 2011.

[164] E. Candes, X. Li, Y. Ma, and J. Wright, "Robust principal component analysis?" *Journal of the ACM*, vol. 58, no. 3, pp. 1–37, 2011.

[165] B. Efron, T. Hastie, I. Johnstone, and R. Tibshirani, "Least angle regression," *The Annals of Statistics*, vol. 32, no. 2, pp. 407–499, 2004.

[166] M. Best, "An algorithm for the solution of the parametric quadratic programming problem," in *Applied Mathematics and Parallel Computing*, Springer, 1996, pp. 57–76.

[167] L. Ghaoui, V. Viallon, and T. Rabbani, "Safe feature elimination in sparse supervised learning," preprint, 2010, http://arXiv:1009.4219.

[168] R. Tibshirani, J. Bien, J. Friedman, T. Hastie, N. Simon, J. Taylor, and R. Tibshirani, "Strong rules for discarding predictors in lasso-type problems," *Journal of the Royal Statistical Society: Series B (Statistical Methodology)*, vol. 74, no. 2, pp. 245–266, 2012.

[169] S. Wright, R. Nowak, and M. Figueiredo, "Sparse reconstruction by separable approximation," *IEEE Transactions on Signal Processing*, vol. 57, no. 7, pp. 2479–2493, July 2009.

[170] Z. Wen, W. Yin, H. Zhang, and D. Goldfarb, "On the convergence of an active set method for l1-minimization," *Optimization Methods and Software*, vol. 27, no. 6, pp. 1127–1146, 2012.

[171] J. Nocedal and S. J. Wright, *Numerical Optimization*, New York, USA: Springer-Verlag, 1999.

[172] J. Tropp and A. Gilbert, "Signal recovery from random measurements via orthogonal matching pursuit," *IEEE Transactions on Information Theory*, vol. 53, no. 12, pp. 4655–4666, December 2007.

[173] D. Donoho, Y. Tsaig, I. Drori, and J.-C. Starck, "Sparse solution of underdetermined linear equations by stagewise orthogonal matching pursuit," submitted to *IEEE Transactions on Information Theory*, 2006.

[174] D. Needell and R. Vershynin, "Signal recovery from incomplete and inaccurate measurements via regularized orthogonal matching pursuit," *IEEE Journal of Selected Topics in Signal Processing*, vol. 4, no. 2, pp. 310–316, April 2010.

[175] W. Dai and O. Milenkovic, "Subspace pursuit for compressive sensing: Closing the gap between performance and complexity," arXiv:0803.0811v1 [cs.NA], 2008.

[176] D. Needell and J. Tropp, "Cosamp: Iterative signal recovery from incomplete and inaccurate samples," *Applied and Computational Harmonic Analysis*, vol. 26, no. 3, pp. 301–321, 2009.

[177] S. Foucart, "Hard thresholding pursuit: An algorithm for compressive sensing," *SIAM Journal on Numerical Analysis*, vol. 49, no. 6, pp. 2543–2563, 2011.

[178] Y. Wang and W. Yin, "Sparse signal reconstruction via iterative support detection," *SIAM Journal on Imaging Sciences*, vol. 3, no. 3, pp. 462–491, 2010.

[179] T. Blumensath and M. Davies, "Iterative hard thresholding for compressed sensing," *Applied and Computational Harmonic Analysis*, vol. 27, no. 3, pp. 265–274, 2009.

[180] ——, "Normalized iterative hard thresholding: Guaranteed stability and performance," *IEEE Journal of Selected Topics in Signal Processing*, vol. 4, no. 2, pp. 298–309, April 2010.

[181] W. Yin, S. Osher, D. Goldfarb, and J. Darbon, "Bregman iterative algorithms for l1-minimization with applications to compressed sensing," *SIAM Journal on Imaging Sciences*, vol. 1, no. 1, pp. 143–168, 2008.

[182] S. Osher, Y. Mao, B. Dong, and W. Yin, "Fast linearized Bregman iteration for compressive sensing and sparse denoising," *Communications in Mathematical Sciences*, vol. 8, no. 1, pp. 93–111, 2010.

[183] K. Toh and S. Yun, "An accelerated proximal gradient algorithm for nuclear norm regularized linear least squares problems," *Pacific Journal of Optimization*, vol. 6, no. 15, pp. 615–640, 2010.

[184] J. Yang and X. Yuan, "Linearized augmented Lagrangian and alternating direction methods for nuclear norm minimization," *Mathematics of Computation*, vol. 82, no. 281, pp. 301–329, 2013.

[185] R. Keshavan, A. Montanari, and S. Oh, "Matrix completion from a few entries," *IEEE Transactions on Information Theory*, vol. 56, no. 6, pp. 2980–2998, June 2010.

[186] Z. Wen, W. Yin, and Y. Zhang, "Solving a low-rank factorization model for matrix completion by a nonlinear successive over-relaxation algorithm," *Mathematical Programming Computation*, vol. 4, no. 4, pp. 333–361, 2012.

[187] R. Baraniuk, V. Cevher, M. Duarte, and C. Hegde, "Model-based compressive sensing," *IEEE Transactions on Information Theory*, vol. 56, no. 4, pp. 1982–2001, April 2010.

[188] D. Wang, Q. Zhang, and J. Liu, "Partial network coding: Theory and application in continuous sensor data collection," in *Proceedings of IEEE Workshop on Quality of Services*, New Haven, CT, USA, June 2006.

[189] B. Chazelle, R. Rubinfeld, and L. Trevisan, "Approximating the minimum spanning tree weight in sublinear time," *SIAM Journal on computing*, vol. 34, no. 6, pp. 1370–1379, July 2005.

[190] Sublinear algorithm surveys, Online: http://people.csail.mit.edu/ronitt/sublinear.html.

[191] M. Mardani, G. Mateos, and G. B. Giannakis, "Subspace learning and imputation for streaming big data matrices and tensors," *IEEE Transactions on Signal Processing*, vol. 63, no. 10, pp. 2663–2677, May 2015.

[192] L. Kuang, F. Hao, L. T. Yang, M. Lin, C. Luo, and G. Min, "A tensor-based approach for big data representation and dimensionality reduction," *IEEE Transactions on Emerging Topics in Computing*, vol. 2, no. 3, pp. 280–291, September 2014.

[193] J. Li, Y. Yan, W. Duan, S. Song, and M. H. Lee, "Tensor decomposition of Toeplitz jacket matrices for big data processing," in *International Conference on Big Data and Smart Computing*, Jeju Island, Korea, February 2015, pp. 11–14.

[194] N. Vervliet, O. Debals, L. Sorber, and L. D. Lathauwer, "Breaking the curse of dimensionality using decompositions of incomplete tensors: Tensor-based scientific computing in big data analysis," *IEEE Signal Processing Magazine*, vol. 31, no. 5, pp. 71–79, September 2014.

[195] N. D. Sidiropoulos, E. E. Papalexakis, and C. Faloutsos, "Parallel randomly compressed cubes : A scalable distributed architecture for big tensor decomposition," *IEEE Signal Processing Magazine*, vol. 31, no. 5, pp. 57–70, September 2014.

[196] T. G. Kolda and B. W. Bader, "Tensor decompositions and applications," *SIAM Review*, vol. 51, no. 3, pp. 455–500, 2009.

[197] J. D. Carroll and J.-J. Chang, "Analysis of individual differences in multidimensional scaling via an N-way generalization of 'Eckart-Young' decomposition," *Psychometrika*, vol. 35, no. 3, pp. 283–319, 1970.

[198] R. A. Harshman, "Parafac2: Mathematical and technical notes," *UCLA Working Papers in Phonetics*, vol. 22, no. 3044, p. 122215, 1972.

[199] J. Douglas Carroll, S. Pruzansky, and J. B. Kruskal, "Candelinc: A general approach to multidimensional analysis of many-way arrays with linear constraints on parameters," *Psychometrika*, vol. 45, no. 1, pp. 3–24, 1980.

[200] R. A. Harshman, "Models for analysis of asymmetrical relationships among N objects or stimuli," in *First Joint Meeting of the Psychometric Society and the Society for Mathematical Psychology*, McMaster University, Hamilton, Ontario, vol. 5, 1978.

[201] R. A. Harshman and M. E. Lundy, "Uniqueness proof for a family of models sharing features of tucker's three-mode factor analysis and parafac/candecomp," *Psychometrika*, vol. 61, no. 1, pp. 133–154, 1996.

[202] A. Huy Phan and A. Cichocki, "Parafac algorithms for large-scale problems," *Neurocomput.*, vol. 74, no. 11, pp. 1970–1984, May 2011, http://dx.doi.org/10.1016/j.neucom.2010.06.030.

[203] A. L. F. De Almeida, G. Favier, and J. C. Mota, "The constrained block-PARAFAC decomposition," in *Conference of ThRee-Way Methods in Chemistry and Psychology*, Chania, Greece, June 2006.

[204] "COMFAC: Matlab code for LS fitting of the complex PARAFAC model in 3-D," http://people.ece.umn.edu/ñikos/comfac.m.

[205] "Factoring tensors in the cloud: A tutorial on big tensor data analytics," www.cs.cmu.edu/~epapalex/tutorials/icassp14.html.

[206] N. D. Sidiropoulos and A. Kyrillidis, "Multi-way compressed sensing for sparse low-rank tensors," *IEEE Signal Processing Letters*, vol. 19, no. 11, pp. 757–760, August 2012.

[207] L. Chiantini and G. Ottaviani, "On generic identifiability of 3-tensors of small rank," *SIAM Journal on Matrix Analysis and Applications*, vol. 33, no. 3, pp. 1018–1037, 2012.

[208] J. Liu, P. Musialski, P. Wonka, and J. Ye, "Tensor completion for estimating missing values in visual data," *IEEE Transactions on Pattern Analysis and Machine Intelligence*, vol. 35, no. 1, pp. 208–220, January 2013.

[209] S. Gandy, B. Recht, and I. Yamada, "Tensor completion and low-n-rank tensor recovery via convex optimization," *Inverse Problems*, vol. 27, no. 2, p. 025010, 2011.

[210] M. Signoretto, R. V. de Plas, B. D. Moor, and J. A. K. Suykens, "Tensor versus matrix completion: A comparison with application to spectral data," *IEEE Signal Processing Letters*, vol. 18, no. 7, pp. 403–406, July 2011.

[211] D. Kressner, M. Steinlechner, and B. Vandereycken, "Low-rank tensor completion by Riemannian optimization," *BIT Numerical Mathematics*, vol. 54, no. 2, pp. 447–468, 2014.

[212] Y. Xu, R. Hao, W. Yin, and Z. Su, "Parallel matrix factorization for low-rank tensor completion," preprint, 2013, http://arXiv:1312.1254.

[213] M. Yuan and C.-H. Zhang, "On tensor completion via nuclear norm minimization," *Foundations of Computational Mathematics*, vol. 16, no. 4, pp. 1031–1068, 2016.

[214] C. T. Zahn, "Graph-theoretical methods for detecting and describing gestalt clusters," *IEEE Transactions on Computers*, vol. C-20, no. 1, pp. 68–86, January 1971.

[215] P. Mordohai and G. Medioni, "Tensor voting: A perceptual organization approach to computer vision and machine learning," *Synthesis Lectures on Image, Video, and Multimedia Processing*, vol. 2, no. 1, pp. 1–136, 2006.

[216] E. Franken, M. van Almsick, P. Rongen, L. Florack, and B. ter Haar Romeny, *An Efficient Method for Tensor Voting Using Steerable Filters*, Berlin, Heidelberg, Germany: Springer Berlin Heidelberg, 2006, pp. 228–240.

[217] W. T. Freeman and E. H. Adelson, "The design and use of steerable filters," *IEEE Transactions on Pattern Analysis and Machine Intelligence*, vol. 13, no. 9, pp. 891–906, September 1991.

[218] E. Pan, M. Pan, Z. Han, and V. Wright, "Mobile trace inference based on tensor voting," in *Proceedings of IEEE Global Communications Conference*, Austin, TX, USA, December 2014, pp. 4891–4897.

[219] B. J. King, "Range data analysis by free-space modeling and tensor voting," PhD dissertation, Troy, NY, USA, 2008.

[220] D. Nion, K. N. Mokios, N. D. Sidiropoulos, and A. Potamianos, "Batch and adaptive parafac-based blind separation of convolutive speech mixtures," *IEEE Transactions on Audio, Speech, and Language Processing*, vol. 18, no. 6, pp. 1193–1207, August 2010.

[221] N. D. Sidiropoulos, G. B. Giannakis, and R. Bro, "Blind parafac receivers for DS-CDMA systems," *IEEE Transactions on Signal Processing*, vol. 48, no. 3, pp. 810–823, March 2000.

[222] N. D. Sidiropoulos, R. Bro, and G. B. Giannakis, "Parallel factor analysis in sensor array processing," *IEEE Transactions on Signal Processing*, vol. 48, no. 8, pp. 2377–2388, August 2000.

[223] E. E. Papalexakis, C. Faloutsos, and N. D. Sidiropoulos, *ParCube: Sparse Parallelizable Tensor Decompositions*. Berlin, Heidelberg, Germany: Springer Berlin Heidelberg, 2012, pp. 521–536.

[224] R. Bro and N. D. Sidiropoulos, "Least squares algorithms under unimodality and non-negativity constraints," *Journal of Chemometrics*, vol. 12, no. 4, pp. 223–247, 1998.

[225] A. Cichocki, D. Mandic, L. D. Lathauwer, G. Zhou, Q. Zhao, C. Caiafa, and H. A. PHAN, "Tensor decompositions for signal processing applications: From two-way to multiway component analysis," *IEEE Signal Processing Magazine*, vol. 32, no. 2, pp. 145–163, March 2015.

[226] L. Li and D. Boulware, "High-order tensor decomposition for large-scale data analysis," in *IEEE International Congress on Big Data*, June 2015, pp. 665–668.

[227] F. Shang, Y. Liu, and J. Cheng, "Generalized higher-order tensor decomposition via parallel ADMM," preprint, 2014, http://arXiv:1407.139.

[228] X. He, D. Cai, and P. Niyogi, "Tensor subspace analysis," in *Advances in Neural Information Processing Systems*, vol. 18, pp. 499–506, 2005.

[229] Y. Xu and W. Yin, "A block coordinate descent method for regularized multiconvex optimization with applications to nonnegative tensor factorization and completion," *SIAM Journal on Imaging Sciences*, vol. 6, no. 3, pp. 1758–1789, 2013.

[230] B. Romera-Paredes and M. Pontil, "A new convex relaxation for tensor completion," preprint, 2013, http://arXiv:1307.4653.

[231] L. Yang, Z.-H. Huang, and Y.-F. Li, "A splitting augmented Lagrangian method for low multilinear-rank tensor recovery," *Asia-Pacific Journal of Operational Research*, vol. 32, no. 1, p. 1540008, 2015.

[232] Y. Xu, "Alternating proximal gradient method for sparse nonnegative tucker decomposition," *Mathematical Programming Computation*, vol. 7, no. 1, pp. 39–70, 2015.

[233] ——, "Block coordinate update method in tensor optimization," *Siam J. Imaging Sciences*, vol. 6, no. 3, pp. 1758–1789, 2014.

[234] Y. Liu and F. Shang, "An efficient matrix factorization method for tensor completion," *IEEE Signal Processing Letters*, vol. 20, no. 4, pp. 307–310, April 2014.

[235] R. Tomioka, T. Suzuki, K. Hayashi, and H. Kashima, "Statistical performance of convex tensor decomposition," *NIPS'11 Proceedings of the 24th International Conference on Neural Information Processing Systems*, pp. 972–980, 2011.

[236] Y. Liu, F. Shang, W. Fan, J. Cheng, and H. Cheng, "Generalized higher order orthogonal iteration for tensor learning and decomposition," *IEEE Transactions on Neural Networks and Learning Systems*, vol. 27, no. 12, pp. 2551–2563, November 2015.

[237] A. Krishnamurthy and A. Singh, "Low-rank matrix and tensor completion via adaptive sampling," in *Advances in Neural Information Processing Systems*, Lake Tohoe, NV, USA, December 2013, pp. 836–844.

[238] Q. Li, A. Prater, L. Shen, and G. Tang, "Overcomplete tensor decomposition via convex optimization," in *IEEE International Workshop on Computational Advances in Multi-Sensor Adaptive Processing*, December 2015, pp. 53–56.

[239] D. Goldfarb and Z. Qin, "Robust low-rank tensor recovery: Models and algorithms," *SIAM Journal on Matrix Analysis and Applications*, vol. 35, no. 1, pp. 225–253, 2014.

[240] J. Brachat, P. Comon, B. Mourrain, and E. P. Tsigaridas, "Symmetric tensor decomposition," in *European Signal Processing Conference*, 2009, pp. 525–529.

[241] B. Ran, H. Tan, Y. Wu, and P. J. Jin, "Tensor based missing traffic data completion with spatial-temporal correlation," *Physica A: Statistical Mechanics and its Applications*, vol. 446, pp. 54 – 63, 2016.

[242] J. Liu, P. Musialski, P. Wonka, and J. Ye, "Tensor completion for estimating missing values in visual data," in *IEEE International Conference on Computer Vision*, September 2009, pp. 2114–2121.

[243] O. H. M. Padilla and J. G. Scott, "Tensor decomposition with generalized lasso penalties," preprint, 2015, http://ArXiv:1502.06930.

[244] B. Jiang, S. Ma, and S. Zhang, "Tensor principal component analysis via convex optimization," *Mathematical Programming*, vol. 150, no. 2, pp. 423–457, 2015.

[245] L. T. Huang, H. C. So, Y. Chen, and W. Q. Wang, "Truncated nuclear norm minimization for tensor completion," in *IEEE Sensor Array and Multichannel Signal Processing Workshop*, June 2014, pp. 417–420.

[246] M. Reisert and H. Burkhardt, "Efficient tensor voting with 3D tensorial harmonics," in *IEEE Computer Society Conference on Computer Vision and Pattern Recognition Workshops*, June 2008, pp. 1–7.

[247] G. Guy and G. Medioni, "Inferring global perceptual contours from local features," in *Proceedings of IEEE Computer Society Conference on Computer Vision and Pattern Recognition*, June 1993, pp. 786–787.

[248] J. Kang, I. Cohen, and G. Medioni, "Continuous multi-views tracking using tensor voting," in *Proceedings of Workshop on Motion and Video Computing, 2002*, December 2002, pp. 181–186.

[249] P. Kornprobst and G. Medioni, "Tracking segmented objects using tensor voting," in *Proceedings of IEEE Conference on Computer Vision and Pattern Recognition*, Hilton Head Island, SC, USA, June 2000, vol. 2.

[250] A. Narayanaswamy, Y. Wang, and B. Roysam, "3-D image pre-processing algorithms for improved automated tracing of neuronal arbors," *Neuroinformatics*, vol. 9, no. 2, pp. 219–231, 2011.

[251] N. Anjum and A. Cavallaro, "Multifeature object trajectory clustering for video analysis," *IEEE Transactions on Circuits and Systems for Video Technology*, vol. 18, no. 11, pp. 1555–1564, November 2008.

[252] J. G. Lee, J. Han, and X. Li, "A unifying framework of mining trajectory patterns of various temporal tightness," *IEEE Transactions on Knowledge and Data Engineering*, vol. 27, no. 6, pp. 1478–1490, June 2015.

[253] D. Mouillot and D. Viale, "Satellite tracking of a fin whale *(Balaenoptera physalus)* in the north-western Mediterranean Sea and fractal analysis of its trajectory," *Hydrobiologia*, vol. 452, no. 1, pp. 163–171, 2001.

[254] D. Zhang and J. P. G. Sterbenz, "Robustness analysis of mobile ad hoc networks using human mobility traces," in *International Conference on Design of Reliable Communication Networks*, Kansas City, MO, USA, March 2015, pp. 125–132.

[255] H. Liu and J. Li, "Unsupervised multi-target trajectory detection, learning and analysis in complicated environments," in *International Conference on Pattern Recognition*, Tsukuba Science City, Japan, November 2012, pp. 3716–3720.

[256] J. G. Ko and J. H. Yoo, "Rectified trajectory analysis based abnormal loitering detection for video surveillance," in *Proceedings of International Conference on Artificial Intelligence, Modelling and Simulation*, Kota Kinabalu, Malaysia, December 2013, pp. 289–293.

[257] "Deep learning wikipedia," https://en.wikipedia.org/wiki/Deep_learning.

[258] A. G. Ivakhnenko and V. G. Lapa, "Cybernetic predicting devices," DTIC Document, Tech. Rep., 1966.

[259] A. G. Ivakhnenko, "Polynomial theory of complex systems," *IEEE Transactions on Systems, Man, and Cybernetics*, vol. SMC-1, no. 4, pp. 364–378, October 1971.

[260] K. Fukushima, "Neocognitron: A self-organizing neural network model for a mechanism of pattern recognition unaffected by shift in position," *Biological Cybernetics*, vol. 36, no. 4, pp. 193–202, 1980.

[261] J. Schmidhuber, "Learning complex, extended sequences using the principle of history compression," *Neural Computation*, vol. 4, no. 2, pp. 234–242, March 1992.

[262] ——, "Deep learning in neural networks: An overview," *Neural Networks*, vol. 61, pp. 85–117, 2015.

[263] Jürgen Schmidhuber, Habilitation thesis, TUM, 1993.

[264] G. E. Hinton, P. Dayan, B. J. Frey, and R. M. Neal, "The 'wake-sleep' algorithm for unsupervised neural networks," *Science*, vol. 268, no. 5214, p. 1158, 1995.

[265] S. Hochreiter, "Untersuchungen zu dynamischen neuronalen netzen," in *Diploma, Technische Universitat Munchen*, p. 91, 1991.

[266] S. Hochreiter, Y. Bengio, P. Frasconi, and J. Schmidhuber, "Gradient flow in recurrent nets: The difficulty of learning long-term dependencies," 2001.

[267] J. Schmidhuber, "Deep Learning," *Scholarpedia*, vol. 10, no. 11, p. 32832, 2015, revision no. 152272.

[268] R. Dechter, "Learning while searching in constraint-satisfaction problems," *Proceedings of the 5th National Conference on Artificial Intelligence*. Philadelphia, PA, August 11–15, *Science*, vol. 1, 1986.

[269] I. Aizenberg, N. N. Aizenberg, and J. P. Vandewalle, *Multi-Valued and Universal Binary Neurons: Theory, Learning and Applications*, Springer Science & Business Media, 2013.

[270] G. E. Hinton, "Learning multiple layers of representation," *Trends in Cognitive Sciences*, vol. 11, no. 10, pp. 428–434, 2007.

[271] S. Hochreiter and J. Schmidhuber, "Long short-term memory," *Neural Computation*, vol. 9, no. 8, pp. 1735–1780, November 1997.

[272] A. Graves, S. Fernández, F. Gomez, and J. Schmidhuber, "Connectionist temporal classification: Labelling unsegmented sequence data with recurrent neural networks," in *Proceedings of the 23rd International Conference on Machine Learning*, ser. ICML '06. New York, NY, USA: ACM, 2006, pp. 369–376.

[273] H. Sak, A. W. Senior, and F. Beaufays, "Long short-term memory based recurrent neural network architectures for large vocabulary speech recognition," preprint, 2014, http://arXiv:1402.1128.

[274] X. Li and X. Wu, "Constructing long short-term memory based deep recurrent neural networks for large vocabulary speech recognition," *IEEE International Conference on Acoustics, Speech and Signal Processing (ICASSP)*, pp. 4520–4524, April 2015.

[275] H. Zen and H. Sak, "Unidirectional long short-term memory recurrent neural network with recurrent output layer for low-latency speech synthesis," in *Proceedings of the IEEE International Conference on Acoustics, Speech and Signal Processing*, Brisbane, Australia, April 2015, pp. 4470–4474.

[276] G. Hinton *et al.*, "Deep neural networks for acoustic modeling in speech recognition: The shared views of four research groups," *IEEE Signal Processing Magazine*, vol. 29, no. 6, pp. 82–97, November 2012.

[277] L. Deng, G. Hinton, and B. Kingsbury, "New types of deep neural network learning for speech recognition and related applications: An overview," in *2013 IEEE International Conference on Acoustics, Speech and Signal Processing*, May 2013, pp. 8599–8603.

[278] L. Deng, J. Li, J. T. Huang, K. Yao, D. Yu, F. Seide, M. Seltzer, G. Zweig, X. He, J. Williams, Y. Gong, and A. Acero, "Recent advances in deep learning for speech research at Microsoft," in *2013 IEEE International Conference on Acoustics, Speech and Signal Processing*, May 2013, pp. 8604–8608.

[279] L. Deng, O. Abdel-Hamid, and D. Yu, "A deep convolutional neural network using heterogeneous pooling for trading acoustic invariance with phonetic confusion," in *2013 IEEE International Conference on Acoustics, Speech and Signal Processing*, May 2013, pp. 6669–6673.

[280] T. N. Sainath, A.-R. Mohamed, B. Kingsbury, and B. Ramabhadran, "Deep convolutional neural networks for LVCSR," in *Proceedings of IEEE International Conference*

*on Acoustics, Speech and Signal Processing*, Vancouver, BC, Canada, May 2013, pp. 8614–8618.

[281] "Slides on deep learning," http://www.cs.nyu.edu/ỹann/talks/lecun-ranzato-icml2013.pdf.

[282] L. Deng and D. Yu, "Deep learning: Methods and applications," *Foundations and Trends in Signal Processing*, vol. 7, no. 34, pp. 197–387, 2014.

[283] D. Yu and L. Deng, *Automatic Speech Recognition: A Deep Learning Approach*, Springer, 2012.

[284] "Deng receives prestigious IEEE technical achievement award," http://blogs.technet.com/b/inside_microsoft_research/archive/2015/12/03/deng-receives-prestigious-ieee-technical-achievement-award.aspx.

[285] K.-S. Oh and K. Jung, "GPU implementation of neural networks," *Pattern Recognition*, vol. 37, no. 6, pp. 1311–1314, 2004.

[286] K. Chellapilla, S. Puri, and P. Simard, "High performance convolutional neural networks for document processing," Suvisoft, 2006.

[287] D. C. Ciresan, U. Meier, L. M. Gambardella, and J. Schmidhuber, "Deep big simple neural nets excel on handwritten digit recognition," preprint, 2010, http://arXiv:1003.0358.

[288] R. Raina, A. Madhavan, and A. Y. Ng, "Large-scale deep unsupervised learning using graphics processors," in *Proceedings of the Annual International Conference on Machine Learning*. New York, NY, USA: ACM, 2009, pp. 873–880.

[289] Y. LeCun, Y. Bengio, and G. Hinton, "Deep learning," *Nature*, vol. 521, no. 7553, pp. 436–444, 2015.

[290] A. Krizhevsky, I. Sutskever, and G. E. Hinton, "Imagenet classification with deep convolutional neural networks," in *Advances in Neural information Processing Systems*, ed. F. Pereira *et al.*, Curran Associates, Inc., 2012, pp. 1097–1105.

[291] I. Sutskever, O. Vinyals, and Q. V. Le, "Sequence to sequence learning with neural networks," in *Advances in Neural information Processing Systems*, ed. F. Pereira *et al.*, North Miami Beach, FL, USA: Curran Associates, Inc., 2014, pp. 3104–3112.

[292] K. Cho, B. van Merrienboer, Ç. Gülçehre, F. Bougares, H. Schwenk, and Y. Bengio, "Learning phrase representations using RNN encoder-decoder for statistical machine translation," preprint, 2014, http://arXiv:1406.1078.

[293] G. E. Hinton, S. Osindero, and Y.-W. Teh, "A fast learning algorithm for deep belief nets," *Neural computation*, vol. 18, no. 7, pp. 1527–1554, 2006.

[294] Y. Bengio, P. Lamblin, D. Popovici, and H. Larochelle, "Greedy layer-wise training of deep networks," in *Advances in Neural information Processing Systems*, vol. 19, 2007, p. 153.

[295] Y. LeCun, B. E. Boser, J. S. Denker, D. Henderson, R. E. Howard, W. E. Hubbard, and L. D. Jackel, "Handwritten digit recognition with a back-propagation network," in *Advances in Neural information Processing Systems*, ed. D. S. Touretzky, Morgan-Kaufmann, 1990, pp. 396–404.

[296] Y. Lecun, L. Bottou, Y. Bengio, and P. Haffner, "Gradient-based learning applied to document recognition," *Proceedings of the IEEE*, vol. 86, no. 11, pp. 2278–2324, November 1998.

[297] Y. Bengio, R. Ducharme, P. Vincent, and C. Janvin, "A neural probabilistic language model," *Journal of Machine Learning Research*, vol. 3, no. February, pp. 1137–1155, 2003.

[298] F. A. Gers, N. N. Schraudolph, and J. Schmidhuber, "Learning precise timing with LSTM recurrent networks," *Journal of Machine Learning Research*, vol. 3, pp. 115–143, August 2002.

[299] A. Graves, D. Eck, N. Beringer, and J. Schmidhuber, "Biologically plausible speech recognition with LSTM neural nets," in *International Workshop on Biologically Inspired Approaches to Advanced Information Technology*, Berlin, Heidelberg: Springer, 2004, pp. 127–136.

[300] S. Fernández, A. Graves, and J. Schmidhuber, "An application of recurrent neural networks to discriminative keyword spotting," in *International Conference on Artificial Neural Networks*, Berlin, Heidelberg: Springer, 2007, pp. 220–229.

[301] R. McMillan, "How Skype used AI to build its amazing new language translator," *Wire*, December 2014.

[302] A. Y. Hannun, C. Case, J. Casper, B. Catanzaro, G. Diamos, E. Elsen, R. Prenger, S. Satheesh, S. Sengupta, A. Coates, and A. Y. Ng, "Deep speech: Scaling up end-to-end speech recognition," preprint, 2014, http://arXiv:1412.5567.

[303] "Plenary speakers," www.icassp2016.org/PlenarySpeakers.asp.

[304] L. Deng, "Achievements and challenges of deep learning," Microsoft, 2015.

[305] D. Ciregan, U. Meier, and J. Schmidhuber, "Multi-column deep neural networks for image classification," in *IEEE Conference on Computer Vision and Pattern Recognition*, Rhode Island, RI, USA, June 2012, pp. 3642–3649.

[306] O. Vinyals, A. Toshev, S. Bengio, and D. Erhan, "Show and tell: A neural image caption generator," in *IEEE Conference on Computer Vision and Pattern Recognition*, Boston, MA, USA, June 2015, pp. 3156–3164.

[307] H. Fang, S. Gupta, F. Iandola, R. K. Srivastava, L. Deng, P. Dollr, J. Gao, X. He, M. Mitchell, J. C. Platt, C. L. Zitnick, and G. Zweig, "From captions to visual concepts and back," in *IEEE Conference on Computer Vision and Pattern Recognition*, Boston, MA, USA, June 2015, pp. 1473–1482.

[308] R. Kiros, R. Salakhutdinov, and R. S. Zemel, "Unifying visual-semantic embeddings with multimodal neural language models," preprint, 2014, http://arXiv:1411.2539.

[309] S.-h. Zhong, Y. Liu, and Y. Liu, "Bilinear deep learning for image classification," in *Proceedings of the 19th ACM International Conference on Multimedia*, New York, NY, USA, 2011, pp. 343–352.

[310] "Nvidia demos a car computer trained with 'deep learning'," www.technology review.com/s/533936/ces-2015-nvidia-demos-a-car-computer-trained-with-deep-learning/.

[311] "What is a driverless car?," www.wisegeek.com/what-is-a-driverless-car.htm.

[312] "Self-driving cars now legal in California," www.cnn.com/2012/09/25/tech/innovation /self-driving-car-california/.

[313] S. Thrun, "Toward robotic cars," *Communications of the ACM*, vol. 53, no. 4, pp. 99–106, 2010.

[314] S. K. Gehrig and F. J. Stein, "Dead reckoning and cartography using stereo vision for an autonomous car," in *Proceedings of International Conference on Intelligent Robots and Systems*, Kyongju, Korea, 1999, vol. 3, pp. 1507–1512.

[315] "The beginning of the end of driving," www.motortrend.com/news/the-beginning-of -the-end-of-driving/.

[316] "European roadmap smart systems for automated driving," www.smart-systems-integr ation.org/public/documents/publications/EPoSS%20Roadmap_Smart%20Systems%20for %20Automated%20Driving_2015_V1.pdf.

[317] W. Zhu, J. Miao, J. Hu, and L. Qing, "Vehicle detection in driving simulation using extreme learning machine," *Neurocomputing*, vol. 128, pp. 160–165, 2014.

[318]  F. A. Gers and E. Schmidhuber, "LSTM recurrent networks learn simple context-free and context-sensitive languages," *IEEE Transactions on Neural Networks*, vol. 12, no. 6, pp. 1333–1340, November 2001.

[319]  R. Józefowicz, O. Vinyals, M. Schuster, N. Shazeer, and Y. Wu, "Exploring the limits of language modeling," preprint, 2016, http://arXiv:1602.02410.

[320]  D. Gillick, C. Brunk, O. Vinyals, and A. Subramanya, "Multilingual language processing from bytes," preprint, 2015, http://arXiv:1512.00103.

[321]  R. Socher, J. Bauer, C. D. Manning, and A. Y. Ng, "Parsing with compositional vector grammars," in *ACL*, 2013.

[322]  R. Socher, A. Perelygin, J. Y. Wu, J. Chuang, C. D. Manning, A. Y. Ng, and C. Potts, "Recursive deep models for semantic compositionality over a sentiment treebank," in *Proceedings of the Conference on Empirical Methods in Natural Language Processing*, Citeseer, 2013, vol. 1631, p. 1642.

[323]  Y. Shen, X. He, J. Gao, L. Deng, and G. Mesnil, "A latent semantic model with convolutional-pooling structure for information retrieval," in *Proceedings of the 23rd ACM International Conference on Conference on Information and Knowledge Management*, 2014, pp. 101–110.

[324]  P.-S. Huang, X. He, J. Gao, L. Deng, A. Acero, and L. Heck, "Learning deep structured semantic models for web search using clickthrough data," in *Proceedings of the 22nd ACM International Conference on Conference on Information and Knowledge Management*, San Francisco, CA, USA, October 2013, pp. 2333–2338.

[325]  G. Mesnil, Y. Dauphin, K. Yao, Y. Bengio, L. Deng, D. Hakkani-Tur, X. He, L. Heck, G. Tur, D. Yu, and G. Zweig, "Using recurrent neural networks for slot filling in spoken language understanding," *IEEE/ACM Transactions on Audio, Speech, and Language Processing*, vol. 23, no. 3, pp. 530–539, March 2015.

[326]  J. Gao, X. He, S. W. tau Yih, and L. Deng, "Learning continuous phrase representations for translation modeling," in *ACL*. Citeseer, 2014.

[327]  J. Gao, P. Pantel, M. Gamon, X. He, and L. Deng, "Modeling interestingness with deep neural networks," Tech. Rep., October 2014, www.microsoft.com/ en-us /research/publication/modeling-interestingness-with-deep-neural-networks/.

[328]  X. He, J. Gao, and L. Deng, "Deep learning for natural language processing: Theory and practice (tutorial)," in *Proceedings of the 23rd ACM International Conference on Information and Knowledge Management*, Shanghai, China, November 2014.

[329]  "Merck molecular activity challenge," www.kaggle.com/c/MerckActivity/details/winners.

[330]  G. E. Dahl, N. Jaitly, and R. Salakhutdinov, "Multi-task neural networks for QSAR predictions," preprint, 2014, http://arXiv:1406.1231.

[331]  "Merck molecular activity challenge," https://tripod.nih.gov/tox21/challenge/leader board.jsp.

[332]  "NCATS announces Tox21 Data Challenge Winners," https://tripod.nih.gov/tox21/chal lenge/leaderboard.jsp.

[333]  I. Wallach, M. Dzamba, and A. Heifets, "Atomnet: A deep convolutional neural network for bioactivity prediction in structure-based drug discovery," preprint, 2015, http://arXiv:1510.02855.

[334]  Y. Tkachenko, "Autonomous CRM control via CLV approximation with deep reinforcement learning in discrete and continuous action space," preprint, 2015, http://arXiv:1504.01840.

[335] A. van den Oord, S. Dieleman, and B. Schrauwen, "Deep content-based music recommendation," in *Advances in Neural Information Processing Systems*, Lake Tahoe, CA, USA, December 2013, pp. 2643–2651.

[336] A. M. Elkahky, Y. Song, and X. He, "A multi-view deep learning approach for cross domain user modeling in recommendation systems," in *Proceedings of the 24th International Conference on World Wide Web*, ser. WWW '15. New York, NY, USA: ACM, 2015, pp. 278–288, http://doi.acm.org/10.1145/2736277.2741667.

[337] D. Chicco, P. Sadowski, and P. Baldi, "Deep autoencoder neural networks for gene ontology annotation predictions," in *Proceedings of the 5th ACM Conference on Bioinformatics, Computational Biology, and Health Informatics*, ACM, Newport Beach, CA, USA, September 2014, pp. 533–540.

[338] Y. Bengio, A. Courville, and P. Vincent, "Representation learning: A review and new perspectives," *IEEE Transactions on Pattern Analysis and Machine Intelligence*, vol. 35, no. 8, pp. 1798–1828, August 2013.

[339] G. E. Hinton, "Deep belief networks," *Scholarpedia*, vol. 4, no. 5, p. 5947, 2009.

[340] M. A. Carreira-Perpinan and G. Hinton, "On contrastive divergence learning," in *Artificial Intelligence and Statistics Conference*, vol. 10, pp. 33–40, 2005.

[341] G. E. Hinton, *A Practical Guide to Training Restricted Boltzmann Machines*, Springer, 2012, pp. 599–619.

[342] ——, "Products of experts," in *Ninth International Conference on Artificial Neural Networks*, vol. 1. Edinburgh, UK: IET, September 1999, pp. 1–6.

[343] ——, "Training products of experts by minimizing contrastive divergence," *Neural Computation*, vol. 14, no. 8, pp. 1771–1800, 2002.

[344] Y. Bengio, "Learning deep architectures for AI," *Foundations and trends® in Machine Learning*, vol. 2, no. 1, pp. 1–127, 2009.

[345] C. Szegedy, A. Toshev, and D. Erhan, "Deep neural networks for object detection," in *Advances in Neural Information Processing Systems*, Lake Tahoe, CA, USA, December 2013, pp. 2553–2561.

[346] H. Larochelle, D. Erhan, A. Courville, J. Bergstra, and Y. Bengio, "An empirical evaluation of deep architectures on problems with many factors of variation," in *Proceedings of the International Conference on Machine Learning*, ACM, Corvallis, OR, USA, June 2007, pp. 473–480.

[347] A. Fischer and C. Igel, "Training restricted Boltzmann machines: An introduction," *Pattern Recognition*, vol. 47, no. 1, pp. 25–39, 2014.

[348] Y. LeCun, B. Boser, J. S. Denker, D. Henderson, R. E. Howard, W. Hubbard, and L. D. Jackel, "Backpropagation applied to handwritten zip code recognition," *Neural Computation*, vol. 1, no. 4, pp. 541–551, December 1989.

[349] J. J. Weng, N. Ahuja, and T. S. Huang, "Learning recognition and segmentation of 3-D objects from 2-D images," in *Proceedings of Fourth International Conference on Computer Vision, 1993*. Berlin, Germany, May 1993, pp. 121–128.

[350] "Convolutional neural network," http://ufldl.stanford.edu/tutorial/supervised/Convolutiona lNeuralNetwork/.

[351] C. Szegedy, W. Liu, Y. Jia, P. Sermanet, S. Reed, D. Anguelov, D. Erhan, V. Vanhoucke, and A. Rabinovich, "Going deeper with convolutions," in *IEEE Conference on Computer Vision and Pattern Recognition*, June 2015, pp. 1–9.

[352] A. Krizhevsky, "Convolutional deep belief networks on cifar-10," www.cs.toronto .edu/k̃riz/conv-cifar10-aug2010.pdf, not published, 2010.

[353] H. Lee, R. Grosse, R. Ranganath, and A. Y. Ng, "Convolutional deep belief networks for scalable unsupervised learning of hierarchical representations," in *Proceedings of the Annual International Conference on Machine Learning*, ser. ICML '09. New York, NY, USA: ACM, 2009, pp. 609–616.

[354] A. Graves, M. Liwicki, S. Fernndez, R. Bertolami, H. Bunke, and J. Schmidhuber, "A novel connectionist system for unconstrained handwriting recognition," *IEEE Transactions on Pattern Analysis and Machine Intelligence*, vol. 31, no. 5, pp. 855–868, May 2009.

[355] J. Bayer, D. Wierstra, J. Togelius, and J. Schmidhuber, "Evolving memory cell structures for sequence learning," in *International Conference on Artificial Neural Networks*, Springer, 2009, pp. 755–764.

[356] S. Fernndez, A. Graves, and J. Schmidhuber, "Sequence labelling in structured domains with hierarchical recurrent neural networks," in *Proceedings International Joint Conference on Artificial Intelligence*, Hyderabad, India, January 2007, pp. 774–779.

[357] A. Graves and J. Schmidhuber, "Offline handwriting recognition with multidimensional recurrent neural networks," in *Advances in Neural Information Processing Systems 21*, North Miami Beach, FL, USA: Curran Associates, Inc., 2009, pp. 545–552.

[358] B. Fan, L. Wang, F. K. Soong, and L. Xie, "Photo-real talking head with deep bidirectional LSTM," in *IEEE International Conference on Acoustics, Speech and Signal Processing*, April 2015, pp. 4884–4888.

[359] "Google voice search: Faster and more accurate," https://research.googleblog.com/2015/09/google-voice-search-faster-and-more.html.

[360] Cisco, "Cisco visual networking index: Global mobile data traffic forecast update 2015–2020," *White Paper*, 2016.

[361] Apache Spark, "Apache Spark: Lightning-fast cluster computing," 2016, http://spark.apache.org.

[362] O. D. Lara and M. A. Labrador, "A survey on human activity recognition using wearable sensors," *IEEE Communications Surveys & Tutorials*, vol. 15, no. 3, pp. 1192–1209, 2013.

[363] G. M. Weiss and J. W. Lockhart, "The impact of personalization on smartphone-based activity recognition," in *AAAI Workshop on Activity Context Representation: Techniques and Languages*, Palo Alto, CA, USA, 2012.

[364] C. Perera, A. Zaslavsky, P. Christen, and D. Georgakopoulos, "Context aware computing for the Internet of things: A survey," *IEEE Communications Surveys & Tutorials*, vol. 16, no. 1, pp. 414–454, 2014.

[365] P. Vincent, H. Larochelle, I. Lajoie, Y. Bengio, and P.-A. Manzagol, "Stacked denoising autoencoders: Learning useful representations in a deep network with a local denoising criterion," *The Journal of Machine Learning Research*, vol. 11, pp. 3371–3408, 2010.

[366] X. Wang, L. Gao, S. Mao, and S. Pandey, "DeepFi: Deep learning for indoor fingerprinting using channel state information," in *IEEE Wireless Communications and Networking Conference*, March 2015, pp. 1666–1671.

[367] N. D. Lane and P. Georgiev, "Can deep learning revolutionize mobile sensing?" in *Proceedings of the 16th International Workshop on Mobile Computing Systems and Applications*, ACM, 2015, pp. 117–122.

[368] J. Ngiam, A. Khosla, M. Kim, J. Nam, H. Lee, and A. Y. Ng, "Multimodal deep learning," in *Proceedings of the International Conference on Machine Learning*, 2011, pp. 689–696.

[369] J. Dean *et al.*, "Large scale distributed deep networks," in *Advances in Neural Information Processing Systems*, Lake Tahoe, NV, USA, December 2012, pp. 1223–1231.

[370] K. Zhang and X.-w. Chen, "Large-scale deep belief nets with MapReduce," *IEEE Access*, vol. 2, pp. 395–403, 2014.

[371] J. W. Lockhart, G. M. Weiss, J. C. Xue, S. T. Gallagher, A. B. Grosner, and T. T. Pulickal, "Design considerations for the WISDM smart phone-based sensor mining architecture," in *Proceedings of the 5th International Workshop on Knowledge Discovery from Sensor Data*, ACM, 2011, pp. 25–33.

[372] L. von Ahn, B. Maurer, C. McMillen, D. Abraham, and M. Blum, "reCAPTCHA: Human-based character recognition via web security measures," *Science*, vol. 321, no. 5895, pp. 1465–1468, 2008.

[373] P. Klemperer, *Auctions: Theory and Practice*, ser. Princeton, NJ, USA: Princeton University Press, 2004.

[374] H. Abu-Ghazaleh and A. S. Alfa, "Application of mobility prediction in wireless networks using markov renewal theory," *IEEE Transactions on Vehicular Technology*, vol. 59, no. 2, pp. 788–802, February 2010.

[375] D. Katsaros and Y. Manolopoulos, "Prediction in wireless networks by markov chains," *IEEE Wireless Communications*, vol. 16, no. 2, pp. 56–64, April 2009.

[376] J.-K. Lee and J. C. Hou, "Modeling steady-state and transient behaviors of user mobility: Formulation, analysis, and application," in *Proceedings of the 7th ACM International Symposium on Mobile Ad Hoc Networking and Computing*, ser. MobiHoc '06. New York, NY, USA: ACM, 2006, pp. 85–96.

[377] B. P. Clarkson, "Life patterns: Structure from wearable sensors," https://dspace.mit .edu/handle/1721.1/8030, 2002.

[378] N. Eagle and A. S. Pentland, "Eigenbehaviors: Identifying structure in routine," *Behavioral Ecology and Sociobiology*, vol. 63, no. 7, pp. 1057–1066, 2009.

[379] W.-C. Peng and M.-S. Chen, "Mining user moving patterns for personal data allocation in a mobile computing system," in *Proceedings of the 2000 International Conference on Parallel Processing*, August 2000, pp. 573–580.

[380] J. Chung, O. Paek, J. Lee, and K. Ryu, "Temporal pattern mining of moving objects for location-based service," in *Database and Expert Systems Applications*, Springer, 2002, pp. 331–340.

[381] J. Reades, F. Calabrese, and C. Ratti, "Eigenplaces: Analysing cities using the space–time structure of the mobile phone network," *Environment and Planning B: Planning and Design*, vol. 36, no. 5, pp. 824–836, 2009.

[382] F. Calabrese, J. Reades, and C. Ratti, "Eigenplaces: Segmenting space through digital signatures," *IEEE Pervasive Computing*, vol. 9, no. 1, pp. 78–84, January 2010.

[383] I. Arel, D. C. Rose, and T. P. Karnowski, "Deep machine learning: A new frontier in artificial intelligence research [research frontier]," *IEEE Computational Intelligence Magazine*, vol. 5, no. 4, pp. 13–18, November 2010.

[384] G. E. Hinton and R. R. Salakhutdinov, "Reducing the dimensionality of data with neural networks," *Science*, vol. 313, no. 5786, pp. 504–507, 2006.

[385] Y. W. Teh and M. I. Jordan, "Hierarchical Bayesian nonparametric models with applications," in *Bayesian Nonparametrics: Principles and Practice*, Cambridge, UK: Cambridge University Press, 2010.

[386] R. Thibaux and M. I. Jordan, "Hierarchical beta processes and the Indian buffet process," in *Artificial Intelligence and Statistics Conference*, vol. 2, pp. 564–571, 2007.

[387] C. E. Rasmussen, "The infinite Gaussian mixture model," in *Advances in Neural Information Processing Systems*, Cambridge, MA, USA: MIT Press, vol. 12., 2000, pp. 554–560.

[388] B. Chen, G. Polatkan, G. Sapiro, L. Carin, and D. B. Dunson, "The hierarchical beta process for convolutional factor analysis and deep learning," in *Proceedings of the International Conference on Machine Learning*, New York, NY, USA: ACM, 2011, pp. 361–368.

[389] F. Wood, "A non-parametric Bayesian method for inferring hidden causes," in *Proceedings of the Twenty-Second Conference on Uncertainty in Artificial Intelligence*, AUAI Press, 2006, pp. 536–543.

[390] D. Knowles and Z. Ghahramani, "Infinite sparse factor analysis and infinite independent components analysis," in *International Conference on Independent Component Analysis and Signal Separation*, Springer Berlin Heidelberg, 2007, pp. 381–388.

[391] T. L. Griffiths and Z. Ghahramani, "The Indian buffet process: An introduction and review," *Journal of Machine Learning Research*, vol. 12, pp. 1185–1224, July 2011.

[392] M. A. Carreira-Perpiñ and G. Hinton, "On contrastive divergence learning," in *Artificial Intelligence and Statistics Conference*, vol. 10, pp. 33–40, 2005.

[393] E. J. Candès, J. Romberg, and T. Tao, "Robust uncertainty principles: Exact signal reconstruction from highly incomplete frequency information," *IEEE Transactions on Information Theory*, vol. 52, no. 2, pp. 489–509, February 2006.

[394] E. Candes, J. Romberg, and T. Tao, "Stable signal recovery from incomplete and inaccurate information," *Communications on Pure and Applied Mathematics*, vol. 2005, no. 59, pp. 1207–1233, 2005.

[395] E. J. Candès and T. Tao, "Near optimal signal recovery from random projections: Universal encoding strategies?" *IEEE Transactions on Information Theory*, vol. 52, no. 12, pp. 5406–5425, December 2006.

[396] D. Donoho, "Compressed sensing," *IEEE Transactions on Information Theory*, vol. 52, no. 4, pp. 1289–1306, April 2006.

[397] H. Nyquist, "Certain topics in telegraph transmission theory," *Transactions of the American Institute of Electrical Engineers*, vol. 47, no. 2, pp. 617–644, April 1928.

[398] C. Shannon, "Communication in the presence of noise," *Proc. Institute of Radio Engineers*, vol. 37, no. 1, pp. 10–21, 1949.

[399] B. K. Natarajan, "Sparse approximate solutions to linear systems," *SIAM Journal on Computing*, vol. 24, no. 2, pp. 227–234, 1995.

[400] M. Elad, *Sparse and Redundant Representations: From Theory to Applications in Signal and Image Processing*, Springer Verlag, 2010.

[401] J. Starck, F. Murtagh, and J. Fadili, *Sparse Image and Signal Processing: Wavelets, Curvelets, Morphological Diversity*, Cambridge, UK: Cambridge University Press, 2010.

[402] J. Starck, E. Candes, and D. Donoho, "The curvelet transform for image denoising," *IEEE Transactions on Image Processing*, vol. 11, no. 6, pp. 670–684, June 2002.

[403] B. A. Olshausen and D. J. Field, "Emergence of simple-cell receptive field properties by learning a sparse code for natural images," *Nature*, vol. 381, no. 6583, pp. 607–609, 1996.

[404] K. Engan, S. Aase, and J. Husoy, "Multi-frame compression: Theory and design," *Signal Processing*, vol. 80, no. 10, pp. 2121–2140, 2000.

[405] M. Aharon, M. Elad, and A. Bruckstein, "K-SVD: An algorithm for designing overcomplete dictionaries for sparse representation," *IEEE Transactions on Signal Processing*, vol. 54, no. 11, pp. 4311–4322, November 2006.

[406] M. Yuan and Y. Lin, "Model selection and estimation in regression with grouped variables," *Journal of the Royal Statistical Society: Series B*, vol. 68, no. 1, pp. 49–67, April 2006.

[407] J. Chen and X. Huo, "Theoretical results on sparse representations of multiple-measurement vectors," *IEEE Transactions on Signal Processing*, vol. 54, no. 12, pp. 4634–4643, December 2006.

[408] F. Bach, "Consistency of the group lasso and multiple kernel learning," *The Journal of Machine Learning Research*, vol. 9, pp. 1179–1225, 2008.

[409] D. Malioutov, M. Cetin, and A. Willsky, "A sparse signal reconstruction perspective for source localization with sensor arrays," *IEEE Transactions on Signal Processing*, vol. 53, no. 8, pp. 3010–3022, August 2005.

[410] S. Cotter, B. Rao, K. Engan, and K. Kreutz-Delgado, "Sparse solutions to linear inverse problems with multiple measurement vectors," *IEEE Transactions on Signal Processing*, vol. 53, no. 7, pp. 2477–2488, July 2005.

[411] J. Meng, W. Yin, H. Li, E. Hossain, and Z. Han, "Collaborative spectrum sensing from sparse observations in cognitive radio networks," *IEEE Journal on Selected Topics on Communications Special Issue on Advances in Cognitive Radio Networking and Communications*, vol. 29, no. 2, pp. 327–337, February 2011.

[412] M. Fazel, "Matrix rank minimization with applications," PhD dissertation, Stanford University, 2002.

[413] E. Candes and B. Recht, "Exact matrix completion via convex optimization," *Foundations of Computational Mathematics*, vol. 9, no. 6, pp. 717–772, 2009.

[414] Z. Liu and L. Vandenberghe, "Interior-point method for nuclear norm approximation with application to system identification," *SIAM Journal on Matrix Analysis and Applications*, vol. 31, no. 3, pp. 1235–1256, 2009.

[415] A. So and Y. Ye, "Theory of semidefinite programming for sensor network localization," *Mathematical Programming*, vol. 109, no. 2, pp. 367–384, 2007.

[416] C. Tomasi and T. Kanade, "Shape and motion from image streams under orthography: A factorization method," *International Journal of Computer Vision*, vol. 9, no. 2, pp. 137–154, 1992.

[417] T. Morita and T. Kanade, "A sequential factorization method for recovering shape and motion from image streams," *IEEE Transactions on Pattern Analysis and Machine Intelligence*, vol. 19, no. 8, pp. 858–867, August 1997.

[418] D. Goldberg, D. Nichols, B. Oki, and D. Terry, "Using collaborative filtering to weave an information tapestry," *Communications of the ACM*, vol. 35, no. 12, pp. 61–70, 1992.

[419] Y. Eldar and M. Mishali, "Robust recovery of signals from a structured union of subspaces," *IEEE Transactions on Information Theory*, vol. 55, no. 11, pp. 5302–5316, November 2009.

[420] Y. Lu and M. Do, "Sampling signals from a union of subspaces," *IEEE Signal Processing Magazine*, vol. 25, no. 2, pp. 41–47, March 2008.

[421] E. Candes and J. Romberg, "Sparsity and incoherence in compressive sampling," *Inverse Problems*, vol. 23, no. 3, pp. 969–985, 2007.

[422] P. Feng and Y. Bresler, "Spectrum-blind minimum-rate sampling and reconstruction of multiband signals," in *Proceedings of IEEE International Conference on Acoustics, Speech, and Signal Processing*, vol. 3, May 1996, pp. 1688–1691.

[423] M. Vetterli, P. Marziliano, and T. Blu, "Sampling signals with finite rate of innovation," *IEEE Transactions on Signal Processing*, vol. 50, no. 6, pp. 1417–1428, June 2002.

[424] E. Candes and T. Tao, "Decoding by linear programming," *IEEE Transactions on Information Theory*, vol. 51, no. 12, pp. 4203–4215, December 2005.

[425] Y. Zhang, "Theory of compressive sensing via 1-minimization: A non-rip analysis and extensions," *Journal of the Operations Research Society of China*, vol. 1, no. 1, pp. 79–105, 2013.

[426] E. Candes and Y. Plan, "A probabilistic and RIPless theory of compressed sensing," *IEEE Transactions on Information Theory*, vol. 57, no. 11, pp. 7235–7254, November 2010.

[427] D. Donoho and X. Huo, "Uncertainty principles and ideal atomic decompositions," *IEEE Transactions on Information Theory*, vol. 47, no. 7, pp. 2845–2862, November 2001.

[428] R. Gribonval and M. Nielsen, "Sparse representations in unions of bases," *IEEE Transactions on Information Theory*, vol. 49, no. 12, pp. 3320–3325, December 2003.

[429] Y. Zhang, "A simple proof for recoverability of $\ell_1$-minimization: Go over or under?" *Rice University CAAM Technical Report TR05-09*, 2005.

[430] A. Cohen, W. Dahmen, and R. A. DeVore, "Compressed sensing and best $k$-term approximation," *Journal of the American Mathematical Society*, vol. 22, no. 1, pp. 211–231, 2009.

[431] E. Candes, "The restricted isometry property and its implications for compressed sensing," *Comptes Rendus Mathematique*, vol. 346, no. 9–10, pp. 589–592, 2008.

[432] S. Foucart and M. Lai, "Sparsest solutions of underdetermined linear systems via $\ell q$-minimization for $0 < q \leq 1$," *Applied and Computational Harmonic Analysis*, vol. 26, no. 3, pp. 395–407, 2009.

[433] S. Foucart, "A note on guaranteed sparse recovery via $\ell_1$-minimization," *Applied and Computational Harmonic Analysis*, vol. 29, no. 1, pp. 97–103, July 2010.

[434] T. Cai, L. Wang, and G. Xu, "Shifting inequality and recovery of sparse signals," *IEEE Transactions on Signal Processing*, vol. 58, no. 3, pp. 1300–1308, March 2010.

[435] Q. Mo and S. Li, "New bounds on the restricted isometry constant $\delta_{2k}$," *Applied and Computational Harmonic Analysis*, vol. 31, no. 3, pp. 460–468, 2011.

[436] M. Davenport, "PhD thesis: Random observations on random observations: Sparse signal acquisition and processing," PhD dissertation, 2010.

[437] R. Baraniuk, M. Davenport, R. Devore, and M. Wakin, "A simple proof of the restricted isometry property for random matrices," *Constructive Approximation*, vol. 28, no. 3, pp. 253–263, 2007.

[438] S. Mendelson, A. Pajor, and N. Tomczak-Jaegermann, "Uniform uncertainty principle for Bernoulli and subgaussian ensembles," *Constructive Approximation*, vol. 28, no. 3, pp. 277–289, 2008.

[439] H. Rauhut, "Compressive sensing and structured random matrices," *Theoretical Foundations and Numerical Methods for Sparse Recovery*, vol. 9, pp. 1–92, 2010.

[440] J. Bourgain, S. Dilworth, K. Ford, S. Konyagin, and D. Kutzarova, "Explicit constructions of RIP matrices and related problems," *Duke Mathematical Journal*, vol. 159, no. 1, pp. 145–185, 2011.

[441] J. Haupt, L. Applebaum, and R. Nowak, "On the restricted isometry of deterministically subsampled Fourier matrices," *Proceedings of the 44th Annual Conference on Information Sciences and Systems*, pp. 1–6, Princeton, NJ, USA, March 2010.

[442] P. Indyk, "Explicit constructions for compressed sensing of sparse signals," *SODA '08 Proceedings of the Nineteenth Annual ACM-SIAM Symposium on Discrete Algorithms*, pp. 30–33, San Francisco, CA, USA, January 20–22, 2008.

[443] S. Vavasis, "Derivation of compressive sensing theorems from the spherical section property," *University of Waterloo, CO*, vol. 769, 2009.

[444] B. S. Kashin, "Diameters of some finite-dimensional sets and classes of smooth functions," *Mathematics of the USSR-Izvestiya*, vol. 11, p. 317, 1977.

[445] A. Garnaev and E. D. Gluskin, "The widths of a Euclidean ball," *Dokl. Akad. Nauk SSSR*, vol. 277, no. 5, pp. 1048–1052, 1984.

[446] D. Du and F. Hwang, *Combinatorial Group Testing and Its Applications*. World Scientific Pub. Co. Inc., 2000.

[447] R. Berinde, A. Gilbert, P. Indyk, H. Karloff, and M. Strauss, "Combining geometry and combinatorics: A unified approach to sparse signal recovery," in *2008 46th Annual Allerton Conference on Communication, Control, and Computing*, September 2008, pp. 798–805.

[448] A. Gilbert and P. Indyk, "Sparse recovery using sparse matrices," *Proceedings of the IEEE*, vol. 98, no. 6, pp. 937–947, June 2010.

[449] A. Gilbert, M. Strauss, J. Tropp, and R. Vershynin, "One sketch for all: Fast algorithms for compressed sensing," 2007, pp. 237–246.

[450] A. C. Gilbert, Y. Li, E. Porat, and M. J. Strauss, "Approximate sparse recovery: Optimizing time and measurements," *SIAM Journal on Computing*, vol. 41, no. 2, pp. 436–453, 2012.

[451] Y. Zhang, "When is missing data recoverable?" *Rice University CAAM Technical Report TR06-15*, 2006.

[452] http://dsp.rice.edu/cs.

[453] J. A. Tropp, J. N. Laska, M. F. Duarte, J. K. Romberg, and R. G. Baraniuk, "Beyond Nyquist: Efficient sampling of sparse bandlimited signals," *IEEE Transactions on Information Theory*, vol. 56, no. 1, pp. 520–544, January 2010.

[454] S. Kirolos, J. Laska, M. Wakin, M. Duarte, D. Baron, T. Ragheb, Y. Massoud, and R. Baraniuk, "Analog-to-information conversion via random demodulation," in *IEEE Dallas Circuits and Systems Workshop*, Dallas, October 2006.

[455] J. N. Laska, S. Kirolos, M. F. Duarte, T. S. Ragheb, R. G. Baraniuk, and Y. Massoud, "Analog-to-information conversion via random demodulation," in *IEEE International Symposium on Circuits and Systems, ISCAS*, New Orleans, LA, USA, May 2007.

[456] M. Mishali and Y. C. Eldar, "From theory to practice: Sub-Nyquist sampling of sparse wideband analog signals," *IEEE Journal of Selected Topics in Signal Processing*, vol. 4, no. 2, pp. 375–391, April 2010.

[457] ——, "Expected rip: Conditioning of the modulated wideband converter," in *2009 IEEE Information Theory Workshop*, Sicily, Italy, October 2009.

[458] M. Mishali, Y. C. Eldar, and J. A. Tropp, "Efficient sampling of sparse wideband analog signals," in *IEEE Convention of Electrical and Electronics Engineers*, Israel, December 2008, pp. 290–294.

[459] M. Mishali, Y. C. Eldar, and A. Elron, "Xampling: Signal acquisition and processing in union of subspaces," *IEEE Transactions on Signal Processing*, vol. 59, no. 10, pp. 4719–4734, October 2011.

[460] T. Michaeli and Y. C. Eldar, "Xampling at the rate of innovation," *IEEE Transactions on Signal Processing*, vol. 60, no. 3, pp. 1121–1133, March 2012.

[461] K. Gedalyahu and Y. Eldar, "Time-delay estimation from low-rate samples: A union of subspaces approach," *IEEE Transactions on Signal Processing*, vol. 58, no. 6, pp. 3017–3031, June 2011.

[462] E. Matusiak and Y. Eldar, "Sub-nyquist sampling of short pulses," *IEEE Transactions on Signal Processing*, vol. 60, no. 3, pp. 1134–1148, March 2012.

[463] M. Mishali and Y. C. Eldar, "Xampling: Compressed sensing of analog signals," *Compressed Sensing Theory and Applications*, Cambridge, UK: Cambridge University Press, 2012.

[464] M. Satyanarayanan, "Pervasive computing: Vision and challenges," *IEEE Personal Communications*, vol. 8, no. 4, pp. 10–17, August 2001.

[465] R. Glidden, C. Bockorick, S. Cooper, C. Diorio, D. Dressler, V. Gutnik, C. Hagen, D. Hara, T. Hass, T. Humes, J. Hyde, R. Oliver, O. Onen, A. Pesavento, K. Sundstrom, and M. Thomas, "Design of ultra-low-cost uhf rfid tags for supply chain applications," *IEEE Communications Magazine*, vol. 42, no. 8, pp. 140–151, August 2004.

[466] L. Mo, Y. He, Y. Liu, J. Zhao, S.-J. Tang, X.-Y. Li, and G. Dai, "Canopy closure estimates with greenorbs: Sustainable sensing in the forest," in *Proceedings of the 7th ACM Conference on Embedded Networked Sensor Systems*, ser. SenSys '09. New York, NY, USA: ACM, 2009, pp. 99–112.

[467] A. Goldsmith, *Wireless Communications*, Cambridge, UK: Cambridge University Press, 2005.

[468] T. S. Rappaport, *Wireless Communications: Principles and Practice* (2nd ed.) USA: Prentice Hall, 2001.

[469] S. S. Chen, D. L. Donoho, and M. A. Saunders, "Atomic decomposition by basis pursuit," *SIAM Journal on Scientific Computing (SISC)*, vol. 20, no. 1, pp. 33–61, 1998.

[470] W. U. Bajwa, J. Haupt, A. M. Sayeed, and R. Nowak, "Compressed channel sensing: A new approach to estimating sparse multipath channels," *Proceedings of the IEEE*, vol. 98, no. 6, pp. 1058–1076, June 2010.

[471] J. L. Paredes, G. R. Arce, and Z. Wang, "Ultra-wideband compressed sensing channel estimation," *IEEE Journal of Selected Topics in Signal Processing*, vol. 1, no. 3, pp. 383–395, October 2007.

[472] C. R. Berger, Z. Wang, Z. Huang, and S. Zhou, "Application of compressive sensing to sparse channel estimation," *IEEE Communications Magazine*, vol. 48, pp. 164–174, November 2010.

[473] P. Zhang, Z. Hu, R. C. Qiu, and B. M. Sadler, "A compressive sensing based ultra-wideband communication system," in *Proceedings of IEEE International Conference on Communications*, 2009, pp. 1–5.

[474] J. Romberg, "Multiple channel estimation using spectrally random probes," in *Proc. SPIE Wavelets XIII*, 2009.

[475] W. U. Bajwa, A. Sayeed, and R. Nowak, "Compressed sensing of wireless channels in time, frequency, and space," in *2008 42nd Asilomar Conference on Signals, Systems and Computers*, October 2008, pp. 2048–2052.

[476] Z. Sahinoglu, S. Gezici, and I. Guvenc, *Ultra-wideband Positioning Systems: Theoretical LImits, Ranging Algorithms and Protocols*, Cambridge, UK: Cambridge University Press, 2011.

[477] G. Staple and K. Werbach, "The end of spectrum scarcity," *IEEE Spectrum Archive*, vol. 41, no. 3, pp. 48–52, March 2004.

[478] S. Haykin, "Cognitive radio: Brain-empowered wireless communications," *IEEE Journal on Selected areas in Communications*, vol. 23, no. 2, pp. 201–220, February 2005.

[479] H. Kim and K. G. Shin, "Efficient discovery of spectrum opportunities with MAC-layer sensing in cognitive radio networks," *IEEE Transactions on Mobile Computing*, vol. 7, no. 5, pp. 533–545, May 2008.

[480] F. C. Commission, "Longley-Rice methodology for evaluating TV coverage and interference," *Office of Engineering and Technology Bulletin*, no. 69, 2004.

[481] S. M. Mishra, A. Sahai, and R. Brodersen, "Cooperative sensing among cognitive radios," in *Proceedings of IEEE International Conference on Communications*, Istanbul, Turkey, June 2006, pp. 1658–1663.

[482] A. Ghasemi and E. S. Sousa, "Collaborative spectrum sensing for opportunistic access in fading environments," in *Proceedings of IEEE International Symposium on New Frontiers in Dynamic Spectrum Access Networks*, Baltimore, MD, USA, November 2005, pp. 131–136.

[483] ——, "Opportunistic spectrum access in fading channels through collaborative sensing," *Journal of Communications*, vol. 2, no. 2, pp. 71–82, March 2007.

[484] W. Saad, Z. Han, M. Debbah, A. Hjørungnes, and T. Başar, "Coalitional games for distributed collaborative spectrum sensing in cognitive radio networks," in *Proceedings of IEEE Conference on Computer Communications*, Rio de Janeiro, Brazil, April 2009, pp. 2114–2122.

[485] G. Ghurumuruhan and Y. Li, "Cooperative spectrum sensing in cognitive radio: Part I: Two user networks," *IEEE Transactions on Wireless Communications*, vol. 6, no. 6, pp. 2204–2213, June 2007.

[486] ——, "Cooperative spectrum sensing in cognitive radio: Part II: Multiuser networks," *IEEE Transactions on Wireless Communications*, vol. 6, no. 6, pp. 2214–2222, June 2007.

[487] J. Unnikrishnan and V. V. Veeravalli, "Cooperative sensing for primary detection in cognitive radio," *IEEE Journal of Selected Topics in Signal Processing*, vol. 2, no. 1, pp. 18–27, February 2008.

[488] S. Cui, Z. Quan, and A. Sayed, "Optimal linear cooperation for spectrum sensing in cognitive radio networks," *IEEE Journal of Selected Topics in Signal Processing*, vol. 2, no. 1, pp. 28–40, February 2008.

[489] W. Zhang, C. Sun, and K. B. Letaief, "Cluster-based cooperative spectrum sensing in cognitive radio systems," in *Proceedings of International Conference on Communications*, Glasgow, Scotland, June 2007, pp. 2511–2515.

[490] ——, "Cooperative spectrum sensing for cognitive radios under bandwidth constraints," in *Proceedings of IEEE Wireless Communications and Networking Conference*, Hong Kong, China, February 2007, pp. 25–30.

[491] C. H. Lee and W. Wolf, "Energy efficient techniques for cooperative spectrum sensing in cognitive radios," in *Proceedings of IEEE Consumer Communications and Networking Conference*, Las Vegas, NV, USA, January 2008, pp. 968–972.

[492] A. Plaza, J. A. Benediktsson, J. Boardman, J. Brazile, L. Bruzzone, G. Camps-Valls, J. Chanussot, M. Fauvel, P. Gamba, J. Gualtieri, M. Marconcini, J. C. Tilton, and G. Trianni, "Recent advances in techniques for hyperspectral image processing," *Remote Sens. Environment*, vol. 113, no. 1, pp. 110–122, 2009.

[493] J. M. Bioucas-Dias, A. Plaza, G. Camps-Valls, P. Scheunders, N. M. Nasrabadi, and J. Chanussot, "Hyperspectral remote sensing data analysis and future challenges," *IEEE Geoscience and Remote Sensing Magazine*, vol. 1, no. 2, pp. 6–36, June 2013.

[494] E. J. Candès and M. B. Wakin, "An introduction to compressive sampling," *IEEE Signal Processing Magazine*, vol. 25, no. 2, pp. 21–30, March 2008.

[495] R. M. Willett, M. F. Duarte, M. A. Davenport, and R. G. Baraniuk, "Sparsity and structure in hyperspectral imaging: Sensing, reconstruction, and target detection," *IEEE Signal Processing Magazine*, vol. 31, no. 1, pp. 116–126, January 2014.

[496] J. E. Fowler, "Compressive pushbroom and whiskbroom sensing for hyperspectral remote-sensing imaging," in *Proceedings of IEEE International Conference on Image Processing*, Paris, France, October 2014, pp. 684–688.

[497] M. F. Duarte, M. A. Davenport, D. Takhar, J. N. Laska, T. Sun, K. E. Kelly, and R. G. Baraniuk, "Single-pixel imaging via compressive sampling," *IEEE Signal Processing Magazine*, vol. 25, no. 2, pp. 83–91, March 2008.

[498] T. S. C. Li, K. F. Kelly, and Y. Zhang, "A compressive sensing and unmixing scheme for hyperspectral data processing," *IEEE Transactions on Image Processing*, vol. 21, no. 3, pp. 1200–1210, March 2012.

[499] M. Golbabaee and P. Vandergheynst, "Hyperspectral image compressed sensing via low-rank and joint-sparse matrix recovery," in *Proceedings of IEEE International Conference on Acoustic, Speech and Signal Processing*, Kyoto, Japan, March 2012, pp. 2741–2744.

[500] G. Martín, J. M. Bioucas-Dias, and A. Plaza, "HYCA: A new technique for hyperspectral compressive sensing," *IEEE Transactions on Geoscience and Remote Sensing*, vol. 53, no. 5, pp. 2819–2831, May 2015.

[501] H. Ren and C. Chang, "Automatic spectral target recognition in hyperspectral imagery," *IEEE Transactions on Aerospace and Electronic Systems*, vol. 39, no. 4, pp. 1232–1249, October 2003.

[502] Y. Chen, N. Nasrabadi, and T. Tran, "Sparse representation for target detection in hyperspectral imagery," *IEEE Journal of Selected Topics in Signal Processing*, vol. 5, no. 3, pp. 629–640, June 2011.

[503] Y. Chen, N. M. Nasrabadi, and T. D. Tran, "Simultaneous joint sparsity model for target detection in hyperspectral imagery," *IEEE Geoscience and Remote Sensing Letters*, vol. 8, no. 4, pp. 676–680, July 2011.

[504] G. Mateos, J. A. Bazerque, and G. B. Giannakis, "Distributed sparse linear regression," *IEEE Transactions on Signal Processing*, vol. 58, no. 10, pp. 5262–5276, 2010.

[505] J. F. C. Mota, J. M. F. Xavier, P. M. Q. Aguiar, and M. Puschel, "Distributed basis pursuit," *IEEE Transactions on Signal Processing*, vol. 60, no. 4, pp. 1942–1956, April 2012.

[506] I. Foster, Y. Zhao, I. Raicu, and S. Lu, "Large-scale sparse logistic regression," in *Proceedings of ACM International Conference on Knowledge Discovery and Data Mining*, New York, NY, USA, June 2009, pp. 547–556.

[507] S. S. Ram, A. Nedić, and V. V. Veeravalli, "A new class of distributed optimization algorithm: Application of regression of distributed data," *Optimization Methods and Software*, vol. 27, no. 1, pp. 71–88, 2012.

[508] D. Kempe, A. Dobra, and J. Gehrke, "Gossip-based computation of aggregate information," in *Proceedings of Annual IEEE Symposium on Foundations of Computer Sciences*, Cambridge, MA, USA, October 2003, pp. 482–491.

[509] Q. Ling, Y. Xu, W. Yin, and Z. Wen, "Decentralized low-rank matrix completion," in *Proceedings of IEEE International Conference on Acoustic, Speech and Signal Processing*, Kyoto, Japan, March 2012, pp. 2925–2928.

[510] A. Nedić and A. Ozdaglar, "Distributed subgradient methods for multi-agent optimization," *IEEE Transaction on Automatic Control*, vol. 54, no. 1, pp. 48–61, Jan. 2009.

[511] A. Nedic and A. Ozdaglar, "Cooperative distributed multi-agent optimization," in *Convex Optimization in Signal Processing and Communications*, Cambridge, UK: Cambridge University Press, 2009.

[512] A. Nedic, A. Ozdaglar, and P. Parrilo, "Constrained consensus and optimization in multi-agent networks," *IEEE Transactions on Automatic Control*, vol. 55, no. 4, pp. 922–938, April 2010.

[513] K. Srivastava and A. Nedic, "Distributed asynchronous constrained stochastic optimization," *IEEE Journal of Selected Topics in Signal Processing*, vol. 5, no. 4, pp. 772–790, August 2011.

[514] J. Tsitsiklis, "Problems in decentralized decision making and computation," PhD thesis, MIT, 1984.

[515] K. Yuan, Q. Ling, and W. Yin, "On the convergence of decentralized gradient descent," *SIAM Journal on Optimization*, 2015.

[516] I. Chen, "Fast distributed first-order methods," Master's thesis, MIT, 2012.

[517] K. I. Tsianos and M. G. Rabbat, "Distributed strongly convex optimization," in *50th Annual Allerton Conference on Communication, Control, and Computing*, Monticello, IL, USA, Oct 2012, pp. 593–600.

[518] D. Bertsekas and J. Tsitsiklis, *Parallel and Distributed Computation: Numerical Methods* (2nd ed.) Belmont, MA, USA: Athena Scientific, 1997.

[519] G. B. Giannakis, Q. Ling, G. Mateos, I. D. Schizas, and H. Zhu, "Proximal splitting methods in signal processing," in *Splitting Methods in Communication and Imaging*, New York, USA: Springer, 2015.

[520] I. Schizas, A. Ribeiro, and G. Giannakis, "Consensus in ad hoc WSNs with noisy links - Part I: Distributed estimation of deterministic signals," *IEEE Transactions on Signal Processing*, vol. 56, no. 1, pp. 350–364, 2008.

[521] W. Shi, Q. Ling, K. Yuan, G. Wu, and W. Yin, "On the linear convergence of the ADMM in decentralized consensus optimization," *IEEE Transactions on Signal Processing*, vol. 62, no. 7, pp. 1750–1761, April 2014.

[522] Q. Ling, W. Shi, G. Wu, and A. Ribeiro, "DLM: Decentralized linearized alternating direction method of multipliers," *IEEE Transactions on Signal Processing*, vol. 63, no. 15, pp. 4051–4064, August 2015.

[523] N. Parikh and S. Boyd, "Proximal algorithms," *Foundations and Trends in Optimization*, vol. 1, no. 3, pp. 1–112, 2013.

[524] W. Shi, Q. Ling, G. Wu, and W. Yin, "EXTRA: An exact first-order algorithm for decentralized consensus optimization," *SIAM Journal on Optimization*, vol. 25, no. 2, pp. 944–966, 2014.

[525] J. C. Duchi, A. Agarwal, and M. J. Wainwright, "Dual averaging for distributed optimization: Convergence analysis and network scaling," *IEEE Transactions on Automatic Control*, vol. 57, no. 3, pp. 592–606, March 2012.

[526] Y. Nesterov, "Primal-dual subgradient methods for convex problems," *Mathematical Programming*, vol. 120, no. 1, pp. 261–283, 2009.

[527] M. Hong, Z.-Q. Luo, and M. Razaviyayn, "On the convergence of alternating direction method of mulitpliers for a family of nonconvex problems," in *40th International Conference on Acoustic, Speech and Signal Processing*, Brisbane, Australia, April 2015.

[528] A. Nedic and A. Olshevsky, "Distributed optimization over time-varying directed graphs," *IEEE Transactions on Automatic Control*, vol. 60, no. 3, pp. 601–615, March 2015.

[529] E. Wei and A. Ozdaglar, "On the $O(1/k)$ convergence of asynchronous distributed alternating direction method of multipliers," in *Global Conference on Signal and Information Processing*, Austin, TX, USA, December 2013.

[530] D. Bertsekas, "Incremental gradient, subgradient, and proximal methods for convex optimization: A survey," *Optimization for Machine Learning*, vol. 4, pp. 85–119, 2012.

[531] Z.-Q. Luo, "On the convergence of the LMS algorithm with adaptive learning rate for linear feedforward networks," *Neural Computation*, vol. 3, no. 2, pp. 226–245, 1991.

[532] D. Blatt, A. O. Hero, and H. Gauchman, "A convergent incremental gradient method with a constant step size," *SIAM Journal on Optimization*, vol. 18, no. 1, pp. 29–51, 2007.

[533] M. Gurbuzbalaban, A. Ozdaglar, and P. Parrilo, "Convergence rate of incremental aggregated gradient algorithms," *IEEE Transactions on Signal Processing*, preprint, 2015.

[534] M. L. Roux, M. Schmidt, and F. Bach, "A stochastic gradient method with an exponential convergence rate for strongly-convex optimization with finite training sets," in *Proceedings of the Annual Conference on Neural Information Processing Systems*, Lake Tahoe, NV, USA, December 2012.

[535] A. Defazio, F. Bach, and S. Lacoste-Julien, "Saga: A fast incremental gradient method with support for non-strongly convex composite objectives," in *Proceeding of Annual Conference on Neural Information Processing Systems*, Montreal, Canada, December 2014.

[536] R. Johnson and T. Zhang, "Accelerating stochastic gradient descent using predictive variance reduction," in *Proceedings of the Annual Conference on Neural Information Processing Systems*, Lake Tahoe, NV, USA, December 2013.

[537] L. Xiao and T. Zhang, "A proximal stochastic gradient method with progressive variance reduction," *Siam Journal on Optimization*, vol. 24, pp. 2057–2075, 2014.

[538] J. Konecny, Z. Qu, and P. Richtarik, "Semi-stochastic coordinate descent," preprint, 2014, http://arXiv:1412.6293.

[539] S. Shalev-Shwartz and T. Zhang, "Stochastic dual coordinate ascent methods for regularized loss minimization," *Journal of Machine Learning Research*, vol. 14, pp. 567–599, 2013.

[540] A. Agarwal and L. Bottou, "A lower bound for the optimization of finite sums," in *Proceedings of the International Conference on Machine Learning*, Lille, France, June 2015.

[541] A. Nemirovsky and D. Yudin, "Problem complexity and method efficiency in optimization," in *Interscience Series in Discrete Mathematics*, Wiley, 1983.

[542] G. Lan, "An optimal randomized incremental gradient method," preprint, 2015, https://arxiv.org/abs/1507.02000.

[543] Y. Li and S. Osher, "Coordinate descent optimization for L1 minimization with applications to compressed sensing: A greedy algorithm," *Inverse Problems and Imaging*, vol. 3, no. 3, pp. 487–503, 2009.

[544] L. Bottou and O. Bousquet, "The tradeoffs of large scale learning," in *Advances in Neural Information Processing Systems*, Vancouver, Canada, December 2008.

[545] M. Zinkevich, M. Weimer, A. Smola, and L. Li, "Parallelized stochastic gradient descent," in *Advances in Neural Information Processing Systems*, Vancouver, Canada, December 2010.

[546] F. Niu, B. Recht, C. Re, and S. J. Wright, "Hogwild: A lock-free approach to parallelizing stochastic gradient descent," in *Advances in Neural Information Processing Systems*, Granada, Spain, December 2011.

[547] C. J. Hsieh, K. W. Chang, C. J. Lin, S. S. Keerthi, and S. Sundararajan, "A dual coordinate descent method for large-scale linear SVM," in *International Conference on Machine Learning*, Helsinki, Finland, July 2008.

[548] S. Boyd, N. Parikh, E. Chu, B. Peleato, and J. Eckstein, "Distributed optimization and statistical learning via the alternating direction method of multipliers," *Foundation and Trends in Machine Learning*, vol. 3, no. 1, pp. 1–122, November 2010.

[549] R. M. Freund and P. Grigas, "New analysis and results for the Frankwolfe method," http://arXiv.org/abs/1307.0873, 2014.

[550] S. Lacoste-Julien, M. Jaggi, M. Schmidt, and P. Pletscher, "Block-coordinate Frank-Wolfe optimization for structural SVMs," in *International Conference on Machine Learning*, Atlanta, GA, USA, June 2013.

[551] M. Grant and S. Boyd, "CVX: Matlab software for disciplined convex programming, Version 2.1," http://cvxr.com/cvx, March 2014.

[552] B. Chun, S. Ihm, P. Maniatis, M. Naik, and A. Patti, "CloneCloud: Elastic execution between mobile device and cloud," in *Conference on Computer Systems*, Salzburg, Austria, April 2011.

[553] M. Chiang, S. H. Low, A. R. Calderbank, and J. C. Doyle, "Layering as optimization decomposition: A mathematical theory of network architectures," *Proceedings of the IEEE*, vol. 95, no. 1, pp. 255–312, January 2007.

[554] S. Kosta, A. Aucinas, P. Hui, R. Mortier, and X. Zhang, "ThinkAir: Dynamic resource allocation and parallel execution in the cloud for mobile code offloading," in *Proceedings of IEEE International Conference on Computer Communications*, Orlando, FL, USA, March 2012.

[555] R. Bifulco, M. Brunner, R. Canonico, R. Hasselmeyer, and F. Mir, "Scalability of a mobile cloud management system," in *ACM Special Interest Group on Data Communications Workshop on Mobile Cloud Computing*, Helsinki, Finland, August 2012.

[556] P. Wendell, J. W. Jiang, M. J. Freedman, and J. Rexford, "Donar: Decentralized sever selection for cloud services," in *Proceedings of the ACM Special Interest Group on Data Communications*, New Delhi, India, August 2010.

[557] Z. Zhang, M. Zhang, A. Greenberg, Y. C. Hu, R. Mahajan, and B. Christian, "Optimizing cost and performance in online service provider networks," in *Proceedings of USENIX Symposium on Networked Sytems Design and Implementation*, San Jose, CA, USA, April 2010.

[558] D. K. Goldenberg, L. Qiu, H. Xie, Y. R. Yang, and Y. Zhang, "Optimizing cost and performance for multihoming," in *Proceedings of ACM Special Interest Group on Data*

*Communications Workshop on Mobile Cloud Computing*, Portland, OR, USA, August 2002.

[559] H. Xu, C. Feng, and B. Li, "Temperature aware workload management in geo-distributed datacenters," in *Proceedings of USENIX International Conference on Autonomic Computing*, San Jose, CA, USA, June 2013.

[560] J. W. Jiang, R. Zhang-Shen, J. Rexford, and M. Chiang, "Cooperative content distribution and traffic engineering in an ISP network," in *Proceedings of ACM Special Interest Group on Measurement and Evaluation*, Seattle, WA, USA, June 2009.

[561] S. Narayana, J. W. Jiang, J. Rexford, and M. Chiang, "To coordinate or not to coordinate? Wide-area traffic management for data centers," in *Proceedings of ACM International Conference on emerging Networking Experiments and Technologies*, Nice, France, December 2012.

[562] H. Xu and B. Li, "Joint request mapping and response routing for geo-distributed cloud services," in *Proceedings of IEEE International Conference on Computer Communications*, Turin, Italy, April 2013.

[563] M. Satyanarayanan, P. Bahl, R. Caceres, and N. Davies, "The case for VM-based cloudlets in mobile computing," *IEEE Pervasive Computing*, vol. 8, no. 4, pp. 14–23, October 2009.

[564] N. McKeown, T. Anderson, H. Balakrishnan, G. Parulkar, L. Peterson, J. Rexford, S. Shenker, and J. Turner, "Openflow: Enabling innovation in campus networks," in *Proceedings of the ACM Special Interest Group on Data Communications*, Seattle, WA, USA, August 2008.

[565] M. Jeonghoon and J. Walrand, "Fair end-to-end window-based congestion control," *IEEE/ACM Transactions on Networking*, vol. 8, no. 6, pp. 556–567, 2000.

[566] F. P. Kelly, A. Maulloo, and D. Tan, "Rate control for communication networks: Shadow price, proportional fairness and stability," *Journal of Operational Research Society*, vol. 49, pp. 237–252, March 1998.

[567] L. Rao, X. Liu, L. Xie, and W. Liu, "Minimizing electricity cost: Optimization of distributed internet data centers in a multi-electricity-market environment," in *Proceedings of the IEEE International Conference on Computer Communications*, San Diego, CA, USA, March 2010.

[568] A. Greenberg, J. Hamilton, D. A. Maltz, and P. Patel, "The cost of a cloud: Research problems in data center networks," *ACM Special Interest Group on Data Communications Computer Communication Review*, vol. 39, no. 1, pp. 68–73, January 2009.

[569] X. Fan, W. Weber, and L. A. Barroso, "Power provisioning for a warehouse-size computer," in *Proceedings of the ACM International Symposium on Computer Architecture*, San Diego, CA, USA, June 2007.

[570] H. V. Madhyastha, T. Isdal, M. Piatek, C. Dixon, T. Anderson, A. Krishnamurthy, and A. Venkataramani, "iPlane: An information plane for distributed services," in *Proceedings of USENIX Symposium on Networked Systems Design and Implementation*, Seattle, WA, USA, November 2006.

[571] "AT&T 2012 Sustainability Report: Water," www.att.com/gen/landing-pages?pid=24188.

[572] "Water use hints at problems at Utah Data Center," www.sltrib.com/.

[573] Office of governor in California, "Governor brown declares drought state of emergency," 2014, http://gov.ca.gov/news.php?id=18368.

[574] Z. Liu, Y. Chen, C. Bash, A. Wierman, D. Gmach, Z. Wang, M. Marwab, and C. Hyser, "Renewable and cooling aware workload management for sustainable data centers," in

*Proceedings of Special Interest Group on Performance Evaluation*, London, UK, June 2012.

[575] E. Frachtenberg, "Holistic datacenter design in the open compute project," *Computer*, vol. 45, no. 7, pp. 83–85, 2012.

[576] D. Alger, *Grow a Greener Data Center*, New Jersey: Cisco Press, 2009.

[577] S. Ren, "Optimizing water efficiency in distributed data centers," in *Proceedings of International Conference on Cloud and Green Computing*, Karlsruhe, Germany, September 2013.

[578] "Google data centers," www.google.com/about/datacenters/.

[579] U. Institute, "Data center industry survey," 2013, http://uptimeinstitute.com/2013 -survey-results.

[580] eBay. [Online]. Available: http://tech.ebay.com/dashboard.

[581] "Prineville data center," [Online]. Available: www.facebook.com/PrinevilleDataCenter/.

[582] K. Papagiannaki, N. Taft, Z. Zhang, and C. Diot, "Long-term forecasting of internet backbone traffic: Observations and initial methods," in *Proceedings of the IEEE International Conference on Computer Communications*, San Francisco, CA, USA, March 2003.

[583] G. Wang, J. Wu, G. Zhou, and G. Li, "Collision-tolerant media access control for asynchronous users over frequency-selective channels," *IEEE Transactions on Wireless Communications*, vol. 12, no. 10, pp. 5162–5171, 2013.

[584] L. Liu, S. Ren, and Z. Han, "Scalable workload management for water efficiency in data centers," in *IEEE Global Communications Conference*, Austin, TX, USA, December 2014.

[585] C. Liang and F. Yu, "Wireless network virtualization: A survey, some research issues and challenges," *IEEE Communications Surveys Tutorials*, vol. 17, no. 1, pp. 358–380, 2015.

[586] A. Fischer, J. Botero, M. Till Beck, H. de Meer, and X. Hesselbach, "Virtual network embedding: A survey," *IEEE Communications Surveys Tutorials*, vol. 15, no. 4, pp. 1888–1906, 2013.

[587] A. Belbekkouche, M. M. Hasan, and A. Karmouch, "Resource discovery and allocation in network virtualization," *IEEE Communications Surveys Tutorials*, vol. 14, no. 4, pp. 1114–1128, 2012.

[588] N. Feamster, L. Gao, and J. Rexford, "How to lease the internet in your spare time," *Special Interest Group on Data Communications Computer Communication Review*, vol. 37, no. 1, pp. 61–64, January 2007.

[589] N. Chowdhury and R. Boutaba, "Network virtualization: State of the art and research challenges," *IEEE Communications Magazine*, vol. 47, no. 7, pp. 20–26, July 2009.

[590] H. Wen, P. K. Tiwary, and T. Le-Ngoc, *Wireless Virtualization*, ser. Springer Briefs in Computer Science. Springer, 2013.

[591] H. Wen, P. Tiwary, and T. Le-Ngoc, "Current trends and perspectives in wireless virtualization," in *International Conference on Selected Topics: Stochastic Game for Wireless Network Virtualization*, Montreal, Canada, August 2013, pp. 62–67.

[592] F. Fu and U. Kozat, "Stochastic game for wireless network virtualization," *Networking, IEEE/ACM Transactions on Networking*, vol. 21, no. 1, pp. 84–97, February 2013.

[593] Q. Zhu and X. Zhang, "Game-theory based power and spectrum virtualization for maximizing spectrum efficiency over mobile cloud-computing wireless networks," in *Annual Conference on Information Sciences and Systems*, Baltimore, MD, USA, March 2015.

[594] M. Yang, Y. Li, J. Liu, D. Jin, J. Yuan, and L. Zeng, "Opportunistic spectrum sharing for wireless virtualization," in *IEEE Wireless Communications and Networking Conference*, Istanbul, Turkey, April 2014, pp. 1803–1808.

[595] G. Liu, F. Yu, H. Ji, and V. Leung, "Distributed resource allocation in full-duplex relaying networks with wireless virtualization," in *IEEE Global Communications Conference*, Austin, TX, USA, December 2014, pp. 4959–4964.

[596] R. Kokku, R. Mahindra, H. Zhang, and S. Rangarajan, "NVS: A substrate for virtualizing wireless resources in cellular networks," *IEEE/ACM Transactions on Networking*, vol. 20, no. 5, pp. 1333–1346, October 2012.

[597] L. Xiao, M. Johansson, and S. Boyd, "Simultaneous routing and resource allocation via dual decomposition," *IEEE Transactions on Communications*, vol. 52, no. 7, pp. 1136–1144, July 2004.

[598] Y. Chen, S. Zhang, S. Xu, and G. Li, "Fundamental trade-offs on green wireless networks," *IEEE Communications Magazine*, vol. 49, no. 6, pp. 30–37, June 2011.

[599] E. Hossain, Z. Han, and H. V. Poor, *Smart Grid Communications and Networking*, Cambridge, UK: Cambridge University Press, 2012.

[600] U.-C. P. S. O. T. Force, "Final report on the August 14, 2003 blackout in the United States and Canada: Causes and recommendations," Tech. Rep., April 2004.

[601] S. Gorman, "Effect of stealthy bad data injection on network congestion in market based power system," *The Wall Street Journal*, April 2009.

[602] A. Abur and A. G. Exposito, *Power System State Estimation: Theory and Implementation*, New York, USA: Marcel Dekker, Inc., 2004.

[603] Y. Liu, M. K. Reiter, and P. Ning, "False data injection attacks against state estimation in electric power grids," in *Proceedings of 16th ACM Conference on Computer and Communications Security*, Chicago, IL, USA, November 2009.

[604] J. J. Grainger and W. D. S. Jr., *Power System Analysis*, New York, USA: McGraw-Hill, 1994.

[605] L. Xie, Y. Mo, and B. Sinopoli, "False data injection attacks in electricity markets," in *Proceedings of IEEE International Conference on Smart Grid Communications*, Gaithersburg, MD, USA, October 2010.

[606] M. Esmalifalak, Z. Han, and L. Song, "Effect of stealthy bad data injection on network congestion in market based power system," in *Proceedings of IEEE Wireless Communications and Networking Conference*, Paris, France, April 2012.

[607] G. Dán and H. Sandberg, "Stealth attacks and protection schemes for state estimators in power systems," in *Proceedings of IEEE International Conference on Smart Grid Communications*, Gaithersburg, MD, USA, October 2010.

[608] M. Esmalifalak, G. Shi, Z. Han, and L. Song, "Bad data injection attack and defense in electricity market using game theory study," *IEEE Transactions on Smart Grid*, vol. 4, no. 1, pp. 160–169, March 2013.

[609] O. Kousut, L. Jia, R. J. Thomas, and L. Tong, "Malicious data attacks on the smart grid," *IEEE Transactions on Smart Grid*, vol. 2, no. 4, pp. 645–658, December 2011.

[610] L. Liu, M. Esmalifalak, and Z. Han, "Detection of false data injection in power grid exploiting low rank and sparsity," in *IEEE International Conference on Smart Grid Communications*, Budapest, Hungary, June 2013.

[611] T. T. Kim and H. V. Poor, "Strategic protection against data injection attacks on power grids," *IEEE Transactions on Smart Grid*, vol. 2, no. 2, pp. 326–333, June 2011.

[612] S. Cui, Z. Han, S. Kar, T. T. Kim, H. V. Poor, and A. Tajer, "Coordinated data-injection attack and detection in the smart grid: A detailed look at enriching detection

solutions," *IEEE Signal Processing Magazine*, vol. 29, no. 5, pp. 106–115, September 2012.

[613] Y. Zhao, A. Goldsmith, and H. V. Poor, "Fundamental limits of cyber-physical security in smart power grids," in *Proceedings of IEEE 52nd Annual Conference on Decision and Control*, Florence, Italy, December 2013.

[614] M. Shahidehpour, W. F. Tinney, and Y. Fu, "Impact of security on power system operation," *Proceedings of the IEEE*, vol. 93, no. 11, pp. 2013–2025, November 2001.

[615] O. Alsac and B. Scott, "Optimal load flow with steady-state security," *IEEE Transactions on Power Apparatus and System*, vol. 93, no. 3, pp. 745–751, May 1974.

[616] M. V. F. Pereira, A. Monticelli, and L. M. V. G. Pinto, "Security-constrained dispatch with corrective rescheduling," in *Proceedings of IFAC Symposium on Planning and Operation of Electric Energy System*, Rio de Janeiro, Brazil, July 1985.

[617] A. J. Wood and B. F. Wollenberg, *Power Generation Operation and Control*, New York, USA: Wiley, 1996.

[618] A. Monticelli, M. V. F. Pereira, and S. Granville, "Security-constrained optimal power flow with post-contingency corrective rescheduling," *IEEE Transactions on Power Systems*, vol. 2, no. 1, pp. 175–180, February 1987.

[619] F. Capitanescu, J. L. M. Ramos, P. Panciatici, D. Kirschen, A. M. Marcolini, L. Platbrood, and L. Wehenkel, "State-of-the-art, challenges, and future trends in security constrained optimal power flow," *Electric Power System Research*, vol. 81, no. 8, pp. 1731–1741, August 2011.

[620] J. Martínez-Crespo, J. Usaola, and J. L. Fernández, "Security-constrained optimal generation scheduling in large-scale power systems," *IEEE Transactions on Power Systems*, vol. 21, no. 1, pp. 321–332, February 2006.

[621] Y. Fu, M. Shahidehpour, and Z. Li, "AC contingency dispatch based on security-constrained unit commitment," *IEEE Transactions on Power Systems*, vol. 21, no. 2, pp. 897–908, May 2006.

[622] F. Capitanescu and L. Wehenkel, "A new iterative approach to the corrective security-constrained optimal power flow problem," *IEEE Transactions on Power Systems*, vol. 23, no. 4, pp. 1533–1541, November 2008.

[623] Y. Li and J. D. McCalley, "Decomposed scopf for improving efficiency," *IEEE Transactions on Power Systems*, vol. 24, no. 1, pp. 494–495, February 2009.

[624] Z. Han, H. Li, and W. Yin, *Compressive Sensing for Wireless Communication*, Cambridge, UK: Cambridge University Press, 2012.

[625] E. J. Candès, X. Li, Y. Ma, and J. Wright, "Robust principal component analysis?" *Journal of the ACM*, vol. 58, no. 3, pp. 1–37, May 2011.

[626] Z. Lin, M. Chen, L. Wu, and Y. Ma, "The augmented lagrange multiplier method for exact recovery of corrupted low-rank matrices," UIUC, Tech. Rep. UILU-ENG-09-2215, Urbana, FL, USA, 2009.

[627] J. Cai, E. J. Candès, and Z. Shen, "A singular value thresholding algorithm for matrix completion," *SIAM Journal on Optimization*, vol. 20, no. 4, pp. 1956–1982, January 2010.

[628] Y. Shen, Z. Wen, and Y. Zhang, "Augmented Lagrangian alternating direction method for matrix separation based on low-rank factorization," Rice CAAM, Tech. Rep. TR11-02, Houston, TX, USA, 2011.

[629] Z. Wen, W. Yin, and Y. Zhang, "Solving a low-rank factorization model for matrix completion by a nonlinear successive over-relaxation algorithm," Rice CAAM, Tech. Rep. TR10-07, Houston, TX, USA, 2010.

[630]  R. D. Zimmerman, C. E. Murillo-Sánchez, and R. J. Thomas, "MAT-POWER steady-state operations, planning and analysis tools for power systems research and education," *IEEE Transactions on Power Systems*, vol. 26, no. 1, pp. 12–19, February 2011.

[631]  R. Baldick, B. H. Kim, C. Chase, and Y. Luo, "A fast distributed implementation of optimal power flow," *IEEE Transactions on Power Systems*, vol. 14, no. 3, pp. 858–864, August 1989.

[632]  M. Kraning, E. Chu, J. Lavaei, and S. Boyd, "Dynamic network energy management via proximal message passing," *Foundations and Trends in Optimization*, vol. 1, no. 2, pp. 1–54, January 2014.

[633]  W. Deng and W. Yin, "On the global and linear convergence of the generalized alternating direction method of multipliers," Rice CAAM, Tech. Rep. TR12-14, Houston, TX, USA, 2012.

[634]  J. Nocedal and S. J. Wright, *Numerical Optimization* (2nd ed.), New York, USA: Springer, 2006.

[635]  M. Schatz, "Cloudburst: Highly sensitive read mapping with mapreduce," *Bioinformatics*, vol. 25, no. 11, pp. 1363–1369, 2009.

[636]  http://lintool.github.io/Cloud9/.

[637]  J. Stamos and H. Young, "A symmetric fragment and replicate algorithm for distributed joins," *IEEE Transactions on Parallel and Distributed Systems*, vol. 4, no. 12, pp. 1345–1354, December 1993.

[638]  W. Yan and P. Larson, "Eager aggregation and lazy aggregation," in *Proceedings of Very Large Data Base Conference*, Zurich, Switzerland, September 1995.

[639]  S. Ramakrishnan, G. Swart, and A. Urmanov, "Balancing reducer skew in mapreduce workloads using progressive sampling," in *Proceedings of ACM Symposium on Cloud Computing*, San Jose, CA, USA, October 2012.

[640]  M. Englert, D. Ozmen, and M. Westermann, "The power of reordering for online minimum makespan scheduling," in *Proceedings of IEEE Annual Symposium on Foundation on Computer Science*, Philadelphia, PA, USA, 2008.

[641]  J. Kleinberg and E. Tardos, *Algorithm Design*, www.pearsoned.co.in/Web/Home.aspx, Pearson Education India, 2006.

[642]  J. Devore, *Probability & Statistics for Engineering and the Sciences*, CengageBrain.com, 2012.

[643]  "Wikipedia page-to-page link," 2013, http://haselgrove.id.au/wikipedia.htm.

[644]  W. Heinzelman, A. Chandrakasan, and H. Balakrishnan, "Energy-efficient communication protocol for wireless microsensor networks," in *Proceedings of Hawaiian International Conference on Systems Science*, Wailea Maui, HI, USA, January 2000.

[645]  I. F. Akyildiz, W. Su, Y. Sankarusubramanian, and E. Cayirci, "A survey on sensor networks," *IEEE Communications Magazine*, vol. 40, no. 8, pp. 102–114, August 2002.

[646]  D. Estrin, R. Govindan, J. Heidemann, and S. Kumar, "Next century challenges: Scalable coordination in sensor networks," in *Proceedings of ACM Annual Conference on Mobile Computing and Networking*, Seattle, WA, USA, August 1999.

[647]  S. Madden, R. Szewczyk, M. Franklin, and W. Hong, "Supporting aggregate queries over ad-hoc wireless sensor networks," in *Proceedings of IEEE International Workshop on Mobile Computing Systems and Application*, Callicon, NY, USA, June 2002.

[648]  S. Madden, M. Franklin, J. Hellerstein, and W. Hong, "Tag: A tiny aggregation service for ad hoc sensor networks," in *Proceedings of USENIX Operating System Design and Implementation*, Boston, MA, USA, December 2002.

[649] C. Intanagonwiwat, D. Estrin, R. Govindan, and J. Heidemann, "Impact of network density on data aggregation in wireless sensor networks," in *Proceedings of IEEE International Conference on Distributed Computing Systems*, Vienna, Austria, July 2002.

[650] B. Krishnamachari, D. Estrin, and S. Wicker, "The impact of data aggregation in wireless sensor networks," in *Proceedings of IEEE International Conference on Distributed Computing Systems Workshop on Distributed Event-based System*, Vienna, Austria, July 2002.

[651] J. Al-Karaki and A. Kamal, "Routing techniques in wireless sensor networks: A survey," *IEEE Wireless Communications*, vol. 11, no. 6, pp. 6–28, December 2004.

[652] W. Heinzelman, J. Kulik, and H. Balakrishnan, "Adaptive protocols for information dissemination in wireless sensor networks," in *Proceedings of ACM Annual Conference on Mobile Computing and Networking*, Seattle, WA, USA, August 1999.

[653] C. Intanagonwiwat, R. Govindan, and D. Estrin, "Directed diffusion: A scalable and robust communication paradigm for sensor networks," in *Proceedings of ACM Annual Conference on Mobile Computing and Networking*, Boston, MA, USA, August 2000.

[654] D. Braginsky and D. Estrin, "Rumor routing algorithm for sensor networks," in *Proceedings of ACM Workshop on Wireless Sensor Networks and Applications*, Atlanta, GA, USA, September 2002.

[655] M. Chu, H. Haussecker, and F. Zhao, "Scalable information-driven sensor querying and routing for ad hoc heterogeneous sensor networks," *International Journal of High Performance Computing Applications*, vol. 16, no. 3, pp. 293–313, August 2002.

[656] N. Sadagopan, B. Krishnamachari, and A. Helmy, "the acquire mechanism mechanism for efficient querying in sensor networks," in *Proceedings of the IEEE International Workshop on Sensor Network Protocol and Applications*, Seattle, WA, USA, May 2003.

[657] O. Younis and S. Fahmy, "Distributed clustering in ad-hoc sensor networks: A hybrid, energy-efficient approach," in *Proceedings of IEEE International Conference on Computer Communications*, Hong Kong, China, March 2004.

[658] V. Kawadia and P. Kumar, "The power control and clustering in ad-hoc networks," in *Proceedings of IEEE International Conference on Computer Communications*, San Francisco, CA, USA, March 2003.

[659] S. Banerjee and S. Khuller, "A clustering scheme for hierarchical control in multi-hop wireless networks," in *Proceedings of IEEE International Conference on Communications*, Anchorage, AK, USA, April 2001.

[660] K. Yao, D. Estrin, and Y. H. Hu, eds., *Special Issue on Sensor Networks, EURASIP Journal on Applied Signal Processing*, vol. 2003, no. 4, 2004.

[661] M. Greenwald and S. Khanna, "Power-conserving computation of order-statistics over sensor networks," in *Proceedings of ACM the Symposium on Principles of Database Systems*, Paris, France, June 2004.

[662] C. Buragohain, D. Agrawal, and S. Suri, "Power aware routing for sensor databases," in *Proceedings of IEEE International Conference on Communications*, Miami, FL, USA, March 2005.

[663] W. Yu, W. Rhee, S. Boyd, and J. M. Cioffi, "Iterative water-filling for Gaussian vector multiple-access channels," *IEEE Transactions on Information Theory*, vol. 50, no. 1, pp. 145–152, January 2004.

[664] A. Boukerche, R. Pazzi, and R. Araujo, "A fast and reliable protocol for wireless sensor networks in critical conditions monitoring applications," in *Proceedings of ACM International Conference on Modeling, Analysis and Simulation of Wireless and Mobile Systems*, Venice, Italy, October 2004.

[665] S. Lindsey and C. Raghavendra, "Pegasis: Power-efficient gathering in sensor networks," in *Proceedings of IEEE Aerospace Conference*, vol. 3, 2002.

[666] J. Wieselthier, G. Nguyen, and A. Ephremides, "On the construction of energy-efficient broadcast and multicast trees in wireless networks," in *Proceedings of IEEE International Conference on Computer Communications*, Tel-Aviv, Israel, March 2000.

[667] N. Shrivastava, C. Buragohain, D. Agrawal, and S. Suri, "Medians and beyond: New aggregation techniques for sensor networks," in *Proceedings of ACM Conference on Embedded Networked Sensor Systems*, Baltimore, MD, USA, November 2004.

[668] W. Hoeffding, "Probability inequalities for sums of bounded random variables," *Journal of the American Statistical Association*, vol. 58, no. 301, pp. 13–30, 1963.

# Index